HYDRAULIC BARRIERS IN SOIL AND ROCK

A symposium
sponsored by
ASTM Committee D18
on Soil and Rock in
cooperation with the
United States Committee
on Large Dams of the
International Commission
on Large Dams
Denver, CO, 25 June 1984.

ASTM SPECIAL TECHNICAL PUBLICATION 874
A. Ivan Johnson, Woodward-Clyde Consultants,
Ronald K. Frobel, U.S. Bureau of Reclamation,
Nicholas J. Cavalli, ICOS Corporation of
America, and C. Bernt Pettersson, Brown &
Root, Inc., editors

ASTM Publication Code Number (PCN)
04-874000-38

ASTM

1916 Race Street, Philadelphia, PA 19103

Library of Congress Cataloging-in-Publication Data

Hydraulic barriers in soil and rock.

(ASTM special technical publication; 874)
"Contains papers presented at the Symposium on
Impermeable Barriers for Soil and Rock ... sponsored by
ASTM Committee D18 on Soil and Rock, in cooperation with
the U.S. Committee on Large Dams (UNSCOLD) of the Inter-
national Commission on Large Dams"—Foreword.
"ASTM publication code number (PCN) 04-874000-38."
Includes bibliographies and index.
1. Reservoirs—Lining—Congresses.   2. Slurry
trench construction—Congresses.   3. Soil permeability—
Congresses.   4. Rocks— Permeability—Congresses.
5. Water, Underground—Pollution—Congresses.
6. Groundwater flow—Congresses.   I. Johnson, A. I.
(Arnold Ivan), 1919-   .   II. Symposium on Impermeable
Barriers for Soil and Rock (1984: Denver, Colo.)
III. ASTM Committee D18 on Soil and Rock.   IV. Inter-
national Commission on Large Dams. United States
Committee.   V. Series.
TC167.H93   1985   675   85-13332
ISBN 0-8031-0417-0

NOTE

The Society is not responsible, as a body,
for the statements and opinions
advanced in this publication.

Printed in Baltimore, MD
July 1985

Foreword

This publication, *Hydraulic Barriers in Soil and Rock*, contains papers presented at the Symposium on Impermeable Barriers for Soil and Rock, which was held in Denver on 25 June 1984. The symposium was sponsored by ASTM Committee D18 on Soil and Rock in cooperation with the U.S. Committee on Large Dams (USCOLD) of the International Commission on Large Dams. A. Ivan Johnson, Woodward-Clyde Consultants, and Ronald K. Frobel, U.S. Bureau of Reclamation, presided as symposium cochairmen. They were assisted in developing the program and as session cochairmen by Nicholas J. Cavalli, ICOS Corporation of America, and C. Bernt Pettersson, Brown & Root, Inc. All are coeditors of this publication.

Related
ASTM Publications

Special Procedures for Testing Soil and Rock for Engineering Purposes, STP 479 (1970), 04-479000-38

Dispersive Clays, Related Piping and Erosion in Geotechnical Projects, STP 623 (1977), 04-623000-38

Permeability and Groundwater Contaminant Transport, STP 746 (1981), 04-746000-38

Hazardous Solid Waste Testing: First Conference, STP 760 (1981), 04-760000-16

A Note of Appreciation
to Reviewers

The quality of the papers that appear in this publication reflects not only the obvious efforts of the authors but also the unheralded, though essential, work of the reviewers. On behalf of ASTM we acknowledge with appreciation their dedication to high professional standards and their sacrifice of time and effort.

ASTM Committee on Publications

ASTM Editorial Staff

Contents

GENERAL DISCUSSION

Overview

Reduction of the subsurface movement of fluids, especially hazardous wastes, has been the subject of much research and considerable controversy over the last few years. The characteristics of the contained fluid play a very important role in determining the properties of the soil and rock system—the solid earth environment. In turn, designers and installers of irrigation ditches, canals, and reservoirs, waste ponds and lagoons, mine tailings ponds, solar energy ponds, excavation dewatering systems, and other types of impoundments are interested in reducing to a minimum fluid movement through the soil and rock and retention installations. Thus, ASTM Committee D18 on Soil and Rock has been concerned for some time about the quality and quantity of the ground water or of any other fluids contained in or moving through the soil and rock voids. A mechanism for reducing leakage from such installations is some type of barrier to hydraulic movement through the soil and rock.

Data and information related to the very complicated interaction of the environmental system of soil and rock containment, fluid barrier, and enclosed fluids must be valid, compatible, and comparable to be useful in the development of guidelines for the designer and installer of any impounding installation. Because the need was recognized for the formation of a broad interdisciplinary consensus group to develop standards related to all types of fluid barriers in soil and rock, Subcommittee D18.20 on Impermeable Barriers was formed in 1981. Since its formation, the group has recognized that probably few, if any, completely impermeable barriers exist, so a name change to "Hydraulic Barriers in Soil and Rock" subsequently has been proposed.

Subcommittee D18.20 now provides a common meeting place where technical experts representing a wide variety of disciplines, and including researchers, users, and manufacturers' representatives, can share experiences for the development of high quality, uniformly acceptable, consensus standards. In addition to its primary objective of the development of needed standards for test methods, the subcommittee's objectives include the establishment of standard methods or guidelines for the quality control of liner or slurry wall installations and of the soil and rock containments; the coordination of data and information on hydraulic barriers; the development of basic testing methods for the soil and rock and associated materials as used for hydraulic barriers; and the promotion of needed research and dissemination

1

of data, information, and research results among people involved in the design and installation of hydraulic barriers.

In partial fulfillment of some of its objectives, ASTM Committee D18, through its Subcommittee D18.20 and in cooperation with the U.S. Committee on Large Dams (USCOLD) of the International Commission on Large Dams, organized a one-day symposium on impermeable barriers, held 25 June 1984 in Denver, Colorado. This ASTM special technical publication presents papers from the symposium, but the title has been changed to conform to the proposed new title of the subcommittee.

The symposium consisted of two half-day sessions—the first on slurry walls and the second on clay and soil-admix liners. For the purposes of this publication, these two types of hydraulic barriers are defined as follows:

Slurry walls—nonstructural walls constructed underground as vertical barriers to the lateral flow of water or other fluids. Slurry walls usually are trenches excavated to a horizontal strata of low permeability material and filled with a slurry of either soil-bentonite (SB) or cement-bentonite (CB).

Soil or soil-admix liners—nonstructural liners placed on horizontal or moderately sloping surfaces as barriers primarily to the vertical flow of water or other fluids. These liners either consist of compacted clay soil or soil admixes using bentonite, cement, and asphalt; chemical; or other additives mixed with the natural soil in order to improve hydraulic properties.

As mentioned at the beginning of this overview, the use of hydraulic barriers to impede the flow of fluids in soil and rock has been the subject of considerable controversy, especially for the past decade. The controversy has in part arisen from environmental concerns over barrier integrity and in part from the interest and desire of the geotechnical engineering design profession and construction industry to provide barriers that will fulfill established requirements and regulations. The central issue is whether a hydraulic barrier, whatever material is used for the barrier, will function satisfactorily over the intended lifetime of the facility. Although the function of the barrier may depend upon several factors, preservation of low permeability has become the primary performance criteria. Deterioration of a slurry wall or lining material resulting in the permeability exceeding a prescribed value has become the definition for "failure," that is, the barrier leaks more than the prescribed amount. Accordingly, the determinations of permeability and changes in permeability in the laboratory and in the field have become among the foremost problems requiring investigation in the field of hydraulic barrier design and installation.

Three previous ASTM symposia relating to the general subject of permeability have been sponsored by Committee D18 on Soil and Rock: *Symposium on Permeability of Soils, ASTM STP 163*, in 1955; *Permeability and Capillarity of Soils, ASTM STP 417*, in 1966; and *Permeability and Ground Water Contaminant Transport, ASTM STP 746*, in 1979. The state of the art undoubtedly has advanced over the 30 years spanned by these symposia. Unfortunately, most studies until recent years have been concerned almost exclusively with the permeability to water, primarily relatively pure ground water, whereas much of the interest today is related to the movement of all kinds of fluids, many of them very hazardous, through the soil and rock or as contaminants to the ground water.

A final consensus on the detailed methods for permeability determinations, either in the laboratory or in the field, may not be in hand, but the problems associated with the methods and with the interactions due to the chemistry of various fluids and the chemistry of the soils and rocks are at least finally recognized. Thus, there is now the start of an empirical basis that can be used by experienced investigators in designing reliable and valid testing programs for the wide range of specific soil or rock and fluid situations that may be encountered. Considerations to be made in the design of laboratory or field testing programs range from the selection of the type of apparatus to the determination of the effects of the permeant on the barrier materials and the soil or rock containment. The possible sources of deviations in test results become greatly compounded for testing using fluids other than water. Interactions between the barrier material and the fluid may cause actual time-dependent changes in permeability as well as affect test conditions or even test equipment, such that apparent changes that have no relationship to real field performance may be observed. Thus, not only must the mechanics of testing be mastered, but the physical and geochemical processes must be understood to such a degree that they can be related to actual field conditions and field performance, so that invalid results can be recognized and discarded.

Another problem exists even if the investigator is aware of all these problems, and that problem relates to quality control of the barrier installation. All of the preliminary testing data can be negated by improper installations that provide different conditions than those assumed in the testing program. There seems to be much that remains to be done to develop adequate field installation quality control standards for hydraulic barriers.

The symposium presented in this special technical publication consisted of both invited and offered papers. In order to accommodate more papers during the one-day symposium, a number of papers were presented as posters with the author available for discussions of the posters during the coffee breaks and part of the lunch hour. Full-length versions of most of the brief poster presentations are published in this special technical publication in addition to most of the papers presented orally. Near the end of each half-day session all authors appearing in that session assembled as a panel to answer questions provided on forms submitted from the audience. After the symposium, all questions were sent to all authors so they had the chance to provide answers even to questions that had not been directed to them. The questions and answers are published in this special technical publication following the pertinent paper or at the end of the publication for those questions of a general nature.

The morning session of the symposium was devoted entirely to presentations on slurry walls and their use as hydraulic barriers. It was pointed out that slurry wall technology apparently was first developed in Italy during the late 1940s as a method for construction of a structural-type wall. During the 1950s in the United States, the technology developed further as nonstructural barriers to impede the flow of water. In more recent years, the growing concern and efforts to mitigate pollution of our ground water is requiring accelerated change and developments in the design, construction, and testing of slurry walls.

The seven papers contained in the first part of this publication present some of the latest innovations and advancements in slurry wall technology from design aspects through construction and testing methodology. The papers discuss soil-bentonite walls, cement-bentonite walls, composite walls of clay and high-density polyethylene sheeting, and grouted walls constructed by use of the vibrating beam technique. One paper also compares the conventional slurry wall construction to the vibrating team technique.

The increasing use of slurry walls as barriers to the movement of pollutants or of contaminated ground water led to questions as to the life of such barriers. The lack of information on the compatibility of the slurry wall materials with many types of permeants has resulted in recent research and standard test method development. Several papers in this special technical publication discuss the effects of organic as well as inorganic permeants on bentonite, describing the testing techniques used and drawing conclusions regarding the effects of the particular chemical permeants.

It is apparent that the technology regarding the use of slurry walls to contain hazardous wastes of all types is in its infancy. The papers presented in this publication represent some of the latest advances in both the construction and the testing for slurry walls. However, further advancement is needed in developing tests of slurry wall materials to determine their long-term performance. As a means of providing improved quality control for slurry walls, ASTM Subcommittee D18.20 has already developed three standard test methods related to the basic properties of a slurry, in particular, density, sand content, and filtration properties of bentonitic slurries.

The afternoon session of the symposium was primarily concerned with clay and soil-admix liners used as hydraulic barriers. Of the fourteen papers presented in the latter part of this publication, four were originally presented in briefer form as posters. The papers presented herein describe a variety of laboratory studies and case histories including the following: direct evaluation of permeability tests utilizing both fixed-wall and flexible-wall permeameters; a proposed field permeability test procedure using a large diameter covered ring; an overview on the construction, testing, and research for soil-cement liners; and the potential use of fly ash in reducing the permeability of otherwise permeable soil liners. Other papers examined factors that affect desiccation cracking of compacted soil and clay liners, the effects of brine on an earth lining, and various other case histories on permeant effects on clay linings. Most papers, however, related to the laboratory studies on the effect of a variety of fluids on bentonitic or soil-admix liner materials. This was not by accident but rather due to a concerted effort by the program committee to bring such studies to light. These papers discussed the effects of such permeants as hydrochloric acid and sodium hydroxide, acid liquor from phosphogypsum disposal, organic leachates, paper mill wastes, brines, and acidic uranium mill tailing liquids on liner materials of compacted natural soils and soil-bentonite mixtures.

Some general scientific conclusions, as well as some philosophical postulations, can be developed from the study of the clay and soil-admix papers published in this publication. The overall objective for permeability determinations is to allow valid predictions of any leakage that may take place through a liner and in the case of waste products of the contaminant transport rates and concentrations with a degree of accuracy acceptable for the operating conditions of a facility. It is, therefore, not sufficient to hypothetically account for perceived differences between tests and full-scale field conditions, but instead there is a need for the actual performance of completed linings to be verified whenever possible. Case studies of constructed linings and possible methods to evaluate field performance are presented in this special technical publication. Such case studies, supported by reliable measurements, must be encouraged to provide support for continued studies of such hydraulic barriers. Also, as brought out during discussions, differences in performance between laboratory-prepared samples and lining materials constructed in the field should be studied further, in particular of linings to be permeated by different fluids other than water. Differences in permeabilities as tested in the laboratory and as tested in the field or as represented by performance in the field may not be due to the test apparatus but rather to the different material placement methods used in the field and in the laboratory.

The testing of lining materials undoubtedly presents a challenge to testing laboratories and the geotechnical profession. The quality of the results is established by the performance of the completed lining. If the permeability remains below the prescribed criteria—without remedial work—then the installation and its supporting testing and engineering design have been successful. In support of this effort, a number of subcommittees and task groups of ASTM Committee D18 on Soil and Rock are actively developing standards for the permeability testing of fine-grained soils.

Despite the recognized difficulties, the investigations presented in this publication demonstrate that clay and soil-admix materials can perform well and maintain low permeabilities when subjected to a variety of permeants. In some cases, additional benefits have been observed due to soil mineralogy, natural attenuation, or buffering by the soil materials. However, there is also evidence of incompatibility between the soils intended for use as a lining and the particular liquid to be impounded. In such cases, alternate lining methods should be investigated irregardless of the measured short-term permeability. There seem to be no valid reasons why linings of clay and soil-admix liners should not be considered from the outset on an equal basis with other lining types in the investigation for a particular application. The techniques and results described in this publication should contribute to successful investigations of lining materials when the interpretations and applications are based on the specifics of each situation.

Many unanswered questions still remain, but this seems inevitable at this relatively early stage of development and investigation for the design and installation of hydraulic barriers, especially when used for fluids other than water. However, the editors believe that the symposium met many of the original objectives of Subcommittee D18.20, namely, to develop information usable in developing useful standards, the promotion of needed research, and the dissemination of information and research results among those people who are involved in the design and installation of hydraulic barriers.

Contributions made by the authors, audience, and technical reviewers are gratefully acknowledged. The editors also thank the ASTM staff and officers of Committee D18 for their assistance and support in organizing and publishing the results of the symposium.

A. Ivan Johnson

Woodward-Clyde Consultants, Denver, CO 80111; symposium cochairman and coeditor.

Ronald K. Frobel

U.S. Bureau of Reclamation, Denver, CO 80225; symposium cochairman and coeditor.

Nicholas J. Cavalli

ICOS Corporation of America, New York, NY 10019; session cochairman and coeditor.

C. Bernt Pettersson

Brown and Root, Inc., Houston, TX 77001; session cochairman and coeditor.

Slurry Walls

Christopher R. Ryan

Slurry Cutoff Walls: Applications in the Control of Hazardous Wastes

REFERENCE: Ryan, C. R., "**Slurry Cutoff Walls: Applications in the Control of Hazardous Waste,**" *Hydraulic Barriers in Soil and Rock, ASTM STP 874*, A. I. Johnson, R. K. Frobel, N. J. Cavalli, C. B. Pettersson, Eds., American Society for Testing and Materials, Philadelphia, 1985, pp. 9–23.

ABSTRACT: Slurry cutoff walls are nonstructural barriers constructed to intercept and impede the flow of fluids underground. There are two basic types of slurry cutoff walls, soil-bentonite (SB) and cement-bentonite (CB). Depending on the nature of the project, either method may have some technical or economic advantage over the other. In both cases, a narrow trench is excavated into the ground using a backhoe or other more specialized equipment. The trench is prevented from collapsing by keeping it full at all times with bentonite slurry similar to drilling mud. In the case of SB walls, the trench is subsequently backfilled with a mixture of soil and bentonite slurry that forms the permanent impervious cutoff wall. With the CB method, cement is added to the slurry, which later sets up, forming the permanent seepage barrier.

Slurry cutoff walls are being used in an increasing variety of applications to provide a barrier to the lateral underground flow of various fluids. Principal applications are site dewatering, underground pollution control, and seepage barriers under dams. In this paper, case studies are used to provide examples of recent applications in the control of leaching hazardous wastes.

Projects cited include:
1. Containment of oil seeping through a reservoir abutment.
2. Cleanup of a polychlorinated biphenyl (PCB) contaminated site.
3. Containment of leachates and methane gas migration from a sanitary landfill site.
4. Cleanup of a site with spilled phenols.

All of the examples were selected because of the unusual conditions under which they were constructed or because of the dramatic evidence of results.

KEY WORDS: slurry cutoff walls, bentonite, groundwater barriers, seepage barriers

Several case studies have been selected to show the range of potential slurry cutoff wall applications, construction methods, and typical conditions under which they are installed. Tremendous progress has been made in the last ten years or so in understanding the mechanics and chemistry of flow through

[1]President, Geo-Con, Inc., Pittsburgh, PA 15235.

9

slurry walls. The increase in knowledge has corresponded, from a timing standpoint, with two other phenomena:

1. A tremendous increase in the number of slurry-wall applications to underground pollution control.
2. An increasing public and corporate awareness of the dangers of the effects of buried hazardous wastes.

The federal Environmental Protection Agency (EPA) has been charged with containing the wastes on hundreds of unclaimed sites under the Superfund Act. Meanwhile, thousands of businesses, local governments, and other owners are seeking remedies for underground pollution problems on their property. The slurry cutoff wall has been already applied to several hundred such sites and is contemplated for many more.

Recent advances in the capacity of excavating equipment and refinements in technique have brought the cost of slurry walls down, and they now compete economically on projects where leachate collectors, clay barriers, or sheeting would have previously been used. The types of walls discussed herein are nonstructural; they are relatively impervious but are not capable of supporting bending moments or significant shear stress. Normally, their strength is of the same order as the surrounding ground.

The technique involves excavating a trench which is kept filled with slurry, whose primary ingredients are bentonite clay and water, and whose function is to maintain the trench open with vertical sides, even below the water table. The excavation is carried out through the slurry from the ground surface using any equipment capable of excavating the trench widths and depths required. Once the trench is excavated to its final depth, a mixture of soil and bentonite is placed in the trench, displacing the bentonite slurry. This type of construction is called an SB slurry cutoff wall.

With a variation on the above technique, called a CB slurry trench, cement is added to the bentonite slurry just before it is introduced into the trench. The resultant slurry has properties substantially similar to normal bentonite slurry with respect to maintaining the sides of the trench. However, once excavation is complete, the CB slurry remains in the trench and sets up and forms the permanent watertight wall.

Construction Methods

Types of Cutoff Walls

The SB slurry trench technique has been in use in the United States for almost 40 years. Figure 1 shows the excavation for a SB cutoff. On projects where the material excavated from the trench is suitable for use as backfill, the SB system can be economical because of the minimum amount of materials required. After the trench has been excavated under a bentonite slurry,

FIG. 1—*Excavation for an SB cutoff.*

more slurry is mixed with the soil adjacent to the trench. A bulldozer is used to work the material to a smooth consistency, and it is then pushed into the trench so that the backfill slope displaces the bentonite slurry forward (Fig. 2). Excavation and backfilling are phased to make the operation continuous with relatively small quantities of new slurry required to keep the trench full and to mix backfill.

CB slurry trenches have been in use in Europe for at least 15 years and in the United States for about 10 years. Figure 3 shows a CB batch plant. Since the entire trench must be filled with slurry materials and since a significant

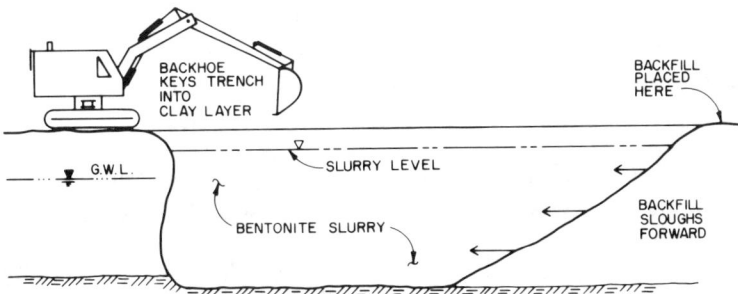

FIG. 2—*Schematic section through an SB cutoff.*

FIG. 3—*CB slurry mixing plant.*

amount of slurry is wasted due to the excavation process and seepage losses through the sides of the trench, the backfill is considerably more expensive than under the SB method. This increased cost is partially offset by the elimination of the backfill mixing operation. However, the CB method can provide the following technical and construction advantages over the SB method:

1. The technique is not dependent on the availability or the quality of soil for backfill.

2. The CB system is more suitable for trenching through areas with difficult access or with not enough room for backfill mixing.

3. Since the trench can be constructed in sections with later sections keyed in by reexcavating a short section, the construction sequence is more flexible to meet site constraints. The long slope of the backfill under the SB system normally requires trenching continuously in one direction.

The SB technique has several advantages over CB, besides lower cost:

1. The resultant wall is generally of lower permeability than CB walls.

2. The backfill can have various materials blended in to suit design conditions.

3. SB backfill is generally more resistant to degradation by most pollutants.

4. Where the excavated material can be mixed as backfill and placed back into the trench, no spoil disposal problem is created.

Given the relative advantages of the two systems, the project requirements should be evaluated to determine the best method to be selected. Where possible, it may be most economical to specify both methods and to allow the contractor to bid with the least expensive system.

Excavating Equipment

The primary requirement for the excavating equipment is the capability to excavate a trench of the design width to the required depths within permissible verticality tolerances. A variety of equipment has in fact been used. In the following paragraphs, the principal types are discussed, along with their relative advantages.

The hydraulic excavator, or backhoe, has been used on many slurry cutoff wall projects in the United States (Fig. 4). The depth limitation of the largest hoes is presently about 20 m (65 ft), but new advances in equipment technology will undoubtedly extend this range. The backhoe, because of its fast cycle time, is the most economical means of excavation. Minimum trench widths are controlled by the thickness of the boom. For large hoes, this can mean 0.9 m (2.5 ft) or more. The thickness of the wall is an important cost factor for CB slurry cutoffs.

The clamshell bucket rigs that were originally developed for cast-in-place concrete slurry walls have been applied to slurry cutoff trenching. These

FIG. 4—*Backhoe excavation.*

buckets may be cable-mounted or attached to a rigid sliding kelly bar (Fig. 5). They may be powered by mechanical means (cables) or by hydraulic cylinders operated by a remote power supply. These rigs have a maximum range up to 76 m (250 ft) and can be used with buckets as thin as 0.6 m (2 ft). Their production is much lower than other methods, so unit costs for excavation are higher.

Another technique more recently introduced into the United States from Europe involves driving a beam into the ground with a vibrating pile-hammer while simultaneously jetting with CB slurry to form a "thin-wall cutoff." The beam is withdrawn while more slurry is injected under pressure. The beam is driven in overlapping imprints to form a continuous wall. The result is a curtain 50 to 100 mm (2 to 4 in) thick with the additional protection of grouting coarse-grained strata with CB slurry. Given the right soil conditions, production is rapid and the thin-wall cutoff uses far less CB slurry than conventional slurry trenching. However, the same narrow width mandates more careful quality control since each square metre of the wall is subjected to one pass of the beam, which does not mix the slurry as in the case of slurry trenching. The principal problem of the vibrated beam has been assuring continuity between adjacent passes at depth. Its range is generally 9 to 15 m (30 to 50 ft), but even within these depths slight deviations may leave "windows" in the wall. Soil profiles with cobbles or boulders are a particular problem and keying into underlying weathered rock or hardpan may not be possible to the extent feasi-

FIG. 5—*Clamshell excavation.*

ble with excavated slurry trenches. The narrow width of wall makes this type of cutoff less suitable for applications in soil where movements due to settlement, subsidence, etc. can be expected later. Design parameters and quality control for thin-wall cutoffs are specialized topics not treated in this paper.

Applications

Because of the range of slurry cutoff wall applications to the control of hazardous waste leachates, it is difficult to pick a few projects to demonstrate the possibilities. Projects completed include, among many, containments for

1. Sanitary landfill leachates.
2. Oil and gas spills.
3. Low-level radioactive waste.
4. Acid mine drainage.
5. Phenols.
6. Polychlorinated Biphenyl (PCB).
7. Trichlorethylene, benzene, and many other organic chemicals.
8. Phosphate mine tailings.
9. Fly ash impoundments.

To illustrate the range of typical applications, three recent projects have been selected. They are

1. Denver, Colorado—Containment of toxic chemicals from a military reservation. This project illustrates use of an SB wall for remedial action.
2. Long Island, New York—Containment of oily wastes seeping from an oil terminal. This project illustrates the use of a CB wall for remedial action.
3. Tampa, Florida—Underseepage containment for new dike construction for a phosphate mine tailings pond. This project illustrates the use of a slurry wall in the construction of a new facility.

Buried Residues from Military Chemicals

The problems at the Rocky Mountain Arsenal near Denver's Stapleton Airport have been the subject of numerous stories in the national press as well as in various technical publications. Apparently, wartime production of chemicals for gas warfare and other uses resulted in the creation of waste products and out-of-spec chemicals. At the time, the most expedient solution was on-site storage in lagoons and buried dumps. Only recently has the extent of the environmental problems emerged. To date, three contracts have been let to create slurry wall containments at various points along the reservation perimeter, in combination with pumping water for treatment and aquifer recharge.

This project was typical in that it involved a length of about 450 m (1500 ft) to depths of as much as 15 m (50 ft). A rock ledge in the soil profile had to be blasted prior to excavation of the slurry trench. Figure 6 shows the slurry field laboratory. Figure 7 shows the trenching in progress, and Figure 8 shows the trench backfilling operation.

The specifications of this particular job required wasting the material excavated from the trench and using imported material in the trench backfill. In general, it is preferable from an economic and technical standpoint to use as much of the material from the trench as possible. If the material is contaminated with chemicals, the addition of a slight amount of bentonite will counteract the effect of the pollution. A series of permeability tests on potential backfill blends is always advisable.

Oily Wastes

The problem at this Long Island oil terminal is typical of problems at many oil tank farms and other similar facilities. Small spills and leaks over the years had created a substantial pool of oily wastes on the groundwater under the terminal, which threatened to seep laterally into the adjacent harbor waters.

The oil pollution problem is one case in which an aquaclude for the slurry wall to key into is not always necessary. It is frequently possible for the wall to

FIG. 6—*Field laboratory for slurry and backfill tests.*

FIG. 7—*Trenching in progress.*

merely intercept the groundwater and literally skim the oily wastes off. The wall is usually built in combination with a central sump to collect the oil contained. There are at least 20 of these types of installations in the United States, and at least 2 are economically as well as environmentally viable; that is, they have paid their own cost in terms of recovered product.

Many of the oily waste applications have been done with the CB technique because the jobs have generally been smaller and have been constructed in areas congested by surface structures and underground lines that would create access problems for the SB technique. CB slurry also has shown none of the detrimental effects when exposed to oily wastes that sometimes occur with other pollutants.

The Long Island case was typical. Figures 9 and 10 show the difficult access and lack of area alongside the trench for backfill mixing. The CB slurry was

FIG. 8—*Backfilling with SB backfill.*

FIG. 9—*CB work in area of tight access.*

FIG. 10—*CB work in area of tight access.*

prepared in an electronically controlled central plant and pumped to the trench. The trenching was carried out much the same as for the SB method (Fig. 11). Numerous underground utilities and other obstructions were successfully passed. After about one day the CB takes its initial set (Fig. 12). Even after full set, the slurry is always soft like a clayey soil, so that it will not impede future underground work at the terminal.

Phosphate Tailings Leachates

Central Florida, in a belt stretching from Tampa to Orlando, is a center for the mining of phosphates. As a part of the process, acidic wastes are created and stored in "gypsum stacks." These impoundments are constructed from gypsum and may be over 30 m (100 ft) high and 500,000 m^2 or more in area. A number of them have experienced dramatic failures due to piping of the dike foundations. Underseepage also creates a nasty environmental problem.

This particular case is the combination of two recent projects in the area. Each job involved the expansion of a gypsum stack into a new area. Both owners wanted to eliminate underseepage for environmental reasons and to ensure the stability of their dikes. Total lineal footage was several kilometers and depths required from 6 to 15 m (20 to 50 ft). Numerous hard ledges of gypsum deposits were penetrated.

FIG. 11—*Trenching under CB slurry.*

FIG. 12—*Setup CB slurry.*

There is one important construction detail that should not be overlooked: the key between the top of the slurry cutoff wall and the overlying dike. To avoid a zone of weakness between the top of the wall and the base of the dam, it is essential to place a cap and to excavate the trench through it. This creates a side contact seal for the slurry cutoff wall and provides a wide area to key the bottom of the dam into. In both cases, gypsum caps were used, although clay would serve as well for a more normal application.

In all other respects, the project was a typical SB slurry cutoff wall. Figures 13, 14, and 15 show slurry mixing, trenching with the extended stick backhoe, and backfilling the trench.

Summary

Slurry cutoff walls have achieved wide recognition in a variety of applications as seepage barriers for pollution control. The two principal techniques, SB and CB, have different relative advantages, but under some conditions are technically interchangeable.

A design for a slurry cutoff wall should take into consideration whether the wall is for permanent or temporary use, the loadings anticipated, and other construction constraints in selecting the technique to be used and the extent to which the work should be controlled by the engineer. Specifications should take account of the built-in safety factors in slurry cutoffs (for example, more

FIG. 13—*Slurry mixing.*

FIG. 14—*Trenching with extended stick.*

FIG. 15—*Backfilling.*

thickness than required in most cases), allow the variability in slurry properties normally experienced during this type of work, and give maximum flexibility to the contractor in selecting materials, equipment, and technique. The
economy, convenience, and positive control of seepage afforded by slurry cutoff walls will bring them acceptance and application on an increasing number
of construction projects in the United States.

DISCUSSION

Y. B. Acar[1] *(written discussion)*—The author proposes to use slurry walls
as a barrier to hazardous wastes. Slurry walls are usually constructed using
high activity clays. Would not such clays be more susceptible to structural
changes due to variations in pore-fluid chemistry? In other words, does the
author use an activity criteria in constructing such walls?

C. R. Ryan (author's closure)—A specific activity criterion is not used in
evaluating potential slurry wall applications. The effects are indirectly measured in the normal compatibility testing that is done on projects with unusual leachates. We have yet to find a leachate whose effect on the SB backfill
cannot be counteracted by relatively minor changes in the constituents. More
bentonite or more fines may be added to accomplish this purpose.

S. B. Ahmed[2] *(written discussion)*—How would you predict long-range
permeability in the field?

C. R. Ryan (author's closure)—The best way to estimate long-term permeability is to perform prejob permeability tests with actual site soils, bentonite,
and leachates. These tests, which can be done in a variety of cells, can usually
be run to stability in a matter of one to two weeks.

S. B. Ahmed (written discussion)—How would you estimate the quantity of
bentonite required say for a slurry cutoff wall in sandy silt with $k = 10^{-4}$
cm/s. Assume the wall is 18 m (60 ft) deep and is keyed into dense stiff impermeable clay?

C. R. Ryan (author's closure)—Between 1 to 3% total bentonite quantity
(by dry weight) should be sufficient in most cases to produce a permeability of
1×10^{-7} cm/s. In all cases where the permeability is critical, prejob permeability testing must be done.

[1]Assistant professor, Department of Civil Engineering, Louisiana State University, Baton
Rouge, LA 70803.
[2]Tennessee Valley Authority, Knoxville, TN 37923.

Gregory W. Druback[1] and Salvatore V. Arlotta, Jr.[1]

Subsurface Pollution Containment Using a Composite System Vertical Cutoff Barrier

REFERENCE: Druback, G. W. and Arlotta, S. V., Jr., **"Subsurface Pollution Containment Using a Composite System Vertical Cutoff Barrier,"** *Hydraulic Barriers in Soil and Rock, ASTM STP 874*, A. I. Johnson, R. K. Frobel, N. J. Cavalli, and C. B. Pettersson, Eds., American Society for Testing and Materials, Philadelphia, 1985, pp. 24–33.

ABSTRACT: A new concept for constructing a vertical impermeable barrier to prevent the migration of polluted groundwater or leachate from a contaminated site or waste disposal area is presented. The composite system is a hybrid cutoff wall constructed with high density polyethylene (HDPE) and sand backfill and is installed using the slurry trench construction method. When installed, a very low permeability, composite, vertical barrier is established with unique engineering properties, including improved chemical resistance, leak detection, and groundwater migration control.

A full-scale construction test project of the system was performed at an existing sanitary landfill in New Jersey in the fall of 1982 to demonstrate the overall fabrication and construction procedures. The test project was conducted in three phases: (1) the excavation of a trench filled with the bentonite slurry mixture through the landfill waste; (2) the cutting and seaming of HDPE material to form an envelope; and (3) the placement of the envelope and subsequent backfilling with sand and water. The success of the operation under often difficult conditions proved that the theory and construction concept of the composite system barrier can be utilized to construct a cutoff wall. Useful information for future applications at contaminated sites was gained.

KEY WORDS: pollution migration, groundwater, vertical barrier, composite system, high density polyethylene membrane, construction technique, slurry trench, monitoring, pumping

The migration of polluted groundwater or leachate from contaminated sites or waste disposal areas, especially abandoned or "orphan" sites, is an environmental problem of national proportions. A new and unique composite construction system has been developed to construct a vertical barrier to prevent this migration.

[1]Senior engineer and vice-president, respectively, Wehran Engineering Corp., Middletown, NY 10940.

The system is a hybrid cutoff wall. It is constructed with high density polyethylene (HDPE) and sand backfill and is installed by using the slurry trench construction method. Once installed, a very low permeability composite barrier is established with several unique engineering properties.

The overall concept and construction methodology was demonstrated during a full-scale construction test at a sanitary landfill in New Brunswick, New Jersey during the fall of 1982. The test was planned to demonstrate the feasibility of the techniques and materials used to construct this composite system under generally difficult (by design) conditions. The test also demonstrated the major advantages of the system over other types of cutoff wall construction.

The composite system essentially functions as a cutoff wall with redundant features for controlling horizontal seepage. A 2.54 mm (100-mil) HDPE sheet is used to form an envelope which lines the walls of an excavated trench. The bentonite slurry used during construction keeps the trench stable for placement of the HDPE sheet and sand backfill and also forms a filter cake on the walls of the trench. This filter cake further decreases the permeability of the composite in-place system. The sand backfill, besides serving as ballast, provides an internal, porous medium for monitoring seepage and, if necessary, for withdrawal of intruding pollutants.

Applications of a Vertical Cutoff Barrier

The prevention of groundwater contamination by the seepage of pollutants is often the major focus of the engineering effort for remedial measures at a chemically contaminated site. Many variables enter into the picture. Some of these variables are (1) quality and direction of the groundwater flow, (2) characteristics of the waste, (3) hydrogeological characteristics of the site and adjacent areas, and (4) human health risks involved.

When the site is a landfill, the prevention of subsurface migration of leachate from the site is usually the goal in order to prevent environmental degradation. A thorough study of the geology, the groundwater regime, and the soil conditions is required to select the best method of preventing subsurface seepage of pollutants or migration of leachate. If leachate enters the groundwater, the now polluted groundwater will migrate, following the hydraulic gradient of the groundwater table. To prevent leachate from moving laterally via the groundwater table, an impermeable barrier or cutoff wall must be constructed perpendicular to the hydraulic gradient of the groundwater. Depending on site stratigraphy, cutoff walls are often keyed into an underlying soil stratum of low permeability to effectively seal off the subsurface horizontal migration of contaminants. This is often done at landfill sites or hazardous waste disposal sites as a remedial measure.

Typically, a certain percentage of precipitation which falls on a landfill eventually ends up percolating through the waste, which is often buried, to form leachate. Any contaminants which are picked up in the leachate through contact with waste will be transported with the leachate.

Some landfill sites contain hazardous, toxic compounds that can be leached out of the waste and that will eventually contaminate the groundwater. Two types of leachate are involved: (1) flowable constituents of the waste; and (2) the flowables generated from infiltration or percolation just mentioned. These leachates are usually referred to as primary and secondary leachate, respectively. The majority of industrial hazardous wastes are produced as liquids and, therefore, can easily become a primary leachate. Both the solute and solvent phases of leachate can affect the permeability of the natural ground, liners, or cutoffs. Wastes can be classified either by source (that is, municipal refuse, industrial and chemical process waste, dredge spoil, etc.) or by physical class (that is, organic, aqueous-inorganic, aqueous-organic, etc.). Physical classification is more useful when assessing likely effects on liners or cutoffs. Organic fluids associated with industrial wastes that are often placed in a disposal facility cover a spectrum of types.

Recent research has indicated possible concerns for soil structures, such as clay dikes, cutoffs, and liners, which are used for containment of wastes. Failure mechanisms fall into four major categories: (1) dissolution, (2) syneresis, (3) piping, and (4) desiccation. It is important to evaluate the materials for containment in relation to the wastes. Generally, strong acids, strong bases, salts, and highly ionic wastes are aggressive toward soils.

Construction, Components, and Operation of the Composite System Vertical Cutoff

The composite system has several advantageous features when used as a vertical barrier to form a cutoff wall. The system, when used for environmental control, combines some of the ideas of clay and of a soil bentonite cutoff wall and adds the capability for monitoring performance and withdrawal of contaminants from within the cutoff wall. The HDPE envelope provides a stable, low permeability (10^{-12} cm/s) membrane and, when installed in the trench, provides improved resistance to permeation and low overall permeability for the system.

The slurry trench construction method allows installation of the composite system in areas with difficult site conditions and to relatively deep levels.

The bentonite slurry used in the construction maintains the stability of the trench during the installation of the HDPE envelope. The slurry also develops a filter cake on the walls of the trench, which provides an additional low permeability bentonite and soil layer.

The composite system is constructed in a similar manner to the soil bentonite cutoff wall. A narrow trench is excavated and filled with bentonite slurry to stabilize the trench walls during construction. The HDPE envelope is fabricated above ground to fit the dimensions of the trench. Sheets are cut

and seamed to form a U-shaped envelope and set with ballast weight to provide an initial submergence into the trench. The envelope is lifted and set in
the trench. Water and sand are pumped into the envelope to develop negative
buoyancy and to form the HDPE to the shape of the trench (Fig. 1).

The sand also provides a medium for placing a system of monitoring or
withdrawal wells within the cutoff wall envelope (Fig. 2). Sand backfill is an
important component of the composite system. Water quality within the wall
can be monitored through the sand on a long-term basis. If leakage is ever
noted by detection of contamination in a monitoring well, the porous sand
medium can be pumped to evacuate and control the contaminated water and
prevent further migration beyond the cutoff.

The sand is graded to provide a high permeability medium within the envelope, ideally with a higher permeability than the surrounding soil in which the
cutoff wall is excavated. Small diameter piezometers or monitoring wells are
placed in the sand backfill. Piezometric head can be routinely observed and
water quality samples extracted for testing.

When piezometers are installed within the homogeneous sand medium
contained in the envelope, continuous monitoring of piezometric head would
indicate the condition of the liner. This would require that the water level
within the envelope be adjusted initially to an elevation below that of the site.
In this way, a net positive head into the composite system is established (see
Fig. 2).

If leakage of the HDPE membrane was detected, the gradient would be
into the envelope and a rise in piezometric head would occur, providing a
simple indicator of membrane integrity.

Water quality monitoring could be also undertaken to determine the type
and characteristics of any substances entering the envelope. If the envelope

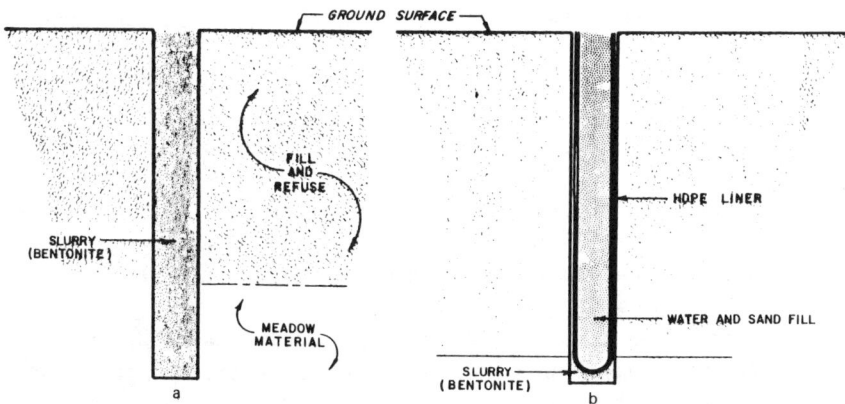

FIG. 1—(a) Excavated trench before placement of the liner; (b) Cross section of the trench
with liner.

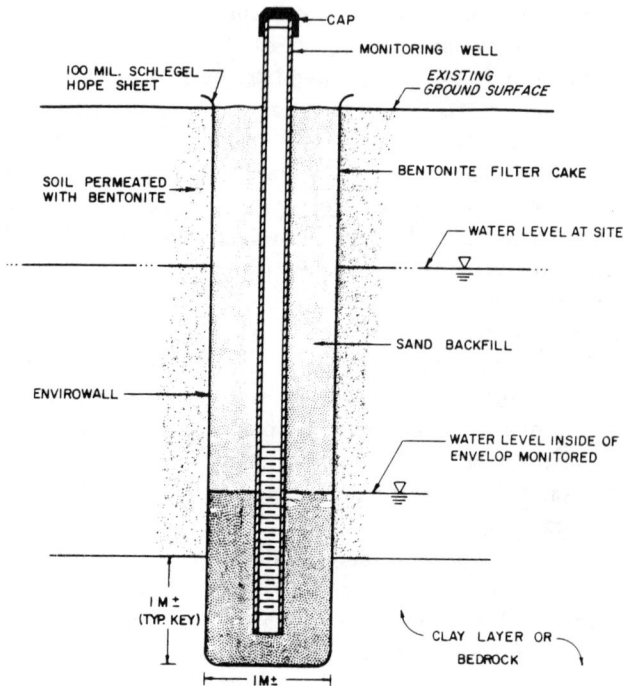

FIG. 2—*Piezometer in sand (not to scale).*

were damaged such that a high rate of leakage occurred, the high permeability sand medium could be also pumped. A series of withdrawal wells or well points, likewise within the sand, could be used to extract any contaminated liquid entering and to maintain a net gradient into the liner. This provides an additional safety feature beyond the initial establishment of a low permeability barrier (Fig. 3).

In the case where the waste is very aggressive toward the bentonite filter cake and HDPE so that the integrity of the membrane is reduced considerably, the sand medium with withdrawal wells or well points simulates a control structure for dewatering. This is especially beneficial where existing soil conditions at the site are not conducive for establishing an efficient pumping system for withdrawal. Collected effluent would then be discharged for treatment. The effectiveness will depend on the spacing of the wells within the sand.

Where conditions are such that the sand and withdrawal wells are being used to control seepage from a site, the composite system will approximate a fully penetrating slot drain or line sink, and a solution for drawdown can be obtained using equations based on this type of drainage concept.

Assuming equilibrium conditions under unconfined flow from a line source

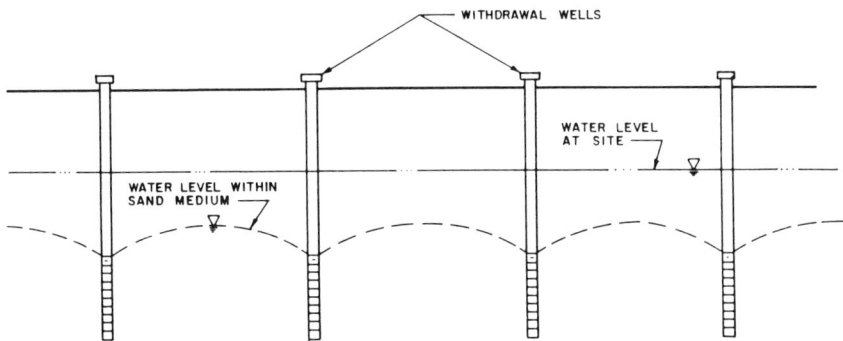

FIG. 3—*Withdrawal well operation (not to scale).*

and with a vertical slot penetrating a homogeneous isotropic stratum and bounded at the base by a horizontal impervious stratum, and further assuming that the Dupuit-Forchheimer conditions apply to the hydraulic gradient, then the flow through a vertical element to the sink can be described by Darcy's law

$$Q = kiA \qquad (1)$$

where

$i = dh/dy$,
$A = hx$, and
$Q = khx\, dh/dy$.

This describes the basic relationship for drawdown and flow. This flow (Q) would be equal to the withdrawal rate within the cutoff (Fig. 4).

Test Project

A full-scale construction test project of the composite system was performed at an existing sanitary landfill in New Jersey in the fall of 1982 to demonstrate the overall construction procedure for fabricating the HDPE envelope and placing it into a trench filled with slurry by backfilling with sand. The project was conducted in three phases:

1. The excavation of a triangular-shaped trench filled with the bentonite slurry mixture through the landfill waste (Fig. 5).
2. The cutting and seaming of HDPE material to form an envelope with the prescribed geometry.
3. The placement of this HDPE envelope in the trench and subsequent backfilling with sand and water.

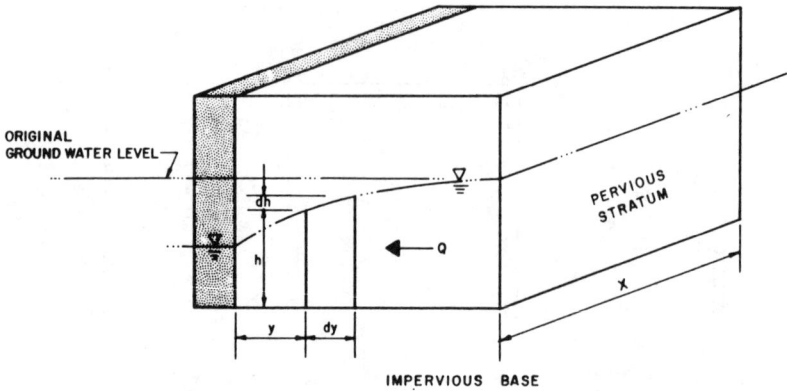

FIG. 4—*Idealized flow from a line source to a fully penetrating slot.*

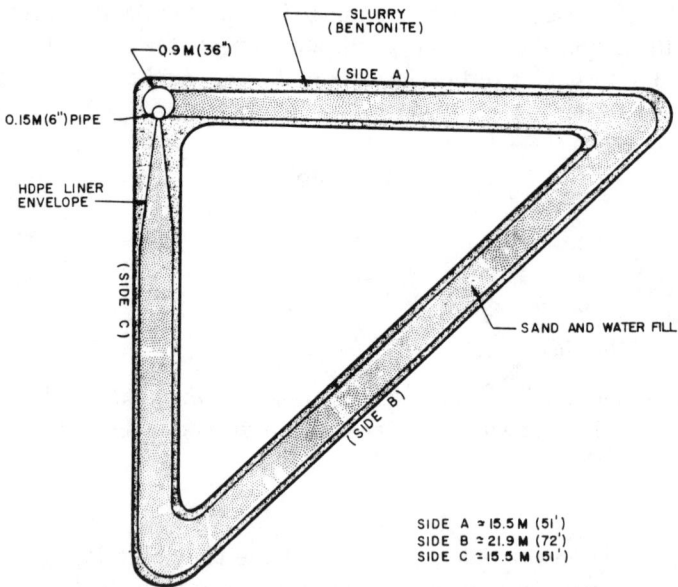

FIG. 5—*Plan view of the HDPE liner in place (not to scale).*

The HDPE envelope was fabricated by Schlegel Lining Technology, Inc. and later submerged by ICOS Corporation of America into a trench excavated to a triangular-shaped plan. This full-scale test introduced several potential advantages of the composite system over other types of cutoff walls and demonstrated the techniques and materials which could be used for construction under realistic conditions.

Indeed, the test site was an inactive portion of a working landfill. The trench was excavated approximately 6.7 m (22 ft) passing through approximately 3 m (10 ft) of previously deposited solid waste and extending down into the underlying foundation of meadow mat (organic peat) which generally occupied the landfill site.

The trench was excavated by a backhoe in triangular plan, creating sharp corners to check the workability of the 2.54 mm (100-mil) HDPE membrane to accommodate very angular corners. Each side of the trench was excavated about 15.5 m (51 ft) long. The triangular configuration used in the test project presented an example of extreme handling conditions and provided a good idea of what problems are presented in this new type of construction. The entire envelope was prefabricated from 10-m (33-ft)-wide rolls of HDPE above ground to conform to the triangular-shaped plan.

Various shapes and sizes were cut and seamed for the one-piece envelope using the Schlegel extrusion welder. Two corners of the envelope were preconstructed using a corner jig. The third corner of the triangle was a closure system which consisted of two interlocking HDPE pipes. Experience indicated that it was beneficial to perform the fabrication as close to the trench as possible to minimize excessive movement of the liner. This reduced the potential for abrasion and handling damage.

The trench was filled with sodium-montmorillonite-bentonite slurry having a viscosity of approximately 40 s by Marsh funnel. The slurry stabilized the walls of the trench until the prefabricated HDPE envelope was ready to be placed.

A 0.9-m (36-in.)-diameter HDPE pipe and a 0.15-m (6-in.)-diameter HDPE pipe were welded on each of the two outer edges of the envelope. These two pipes were used to form an interlocking joint at the third corner of the triangle. The 0.9-m (36-in.)-diameter pipe was slotted as the envelope was submerged into the trench on the last triangular leg. This pipe joint would be bonded later by pressure grouting.

A crane was used to move the 2.54 mm (100-mil) HDPE into position for seaming and fabrication. Continuous support was provided for the envelope with lifting bars and battens to keep it from bending and wrinkling.

The entire fabrication process required several days for this first time installation. As fabrication of the envelope neared completion and was made ready for installation, the trench was checked and cleaned of any debris which might have inhibited the installation. A clam shell was used to remove any objects that might be protruding from the walls or collecting at the bottom of the trench. Cranes were again used for the lift of the liner into the trench.

Using the spreader bars to lift each envelope side and keep them straight, the cranes worked together to set the membrane envelope in place above the trench. The side with the 0.9-m (36-in.) pipe attached was lowered into the trench first; water was pumped into the envelope so that this initial side was lowered sufficiently to allow a front-end loader to begin filling with sand. An overflow trench was prepared in one corner of the trench to handle the slurry that was displaced as the envelope was submerged. The pipe joint was then made by sleeving the 0.15-m (6-in.)-diameter pipe into the 0.9-m (36-in.)-diameter pipe, making the envelope continuous.

Sand continued to be dumped into the envelope at a rapid pace until it was completely submerged. The membrane expanded well to conform to the shape of the trench. The test program was terminated at this point. However, at a remedial site, monitoring wells and piezometers would be installed and the sleeved pipe joint would be pressure grouted to complete the installation. Also, the excess membrane above grade would be welded to completely enclose the sand medium within the envelope, thereby protecting it from the weather or accidental contamination from the just-mentioned grade. If a site is scheduled to be capped with an HDPE membrane, the butting outside edges of the capping membrane and the top of the envelope can be jointed by welding to complete the enclosure of the site.

Conclusion

The test project made the composite-system vertical-barrier concept a reality and demonstrated the feasibility of installing a composite cutoff wall at a landfill site. It appears from the test project that larger, longer installations will be easier to install and will have cheaper unit costs than the test project.

DISCUSSION

K. Ferguson[1] *(written discussion)*—How do you prevent piping of contaminated water around trench HDPE interface?

G. W. Druback and S. V. Arlotta (authors' closure)—The trench HDPE interface should not be a location for piping because the pressure of the backfill will force the liner against the sidewall and bottom of the trench. At this interface the presence of the bentonite filter cake and the trench sidewalls permeated with bentonite slurry would prevent piping from occurring. Also, where localized sidewall excavation surfaces may vary to an extent that the 2.54-mm (100-mil) HDPE cannot follow exactly, these small areas will be filled with undisplaced bentonite slurry that would prevent further piping.

[1]Project engineer, Steffen Robertson & Kirsten Colorado, Inc., Lakewood, CO.

S. Hansen[2] *(written discussion)*—Given the greater density of the slurry, how do you seat the plastic membrane against the wall with a less dense backfill material?

G. W. Druback and S. V. Arlotta (authors' closure)—The bentonite slurry mixture utilized to maintain a stable vertical sidewall in the trench has a density slightly greater than water or about 1025 kg/m^3 (64 lb/ft^3). The sand backfill would have a density in the 1441 to 1762 kg/m^3 (90 to 110 lb/ft^3) range. The difference in density is enough to displace the slurry and cause sufficient pressure to seat the membrane against the bentonite filter cake and the sidewall permeated with bentonite slurry.

[2]U.S. Bureau of Reclamation, Durango, CO.

G. A. Leonards,[1] *Fred Schmednecht,*[2] *Jean-Lou Chameau,*[1] *and Sidney Diamond*[1]

Thin Slurry Cutoff Walls Installed by the Vibrated Beam Method

REFERENCE: Leonards, G. A., Schmednecht, F., Chameau, J.-L., and Diamond, S., **"Thin Slurry Cutoff Walls Installed by the Vibrated Beam Method,"** *Hydraulic Barriers in Soil and Rock, ASTM STP 874,* A. I. Johnson, R. K. Frobel, N. J. Cavalli, and C. B. Pettersson, Eds., American Society for Testing and Materials, Philadelphia, 1985, pp. 34–44.

ABSTRACT: In recent years, thin slurry walls installed by vibrating a beam into the ground and injecting a suitable slurry as it is withdrawn have been used successfully and economically in a variety of applications, most notably as (*a*) permanent seepage cutoffs either in new or existing dams or levees, (*b*) a means of temporarily controlling groundwater during construction, and (*c*) a device for containment of toxic wastes.

This paper describes the equipment and construction techniques for installing the cutoff, and details of various types of slurries suitable for different applications are given. The results of field investigations undertaken to assess the permeability and continuity of vertical and battered (inclined) slurry walls are presented.

KEY WORDS: permeability, slurry walls, slurry, vibrated beam

The vibrating beam technique to install thin slurry walls was introduced in Europe approximately 25 years ago. It was a major improvement over the slow and expensive system of installing a grout or slurry curtain wall by welding grout pipes to individual interlocking sheet piles, conventionally driving the piles into the ground, and then grouting as they were extracted.

The vibratory driver-extractor was developed and entered into the commercial market in the late 1950s, providing a machine that could both drive and extract a pile without changing from a conventional hammer to an extractor. The vibratory driver-extractor permitted reusing the same beam element and overlapping with the previous beam insertion to form a continuous wall. A wide-flange beam was used to provide sufficient bending resistance to a wider beam while maintaining a thin web.

[1]Professor, School of Civil Engineering, Purdue University, West Lafayette, IN 47907.
[2]President, Slurry Systems, Inc., Gary, IN 46406.

The vibrating beam technique has been used successfully to install more than 5 000 000 m² (50 million ft²) of thin slurry wall in Europe with no known failures. In the United States, since 1975, nearly 400 000 m² (4 million ft²) have been installed successfully in a variety of applications, primarily (a) as permanent seepage cutoffs either in new or existing dams or levees, (b) for temporary control of groundwater during construction, and (c) as barriers for the containment of toxic wastes.

This paper describes the equipment and construction techniques for installing the slurry wall and discusses the results of investigations undertaken to assess the permeability and continuity of vertical and battered (inclined) slurry walls.

Equipment

Vibratory Driver-Extractor

Recent developments and improvements in the electrical and mechanical systems have made the driver-extractor a much more versatile and reliable machine. Size and power needed depend on required depth of penetration and on resistance of material being penetrated. As an example, in a recent application at the Wheatfield, Indiana power station, a tandem 4-75 PTC vibrator (distributed by the L. B. Foster Co.) was used. Four 55 930 W (75 hp) electric motors drive its eccentric weights to obtain a moment of 784 N·m (6940 in·lb) at a frequency of approximately 800 cpm. The frequency can be increased up to about 1100 cpm by changing the drive sprocket ratios, thereby obtaining a closer match with the natural frequency of the ground being penetrated.

Crane and Leads

The crawler crane mainly hoists the weight of the vibrator, leads, and beam; hence, the primary emphasis in selection of equipment is on line pull, line speed, and low maintenance.

The vibrator guiding leads are similar to those used for tight tolerance pile driving installations with the following exceptions: (1) the bottom of the lead has a vertical hydraulic support jam with a bearing pad to improve stability, and (2) the spotter is adjustable vertically and laterally. Accurate control of the leads to maintain successive beam penetrations in the same plane is essential to ensure the continuity of the wall.

Injection Beam

The injection beam is a standard wide flange (WF) beam section with sufficient rigidity to maintain vertical straightness without bowing under the dead weight and forces of the vibrator. Wear plates are welded to the tip of the beam to control the width of the slurry wall and to provide resistance to abrasion. For the injection of slurry, full length supply pipes with replaceable nozzles are welded to the beam. Figure 1 shows a sketch of the injection beam and the installation procedure.

Slurry Transfer

Positive displacement pumps can be used effectively to pump slurry directly into the injection beam up to a distance of 762 m (2500 ft). Alternately, slurry

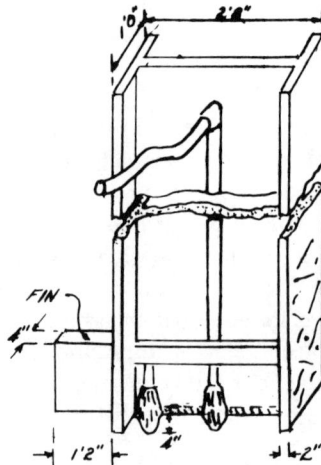

(a) TYPICAL INJECTION CONFIGURATION

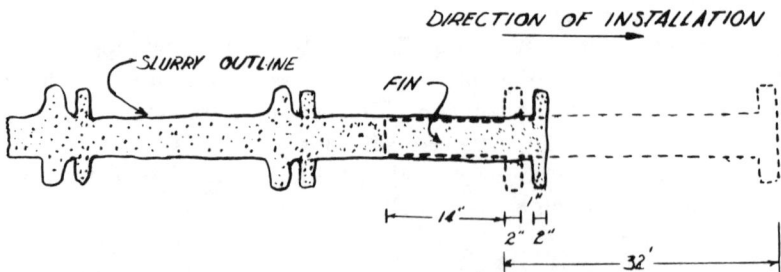

(b) PLAN VIEW OF SLURRY WALL INSTALLATION

FIG. 1—*Installation technique.*

can be pumped from a batch plant to an injection rig storage tank. Pumping directly into the injection beam is generally preferred. However, long pumping distances and extreme cold weather may make it necessary to use the storage tank method.

A sketch of the assembled slurry placement system is shown in Fig. 2.

Slurry Types and Applications

The performance of slurry walls—whether used for permanent seepage cutoff, temporary groundwater control, or waste containment—is controlled by

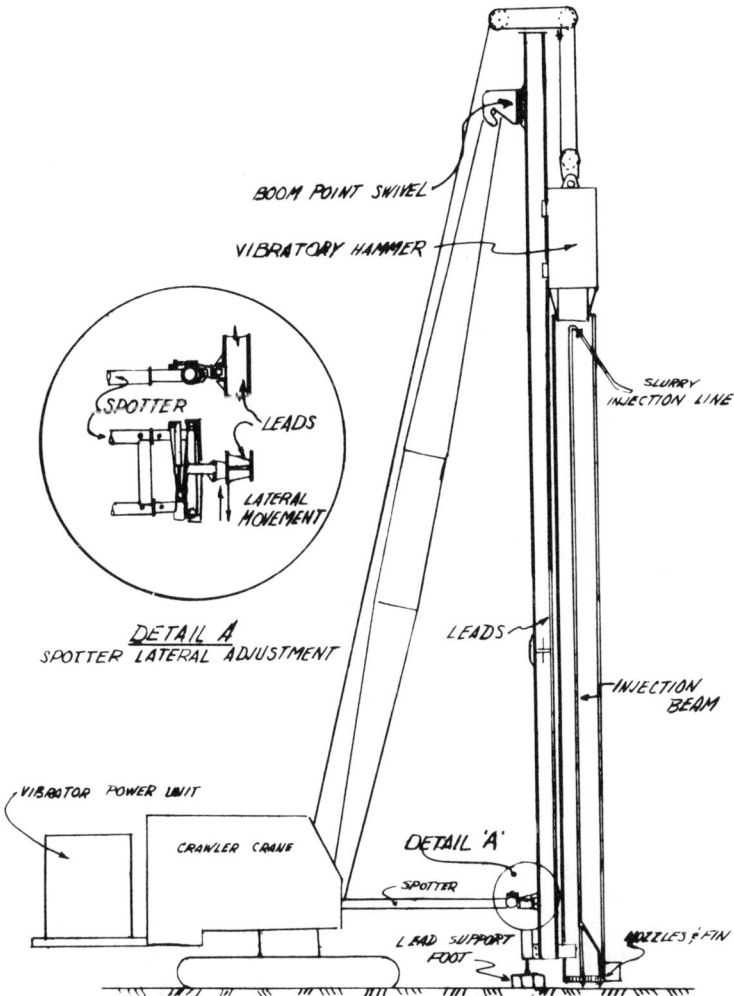

FIG. 2—*Typical vibrated beam injection assembly.*

the permeability of the slurry after it has set up and by the continuity of the wall.

The inherent permeability of the set slurry is a function of the components used, the adequacy of mixing, and the effects of specific chemical admixtures that may be incorporated in the mix.

Two distinct families of slurry materials have been used, based respectively on portland cement-bentonite formulations and on patented formulations involving asphalt emulsions (designated as ASPA mixes). The latter are considerably more expensive on a unit-wall-area basis but are more impermeable and highly resistant to acid and other chemical permeants.

Formulations of portland cement-bentonite slurries usually include sodium carbonate or other dispersant and small percentages of chemical "adducts" that modify the rheological characteristics of the bentonite component. Only high-quality sodium bentonite of high swelling characteristics is suitable for this application. The ratio of cement to bentonite is important: as the cement content increases, the strength and resistance to erosion of the wall increases, but its ability to resist ground displacements without cracking diminishes. Permeability is also adversely affected at higher cement contents. A mix with 5 to 7% bentonite and with a cement/bentonite ratio of approximately 2:1 generally optimizes the requirements for impermeability, flexibility, and resistance to erosion. Lower cement/bentonite ratios are used if impermeability is the primary requirement.

Permeabilities of cement-bentonite slurries vary but are ordinarily expected to be in the range of 0.5 to 3×10^{-6} cm/s. Laboratory-mixed slurries in which oriented boundary layers (cakes) are allowed to form are much less permeable: values on the order of 3×10^{-8} cm/s are usually measured. Thus, cake formation is necessary in installed slurry walls if permeabilities below about 10^{-6} cm/s are required.

Cement-bentonite slurry walls are adversely affected by strongly acid leachates, which tend to dissolve the hydrated cement and flocculate the bentonite. Under such conditions, or where higher degrees of impermeability are required, asphalt-emulsion-based compositions are recommended. Such walls are also resistant to the presence of salts and various other chemical agents. Permeabilities for such systems measured in the laboratory start at about 10^{-8} cm/s, but the slurries rapidly decrease in permeability with loss of fluid and in a few hours approach permeabilities so small they cannot be readily measured (less than 10^{-10} cm/s).

Wall continuity, the equally important requirement for low permeability, is ensured by using a fin welded to the tip of the vibrating beam and by maintaining beam overlap on successive penetrations (Fig. 1b). Past experience in Europe and the United States indicates that the continuity of successive emplacements can be generally insured for vertical walls. Nevertheless, research is underway to monitor successive beam emplacements by remote sensing, both to verify continuity and to assist in controlling the field operations.

Although experience to date is very positive, further information is needed on (1) the "local" permeability of the slurry and (2) the continuity of inclined and intersecting walls. The local permeability is especially important for toxic waste containment. Data on the continuity of inclined walls are needed to assess their suitability as seepage cutoffs under existing canals, when the depth of the impermeable layer renders the use of vertical walls uneconomical.

During the past year, the authors embarked on a field and laboratory testing program with four related goals: (1) to determine the variation in effective permeability of slurry walls installed at sharp inclinations (up to 45°); (2) to check the continuity of vertical and inclined walls, especially at intersections of the wall segments; (3) to determine the variation in thickness of the walls; and (4) to develop expertise in the construction of inclined walls. The results of these efforts follow.

Test Section

A test section with the configuration shown in Fig. 3 was installed by Slurry Systems, Inc. in a silty sand deposit near the Gary, Indiana airport. The test section consists of a trough with battered walls inclined at 45° from the horizontal in the east-west direction and two vertical walls that close the north and south ends of the trough. The walls were installed with a wear plate expected to provide a wall thickness of 100 mm (4 in.). The slurry consisted of 6% bentonite, 12% cement, a small percentage of soda ash, Tamol 850 conditioner, and water. The surface of the trough was covered with Visqueen plastic sheets, which were sealed into a cement-bentonite filled trench surrounding the trough. To minimize the amount of pumping necessary to dewater the sand on the outside of the test section, an outer cutoff wall was installed (Fig. 3). This outer wall penetrated an underlying clay layer at a depth of 11.3 m (37 ft). A dewatering well was installed in the northeast corner of the test area between the trough and the outer wall. Four piezometers were installed to monitor the water levels. Three piezometers (1, 2, and 3) were located inside the trough in the north-south direction. The fourth piezometer was located outside the trough in the southwest corner of the test area.

Pumping Tests

The sand between the trough and the outer wall was dewatered rapidly to a depth well below the apex of the trough, and the water levels in Piezometers 1, 2, and 3 were recorded every 24 h during the following ten days. The results given by the three piezometers are consistent (Table 1) and indicate a fairly uniform drop in water level over the area of the trough. The changes in water

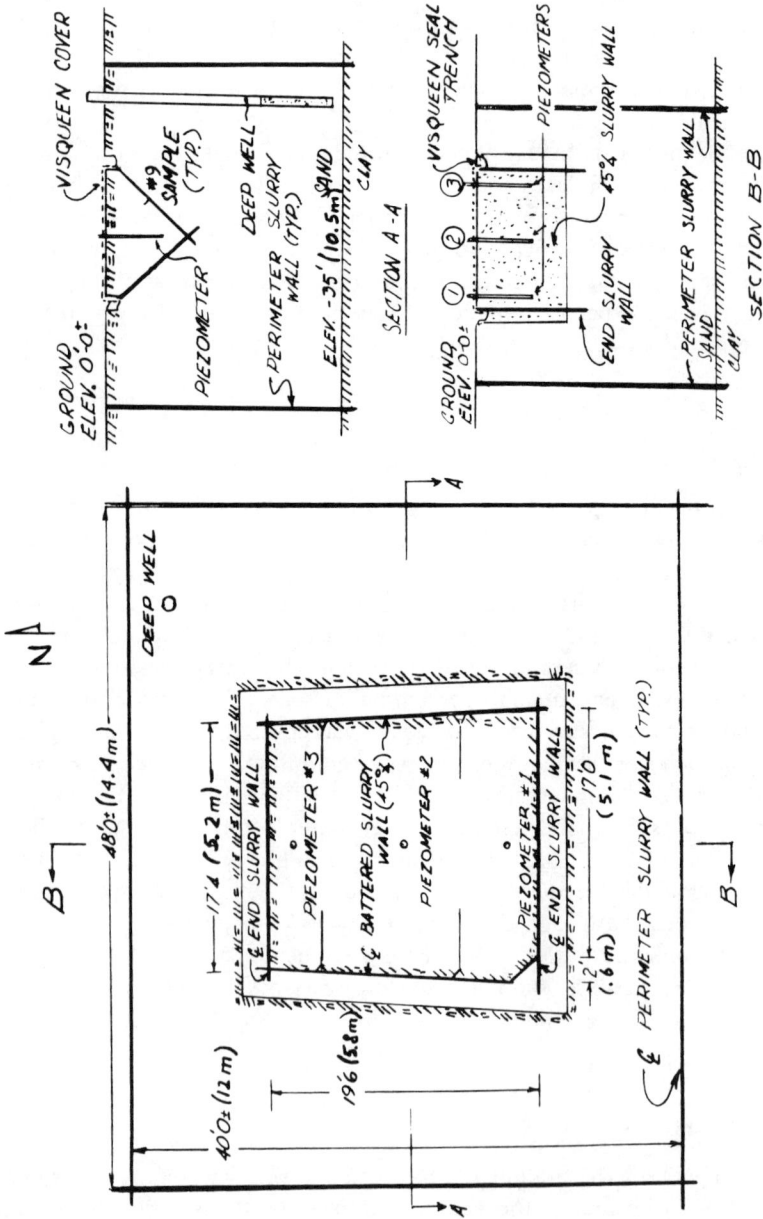

FIG. 3—Plan and sections of field installation.

TABLE 1—*Water level readings.*

Date	Piezometer 1		Piezometer 2		Piezometer 3	
	Water Level	ΔH	Water Level	ΔH	Water Level	ΔH
10/4/83	−3.44	...	−3.50	...	−3.54	...
10/5/83	−4.07	0.63	−4.10	0.60	−4.14	0.60
10/6/83	−4.48	1.04	−4.46	0.96	−4.58	1.04
10/7/83	−4.88	1.44	−4.87	1.37	−4.98	1.44
10/8/83	−5.15	1.71	−5.29	1.79	−4.79	1.25
10/9/83	−5.48	2.04	−5.46	1.96	−5.04	1.50
10/10/83	−5.63	2.19	−5.62	2.12	−5.50	1.96
10/11/83	−5.77	2.33	−5.83	2.33	−5.85	2.31
10/12/83	...	...	...	...	...	...
10/13/83	−6.27	2.83	−6.25	2.75	−6.33	2.79
10/14/83	−6.42	2.98	−6.43	2.93	−6.52	2.98

level observed in Piezometer 2 are plotted versus time in Fig. 4. Also plotted in Fig. 4 are a series of theoretical curves obtained from the equation

$$t = \frac{\sqrt{2}\, n}{(k/d)} \, Ln \, \frac{H_0}{H} \cdot \frac{3\sqrt{2}\, L + 2H}{3\sqrt{2}\, L + 2H_0} \qquad (1)$$

where

t = time,
d = thickness of the slurry wall,
k = coefficient of permeability of the slurry,
n = the effective porosity of the sand inside the trough,
H_0 = height of water above the apex at time $t = 0$,
H = height of water above the apex at time t, and
L = length of the trough.

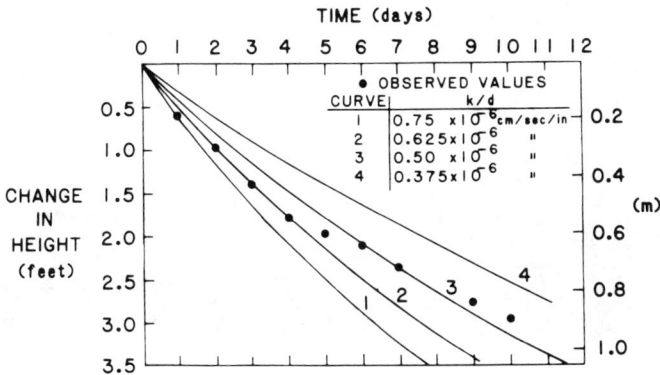

FIG. 4—*Change in height of the water level inside the trough.*

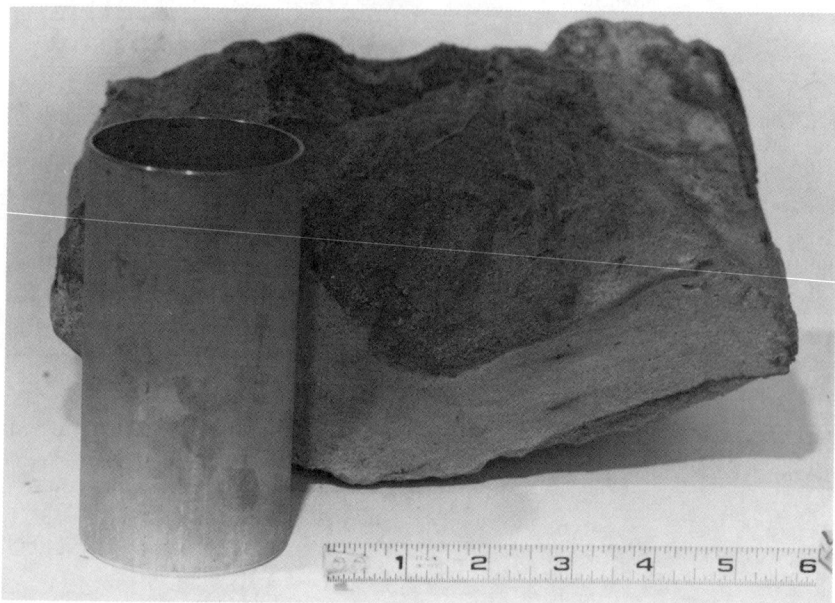

FIG. 5—*Photographs of intersection at apex of trough.*

Eq 1 was derived by considering the head lost across the thickness of the slurry wall to equal the head inside the trough at any particular point and time. Using this relationship, the time corresponding to any selected height, H, can be computed and plotted versus the change in height $\Delta H = H_0 - H$. Four theoretical curves are given in Fig. 4 for four different values of the ratio k/d and a value of the effective porosity of the dune sand (volume of voids drained per unit of total volume) of the order of 0.25. Comparing these curves to the data points indicate that a close match is obtained for a ratio k/d of 0.50×10^{-6} cm/s. The design thickness of the slurry walls is 100 mm (4 in.), hence the overall effective permeability of the slurry walls is estimated to be approximately 2×10^{-6} cm/s. The pumping test was repeated with essentially the same results.

Excavation of the Trough

Following the field pumping tests, the test section was carefully excavated to inspect the vertical and battered slurry walls, as well as their intersections. The thickness of each slurry wall was measured at 0.3-m (1-ft) increments in depth along the centerline of the wall. The thickness of the vertical wall was very consistent over the entire section, ranging from 89 to 100 mm ($3\frac{1}{2}$ to 4 in.). The thickness of the battered wall was less, ranging from 50.8 to 63.5 mm (2 to $2\frac{1}{2}$ in.), with a reduced thickness in the upper 0.6 to 0.9 m (2 to 3 ft). The intersections between the vertical and battered walls and at the apex of the trough were very complete; actually, the thickness usually increased at the intersections to form a bulb of slurry (Fig. 5).

The results indicate that battered slurry walls can be constructed in sandy soil to provide satisfactory seepage barriers. Except for the upper 0.6 to 0.9 m (2 to 3 ft) of wall, a continuous cutoff with an overall effective permeability of the order of 2×10^{-6} cm/s was provided. With further experience this performance can be improved upon; however, qualified personnel and a high level of quality control will always remain a prerequisite for success.

DISCUSSION

S. B. Ahmed[1] (written discussion)—Is it feasible to construct a vibratory beam cutoff wall through loose fly ash like (number of SPT blows, 0 to 4)

[1]Tennessee Valley Authority, Knoxville, TN 37902.

approximately 15.24 m (50 ft) deep? Would not the vibrations cause liquefaction?

G. Leonards, F. Schmednecht, J.-L. Chameau, and S. Diamond (authors' closure)—If local liquefaction occurred, a general mixing of the slurry wall material and the soil would occur and the final result would be a much lower permeability. General soil liquefaction on past jobs has been solved by monitoring and lowering pore pressures by drainage and by minimizing vibration energy.

S. B. Ahmed (written discussion)—Name the projects where inclined (vibratory beam) slurry walls are constructed? What are the advantages over straight walls?

G. Leonards, F. Schmednecht, J.-L. Chameau, and S. Diamond (authors' closure)—The inclined (vibratory beam) wall has been only installed as a test cell to date. Its advantage is to provide an impervious bottom layer when Mother Nature neglected to provide one at a reasonable depth. It may be especially useful as a remedial measure for an existing facility that has developed unacceptable leakage.

B. S. Beattie[2] (written discussion)—How can a hydrocarbon-based asphaltic material offer long-term impermeability in the presence of aromatic hydrocarbon wastes?

G. Leonards, F. Schmednecht, J.-L. Chameau, and S. Diamond (authors' closure)—An asphalt-based ASPA mix would be expected to soften and lose its impermeability as the asphalt dissolved in long-term contact with an aromatic hydrocarbon liquid or a concentrated solution. However, low concentrations of aromatic hydrocarbons dissolved in (or suspended in) water should not influence its integrity, even after long-term contact.

B. S. Beattie (written discussion)—How can the beam method successfully install a continuous wall through hard-rock conditions without the formation of windows in the wall?

G. Leonards, F. Schmednecht, J.-L. Chameau, and S. Diamond (authors' closure)—Like other methods for installing slurry walls, the vibrating beam method is not intended to provide seepage barriers in hard rock. In such cases, injection grouting is commonly used.

Y. A. Acar (written discussion)—Have the authors assessed the compatibility of slurry wall material to toxic wastes? If not, how could they propose the method in containment of such wastes?

G. A. Leonards, F. Schmednecht, J.-L. Chameau, and S. Diamond (authors' closure)—Quality control and the use of slurries with different combinations of materials make the vibratory beam method a good candidate for toxic waste containment. For example, the compatibility of cold asphalt emulsions with aromatics has been tested for approximately one year with excellent results.

---

[2]Federal Bentonite, Montgomery, IL 60538.

Christopher P. Jepsen[1] and Mark Place[2]

Evaluation of Two Methods for Constructing Vertical Cutoff Walls at Waste Containment Sites

REFERENCE: Jepsen, C. P. and Place, M., **"Evaluation of Two Methods for Constructing Vertical Cutoff Walls at Waste Containment Sites,"** *Hydraulic Barriers in Soil and Rock, ASTM STP 874*, A. I. Johnson, R. K. Frobel, N. J. Cavalli, and C. B. Pettersson, Eds., American Society for Testing and Materials, Philadelphia, 1985, pp. 45–63.

ABSTRACT: This paper compares two methods of constructing vertical cutoff walls when considered in the context of controlling the migration of waste materials. It will be argued that the conventional slurry trench method has significant advantages over the vibratory beam injection (VBI) method. The discussion will compare the quality control of each method, relating its importance to the central issue of cutoff wall continuity. Important issues to be discussed in conjunction with quality control considerations include geometry, wall thickness, Darcy's equation, wall composition, and chemical resistance.

KEY WORDS: cutoff wall, slurry, thixotropic, bentonite, cement, filter cake, backfill, asphaltic emulsion, permeability, plumb, deflection

A cutoff wall may be defined as a man-made continuous barrier to groundwater flow that typically exhibits a very low permeability relative to surrounding soils and is comprised of any of a variety of natural and man-made materials including bentonite, soil, cement, fly ash, asphaltic emulsions, and water. In the United States, there are presently two techniques used for constructing cutoff walls: (1) conventional slurry trenching methods that have been in use for over 35 years; and (2) the vibratory beam injection (VBI) method that has been in use for approximately 10 years (Figs. 1 and 2). The former is considered by those in the construction industry to be a slurry trenching method while the latter is considered a grouting method. The two methods are markedly different not only in their approach to the construction of the cutoff wall itself but also in the features of their respective finished products. These differences dictate that a different set of quality control measures apply to each

[1]Technical supervisor, American Colloid Co., Skokie, IL 60077.
[2]Sales representative, John P. Place, Inc., Pittsburgh, PA 15236.

45

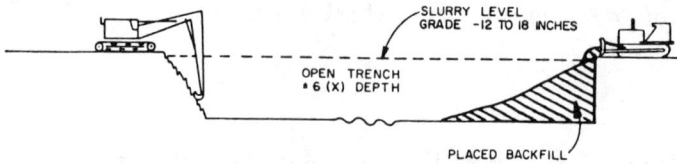

FIG. 1—*Conventional slurry trenching method.*

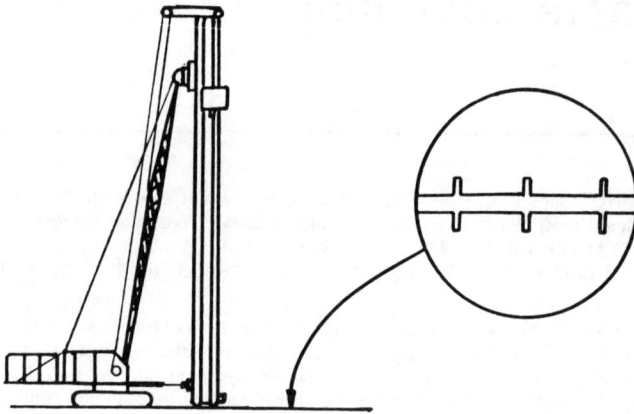

FIG. 2—*Vibratory beam injection method (source: Slurry Systems brochure).*

method if continuity, low permeability, and other important cutoff wall features are to be achieved by each method.

CONVENTIONAL SLURRY TRENCHING METHOD

Conventional slurry trenching is a method by which a continuous trench is excavated under a slurry consisting primarily of sodium bentonite and water. Excavation is accomplished with backhoes to depths of 15 m, with drag lines to depths of 25 m, and with clamshells to depths of over 100 m [1]. The slurry, or colloidal suspension, contains sufficient solids to maintain vertical trench walls in almost any type of soil without any other means of support. Following excavation, the trench is backfilled with a mixture of excavated soil, bentonite, and water to form a continuous soil-bentonite (SB) cutoff wall. In some instances, borrowed soil materials are used to form the backfill. A variation of this technique is the cement-bentonite (CB) slurry trenching method in which a slurry consisting of cement, bentonite, and water is used. This modified slurry initially serves the purpose of maintaining trench wall stability, but, unlike the SB method, there is no backfilling operation because the cement-bentonite slurry hardens in place to form the cutoff wall.

Slurry Trenching Quality Control Parameters

There are essentially three major areas of quality control that govern the slurry trenching method: (1) slurry quality; (2) trench dimensions and continuity; and (3) cutoff wall composition and placement.

A discussion of each of these topics follows.

Slurry Quality

The slurry material utilized in both the SB and CB methods is sodium bentonite, also known as Wyoming bentonite. This highly plastic clay material {Plasticity Limit equal to 40 to 70, according to ASTM Method for Plastic Limit and Plasticity Index of Soils [D 424-59 (1971)]} has the ability to absorb nearly 5 times its weight in water and to swell to 10 to 13 times its dry size upon complete hydration. When a sufficient quantity of water is added to bentonite and mixed vigorously, the result is a stable thixotropic gel that has ideal viscosity, density, and filter loss properties for slurry trenching.

As previously stated, the primary function of the slurry is to maintain trench wall stability. The two primary factors cited for trench wall stability are the hydrostatic force exerted by the slurry and the filter cake formation on the walls. Actually, these two factors work together. The weight of the slurry must exert its force not only on the soil pore fluid but also on the soil particles if it is to support the trench walls. The filter cake (which forms when water is squeezed out of the slurry through the trench walls) allows the hydrostatic force of the slurry to be more directly transferred to the walls of the trench. This is why unstable wall conditions are frequently associated with poor cake formation.

Filter cake formation is a function of many parameters including clay type, clay concentration, makeup water, formation time, level of bentonite contamination, quantity and type of chemical additives, and hydrostatic pressure. The permeabilities of bentonite and cement-bentonite filter cakes have been reported as low as 2.3×10^{-11} m/s and 1.0×10^{-9} m/s, respectively [1,2]. The higher filtrate losses associated with CB filter cakes is due to the calcium in the cement, which causes a partial flocculation and aggregation of the bentonite. It may be noted that filtrate loss is, indeed, a measure of slurry stability because the rate of filter loss is directly proportional to the level of bentonite contamination. High filtrate loss can be controlled in a number of different ways:

1. Utilize specially treated contaminant resistant bentonites which have proven very effective in a number of slurry trenching projects.
2. Utilize chemical treatments at the project site which act either as dispersants or bulking agents in controlling filter loss.
3. Increase the bentonite concentration in the slurry.

4. Waste the contaminated slurry in the trench and replace it with fresh slurry although disposal may pose a significant environmental problem.

Controlling filtrate loss is, indeed, very important in maintaining trench continuity. Filtrate (water) losses through trench walls result in a constant tendency for SB and CB slurries to thicken in the trench as the residual bentonite solid concentration in the slurry increases. It is well known that as the bentonite solids concentration in a slurry increases arithmetically, the attendant slurry viscosity and gelation increase geometrically. Thus, relatively small increases in the bentonite solids concentration of a slurry can result in significant increases in viscosity and gel strength that may hinder both excavating and backfilling operations. Filter loss, however, should not be confused with *slurry loss* that occurs when sudden large voids or very loose, coarse-grained soils are encountered during excavation. Large, sudden losses of slurry can lead to a trench collapse when insufficient hydrostatic pressure is available to keep the trench open. Experienced slurry trench contractors, however, have reserve bentonite mixing and slurry pumping capacity to address most conditions.

In the SB method, the slurry may contain anywhere from 4 to 7% bentonite by weight. An ideal makeup water should be low in total dissolved salts and have a pH in the range of 6.5 to 10. A poor quality makeup water consumes more bentonite than a relatively fresh water to achieve the same viscosity, density, and fluid loss properties. When introduced to the trench, the slurry should be fully hydrated with an apparent viscosity of 0.015 to 0.020 Pa $\cdot$ s (15 to 20 cP) and have a density of approximately 1050 kg/m^3 [3,4]. In the field, these two slurry parameters may be readily checked with a Marsh funnel and a mud balance, respectively. A minimum viscosity of at least 35 Marsh funnel seconds is recommended. Filter loss, which is of secondary importance, should be between 15 and 30 cm^3 at 488 kg/m^2 of pressure.

After a section of the trench has been fully excavated, the slurry is sometimes cleaned to remove gravel or sandy soil particles which have collected in the slurry, especially at the bottom of the trench. The purpose of this is to lower the viscosity of the slurry so that it will be uniformly displaced during the backfilling operation. A clamshell or backhoe bucket is sometimes used as a rough cleanup in bringing larger particles to the surface. A more thorough cleaning can be achieved by airlifting the slurry for circulation through desanding units.

In the CB method, the self-hardening slurry typically contains 4 to 7% bentonite (by weight), 8 to 25% portland cement, and 65 to 88% water [5]. If too little bentonite is used, the cement will settle out, causing excessive water bleeding. If too much bentonite is used, the slurry will be too difficult to work with and the wall may be too weak. If too little cement is used, no setting of the slurry will be achieved. Figures 3 and 4 illustrate some of the important interrelationships between cement, bentonite, and water [6]. The percentage

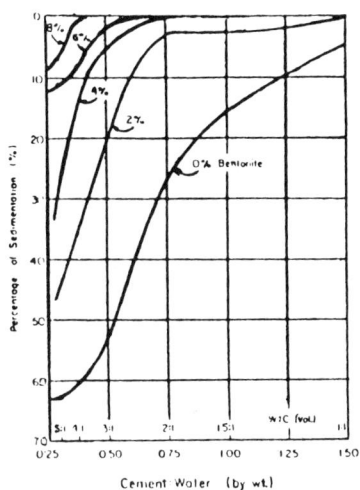

FIG. 3—*Experimental relation of percentage of sedimentation and cement–water ratio by volume (after Burgin [6]).*

of sedimentation referred to in Fig. 3 is determined as follows. First, place a liter of the slurry in a standard 1-L graduated cylinder. At the end of 2 h, the volume of clear liquid formed at the top of the cylinder due to sedimentation of the cement grains is noted. This volume, expressed as a percentage of the total volume, gives the percent sedimentation.

The cement, which acts as a binder for the wall, is always added to a fully hydrated bentonite-water slurry just prior to being introduced in the trench. This requirement, called for in most specifications, minimizes the viscosity of the slurry in the trench during excavation so that excessive suspended solids from the soil will not be entrained in the slurry when it eventually hardens. When the CB slurry is introduced in the trench, it should have a density of approximately 1090 kg/m^3 and a Marsh funnel viscosity ranging from 30 to 50 s [5,7].

Trench Dimensions and Continuity

It is relatively easy to control the dimensions of a slurry trench during excavation so that a continuous trench of uniform dimensions is excavated. The width of a trench is established by the bucket or clamshell of the excavating equipment; the depth is controlled by direct probing or sounding of the trench bottom at maximum 6 m intervals and by observation of materials excavated from the trench; and the longitudinal continuity is ensured by the natural excavating motion of backhoe and draglines or by a quick sideways check with a clamshell. In the case of CB cutoff walls which set up overnight,

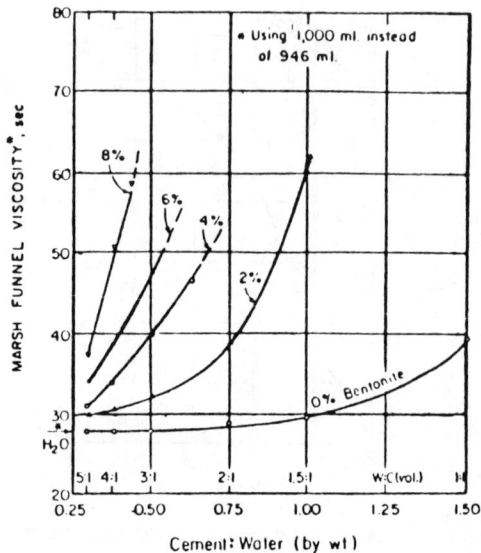

FIG. 4—*Effects of cement–water ratio and percent bentonite on marsh funnel viscosity (after Burgin [6]).*

longitudinal continuity is ensured by keying the beginning excavation of the day 1 to 2 m (depending on depth) into the end of the wall constructed on the previous day. It is for this reason that repairing a section of a CB wall is so easy. To prevent a trench collapse once excavated, a stable slurry should be maintained at all times no more than 0.4 m below the working surface. Thus, slurry quality is of paramount importance for maintaining trench continuity.

Cutoff Wall Composition and Placement

Much has been written about the composition and placement of slurry cut-off walls. A summary of the most important design and construction considerations of SB and CB cutoff walls follows.

Soil-Bentonite Cutoff Walls—As previously stated, most SB cutoff walls are formed with the soil excavated from the trench. However, borrowed soil materials are used in three situations: (1) if the soil is too contaminated to be mixed with bentonite; (2) if the soil is too coarse to meet low permeability requirements at economical bentonite application rates; or (3) if the soil is too fine to meet the necessary strength requirements of the wall. In trenches excavated through soils highly contaminated with volatile organic constituents, there are also air quality considerations that may eliminate the excavated soil as a viable constituent in the backfill mix. The typical SB backfill consists of 2.0 to 4.0% bentonite (by weight of the *total mixture*), 25 to 35% water, with the balance being soil [1].

SB cutoff walls are typically 0.5 to 1.8 m wide. Trench widths narrower than 0.5 m tend to restrict the backfilling of the trench, which could cause bridging of the material. The permeability of SB cutoff walls is in the range of 10^{-9} to 10^{-10} m/s. Research has shown that the permeability of the backfill is dependent on both the gradation of the soil being used as well as the bentonite content. The greater the percentage of fines (soil particles smaller than 7.5×10^{-5} m) in the backfill and the more plastic those fines are, the lower the wall permeability will be (see Fig. 5). In addition, the greater the percentage of bentonite in the backfill, the lower the wall permeability will be (Fig. 6).

Most specifications require that the SB backfill consist of at least 15 to 20% fines and that the bentonite consist of 4 to 7% of the *dry weight of the soil.* Historically, "achieving" these two requirements has been, perhaps, the most questionable aspect of the method, given that soil conditions do change both vertically and longitudinally in a trench. Unless it is known ahead of time that all soil borings at all excavation depths contain the minimum required percentage of fines, certain portions of a cutoff wall could conceivably have "pockets" lacking the necessary fines. Of course, using a borrow material to supplement excavated soil can readily help solve this problem. However, "rules of thumb" and "past experience" without the use of standardized testing are no longer acceptable given the maturity of the technique and the increased understanding of regulatory agencies and engineers that are involved in reviewing and enforcing specifications.

To positively control the just mentioned critical factors, a gradation test [ASTM Particle-Size Analysis of Soils (D 422-63)] and a moisture content test [ASTM Laboratory Determination of Water (Moisture) Content of Soil, Rock, and Soil-Aggregate Mixtures (D 2216-80)] should be made on the SB backfill mixture prior to adding water or slurry. By taking three random sam-

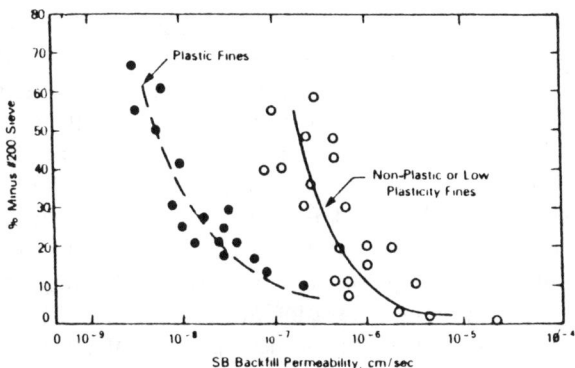

FIG. 5—*Permeability of soil-bentonite backfill related to fines content (after D'Appolonia and Ryan 1979 [1,4,7]).*

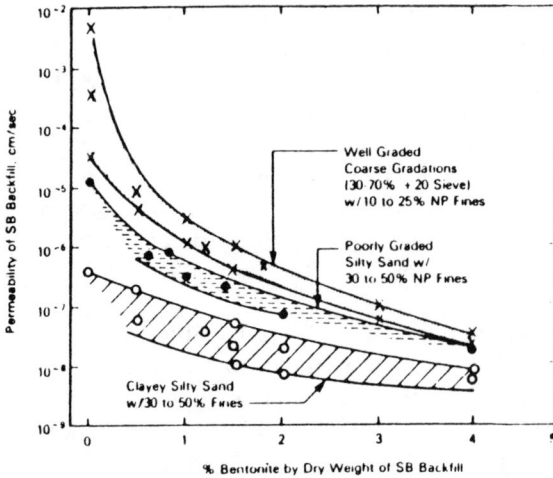

FIG. 6—*Relationship between permeability and quantity of bentonite added to soil-bentonite backfill (after D'Appolonia and Ryan 1979 [1,4,7]).*

ples every 30 m, or more often if required, an average variation in material properties can be determined. Only after such results have been approved should the SB backfill be mixed with water or slurry or both and introduced to the trench. Due to slurry variability in the trench, it is now generally recognized that the proportion of dry bentonite added to the soil should not be adjusted to account for any bentonite content of the slurry but rather should be used as a safety, or insurance, factor. In any event, a safety factor should be incorporated into the percentage of bentonite required in the design mix to account for likely variabilities in the backfill. The magnitude of such a safety factor should be site specific.

Mixing of the backfill is usually accomplished with a bulldozer adjacent to the trench by windrowing, tracking, and blading the mixture. The method or type of mixing equipment is unimportant, provided that a homogenous mix is achieved in which all the soil particles are coated with the bentonite slurry. An ideal backfill should have a smooth semiliquid consistency with a slump in the range of 0.1 to 0.15 m and should stand on a slope of 5 to 10:1. Achieving the proper backfill mix consistency is of paramount importance if trench wall continuity is to be achieved. If the material is too stiff it may form such a steep slope in the trench that it may fold over and trap pockets of slurry that would destroy wall continuity. If the material is too watery, it will form a very flat backfill that not only may separate and disperse into the slurry, but also may interfere with the continuing excavation at the other end of the trench.

As previously stated, controlling the viscosity of the slurry in the trench will also help ensure proper backfill placement. The most common and cost-effective method of control is to use the slurry from the trench itself to form the backfill mix. This makes room for fresh, less viscous slurry to be introduced in the trench. An added benefit of doing this is that the suspended fines in the trench slurry end up in the backfill, which may contribute to lowering its permeability. The rule of thumb is that the backfill will easily displace the slurry if its specific weight is at least 250 kg/m^2 greater than that of the slurry in the trench [1, 7]. Generally, this differential can be readily achieved using the slurry replacement technique just described. Mechanical desanders in general add cost to a project without providing any measurable benefit to the quality of the finished product.

Cement-Bentonite Cutoff Walls—CB cutoff walls, typically 0.3 to 1.0 m thick, are thinner than SB cutoff walls for two reasons: (1) they are more expensive to construct; and (2) there is no backfilling operation. The composition of the slurry that forms the wall has been already discussed. However, a common misperception by engineers unfamiliar with the slurry composition is that the slurry will set up as a rigid wall with the strength of concrete. In reality, CB cutoff walls typically have an unconfined compressive strength of only 1.5×10^4 to 2.8×10^4 kg/m^2 or approximately that of surrounding soils [7].

In addition, the final mix is very plastic and can withstand compressive strains of several percent without cracking. Low strength, high plasticity (compressibility), and low permeability (1×10^{-8} m/s) are all qualities of CB cutoff walls that are associated with the rather large quantities of bentonite in the slurry mix, 4 to 7% by weight.

The quantity of suspended solids that are entrained in the CB slurry should be kept to a minimum. Excessive suspended soil particles (particularly sand) tend to increase the wall permeability and increase wall strength, which can make it more brittle and susceptible to cracking. Tests have shown, however, that suspended solids are not an important factor in the permeability of a CB cutoff wall until the concentration is high enough to permit interparticle contact. This point is reached at percentages (by volume) in the range of 10% [4].

THE VIBRATORY BEAM INJECTION METHOD

History of the Technique [10]

This technique was pioneered by European geotechnical contractors over 25 years ago. Development of this technique was based on a need for effective site dewatering for the construction of underground structures. Principal applications were located in relatively unconsolidated alluvial areas. The original method utilized interlocking sheet piles to which grout pipes were welded. However, this method was slow and expensive and had depth limitations that made it useful in only relatively shallow installations. Later, when vibratory driver-extractor equipment was developed in the early 1960s, the technique of overlapping wide flange beam insertions came into use. In recent years, modifications of the beam, equipment, and slurry mixes have all served to improve the overall quality of the technique. The technique has been used to install approximately 4.6 million m^2 in Europe as of 1974. In the United States, approximately 0.3 million m^2 have been installed. Today, applications have been expanded to include permanent groundwater migration control at disposal sites.

Description of the Method

The VBI method is a technique that involves penetration, to a preselected depth, of a wide flange beam (typical web depth 0.5 to 0.85 m, typical flange width 0.25 to 0.38 m) containing integral grout injection equipment [8]. As the beam is vibrated into position, soil consolidation and displacement occurs. After reaching the desired depth, the beam is vibrated up out of the penetration, leaving a consolidation void which is pressure-grouted full with a grout that hardens in place. Successive beam penetrations are overlapped to form the cutoff wall. The operations involved with the technique follow [9]:

1. A portable mixing plant is generally established at the site.
2. The area along the path of the proposed cutoff wall is made relatively flat.
3. A narrow trench (approximately 0.30 m) is dug along the centerline of the trench and is filled with slurry.
4. The crane-vibratory hammer is centered over the trench and the beam is lowered to the bottom of the trench.
5. The vibratory hammer and pump, which deliver slurry to injection nozzles at the bottom of the beam, are activated. The beam is driven into the earth until it keys into an impervious strata or some predetermined depth.
6. The beam is slowly withdrawn, leaving a consolidated void which is filled by the grout being pumped.
7. The beam is moved ahead in the longitudinal direction of the trench (allowing for a 10% beam depth overlap) and then redriven into the earth.

The net result of this method is a thin curtain wall that is typically 0.025 to 0.100 m wide, depending on the width of web on the beam or the width of the fin welded near the bottom of the beam. A side benefit of the method already mentioned is the densification of soil in the proximity of the beam penetration, which helps reduce nearby soil permeabilities and adds shear strength to the local soil mass.

The VBI method is best suited in areas where saturated loose granular soils predominate. Medium to stiff clays are hard to penetrate with the method, and rocky soils pose beam deflection and alignment problems. It is claimed that this method can economically construct cutoff walls to a depth of 25 m [10].

Vibratory Beam Quality Control Parameters

There are essentially two major areas of quality control that govern the effectiveness of the finished product in the ground:

1. Wall composition, properties, and contaminant resistance.
2. Trench wall continuity.

A discussion of these two areas follows.

Wall Composition, Properties, and Contaminant Resistance

There are a variety of slurry mixes that may be used with this method. Among the most common are cement-bentonite, fly ash-bentonite, kaolin-bentonite, and asphalt emulsions [10]. CB slurries are used when strength is required. Fly ash-bentonite slurries are used in reactive soils [10]. Kaolin-bentonite slurries are used in high saline conditions [10]. Asphalt emulsions, specifically ASPEMIX by Slurry Systems (more expensive than the other slurry mixes), are intended for highly contaminated ground conditions and have been reported to have a very high resistance to attack by acids, brine, and other industrial and chemical wastes. Although the exact proportions of some of these slurries are proprietary, all exhibit thixotropic properties [10]. The permeabilities of the bentonite mixtures vary from 1×10^{-8} to 1×10^{-10} m/s, whereas the asphaltic emulsions undergo internal swelling and seal shut, creating a material with a permeability of less than 1×10^{-14} m/s [11]. In addition to having low permeability, asphaltic emulsions are reported to be highly compressible.

The just mentioned slurry mixtures may, indeed, have a very high contaminant resistance to the wastes mentioned. However, before such products are construed as a panacea specifically for waste containment, the following questions should be thoroughly explored over long-term testing periods.

1. What is the appropriate means of testing the hardened mixtures for compatibility with a specific leachate? Should 2 to 4 pore volume displacements of leachate be passed through the material like a SB or CB cutoff wall sample, or should a material of such low permeability as ASPEMIX be tested by means of an immersion test like a geomembrane?

2. Is there a danger of secondary contamination from the wall itself or from by-products that may form when the wall constituents and the leachate react?

3. Are the various emulsions stable and will they remain somewhat flexible over time, or do they age and crack over time in specifically contaminated environments?

The importance of wall compressibility (plasticity) is particularly critical when considered in the context of hydraulic gradients. Hydraulic gradients are a function of barrier thickness. Thus, the hydraulic gradient through a cutoff wall constructed with the VBI method will, on average, be ten times greater than that of a wall constructed by the SB or CB method given the same hydrogeologic conditions. The possibilities of a piping failure, therefore, are much greater with the VBI method in the event of the wall cracking.

Another potential source of excessive leakage is wall discontinuities caused by vibration of the beam. For instance, gravel can be dislodged and entrained in the slurry as it hardens. Such large particles are undesirable because they contribute to the short circuiting of water through the wall, as water will channel along the surface of such gravel particles, particularly if they protrude through both sides of the wall. The vibrations of the beam also can cause soil adjacent to completed portions of the wall to squeeze, or neck-in, creating a very thin wall. A permanent neck-in of the walls, of course, could result in a localized area of high enough hydraulic gradient to induce failure. An area of total collapse (which is also possible with the conventional slurry trenching method) will provide little if any barrier to groundwater flow. The just mentioned wall discontinuities from the VBI method have been observed by the Army Corps of Engineers at a test site installation for the Corps Mobile, Alabama District Lock "E" Tennessee-Tombighee Project [12,13]. Visual inspection of the wall revealed wall discontinuities with gravels protruding through both sides of the wall [12,13].

Trench Wall Continuity

The question of cutoff wall continuity is perhaps the most controversial quality control question of this method. Unlike the SB or CB slurry trenching methods, there is no visual inspection of excavated material that qualitatively proves that a continuous trench has, indeed, reached a desired aquaclude. Rather, the VBI method must rely on indirect methods for determining both beam alignment and beam penetration into the required aquaclude.

Beam alignment can be controlled in a number of different ways: (1) monitoring of a double guide support system that warns the vibratory hammer crane operator of a beam deflection problem [10]; (2) monitoring of beam alignment with a "Digitilt" inclinometer (a device that is time consuming and slows down product) [11] (3) monitoring with an above-ground laser monitor (the laser system however cannot monitor beam deflections under the ground surface) [11] (4) providing a 10% beam penetration overlap as a factor of safety [10,11].

The vibratory guidance system used by Slurry Systems is very similar to that used for tight tolerance pile driving installations with the following exceptions: (1) the vibrator dimensions are such that it has to be kept external to the lead; (2) the bottom of the lead has a vertical hydraulic support ram with a bearing pad for additional control on plumbness; and (3) the lead has a spotter that can adjust the beam in any direction. According to Slurry Systems, the adjustability for plumbness control is necessary to compensate for crawler crane stability on loose and uneven ground [8].

Determination of beam penetration into a desired aquaclude is claimed to be documented by the following:

1. Beam "communication"—beam penetrations into impervious layers require power surges as recorded in vibrator electric motor amperage combined with an increase in injection pressure of the slurry and a damping of the vibrator amplitude.

2. Confirmation with consultants boring logs and observations of calibrated markings on the beam.

3. Extraction of a bottom sample, although valid only for the location where sampled.

There are essentially two basic ways in which cutoff wall continuity can be affected by beam alignment: (1) beam deflection caused by striking movable or immovable rocks or rock strata; and (2) successive nonplumb beam penetrations.

Beam deflections can potentially affect wall continuity by deflecting either along the wall or perpendicular to it. Computer simulations have been used to predict such deflections in rather simple, assumed geologic formations. One such study predicted the deflection of an A36 W33X 201 beam penetrating approximately 30 m into a saturated, loose sand strata that strikes movable and immovable rocks at different depths [9]. The study reported deviations in wall alignment of up to 0.07 m when a movable rock was struck at the bottom of the guide trench. Immovable rocks near the ground surface were found to cause the worst problems in that they would tear up equipment and buckle the beam before deviations in excess of 0.05 m could occur. The study indicated that beam deflections were not of sufficient magnitude to cause wall discontinuities. However, such results must be evaluated in the context of the assumptions being made, particularly when extrapolating to actual site conditions.

The importance of maintaining verticality cannot be overstated when using the beam method. If each penetration of the beam does not overlap the prior penetration, ungrouted voids or "windows" will result. These windows will allow unimpeded migration of the pollutants/groundwater through the otherwise impermeable wall. Currently, it is claimed that guide leads can maintain plumbness of the beam in the vertical plane of the cutoff wall within the limits of ±1%. Referring to Figs. 7 to 9, the practical results of these limits can be seen [14]. Figure 7 indicates the deviation from vertical that can be found under field conditions at a 1% deviation from plumb. One percent equates to 0.0157 radians (0.9°) measured at the point of entry. From the conditions set forth in Fig. 7, the horizontal displacement is determined by multiplying beam penetration depth by the sine of 0.9°.

According to specifications provided by Slurry Systems (12/83), deviations

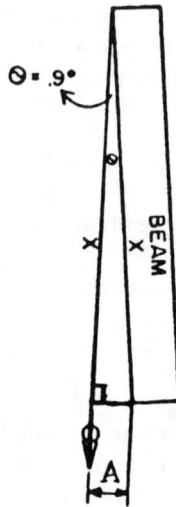

FIG. 7—*Tolerance: 1% of plumb for vibratory beam injection method. One percent of 90° = 0.9°. A = horizontal displacement.*

from plumb are compensated for in two ways: (1) successive beam penetrations are overlapped 10% (0.084 m in the case of an assumed 0.84-m beam depth) as called for in Slurry Systems specifications); and (2) a fin extends the beam overlap with its length of 0.36 m (it is located near the bottom of the beam opposite the direction of travel) [8]. Taking these factors into account, two different field deviations can be defined. A minor field deviation [$\chi \sin\Theta - (8.4 + 36)$] may be defined as the amount of horizontal deviation measured in centimeters at the deepest point of penetration when the first penetration is plumb and the second penetration is out of plumb by 1%. A major field deviation [$2\chi \sin\Theta - (8.4 + 36)$] may be defined as the amount of horizontal deviation measured in centimeters when the first penetration is out of plumb away from the crane 1% and the second penetration is out of plumb in the direction of travel 1%. As illustrated in Figs. 8 and 9, even if a tolerance of 1% is practical, geometry dictates that windows will form in the cutoff wall at certain depths. Combining this phenomenon with the beam deflections (that may occur depending on subsurface conditions), even larger and more frequent "windows" will occur. Of course, geophysical testing methods such as resistivity and sonar sounding can be used as a postconstruction technique to qualitatively identify if windows exist in a cutoff wall. However, such methods are time consuming, costly, and do not provide answers when needed, that is, during the construction process. Thus, given the beam and fin dimensions cited above and assuming no deflections of the assumed beam, it is evident from geometry that present construction practices limit the positive effectiveness of the VBI method to penetration depths of 14.1 m or less.

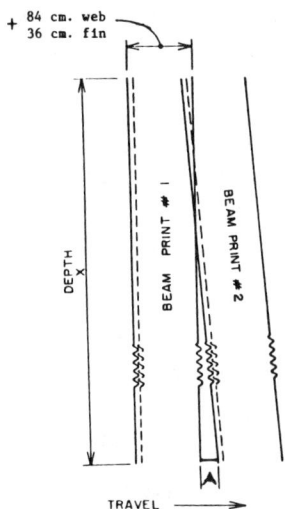

Penetration Depth (meters)	Horizontal Deviation (X sinθ − 44.4) (cm.)
5.0	−36.5*
10.0	−28.7
20.0	−13.0
28.3	0.0
30.0	2.7**

*negative numbers represent the amount of overlap (ie no windows)

**positive numbers represent the width of the window in the cut-off wall

$$\left[+\begin{array}{l}\text{(10\% of beam web, 8,4cm.)} \\ \text{(length of fin, 36cm.)}\end{array}\right] = 44.4$$

FIG. 8—*Minor field deviation ("window" size at "key" elevation).*

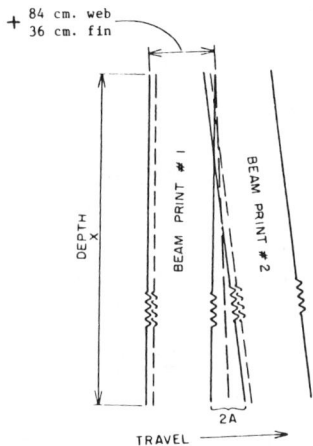

Penetration Depth (meters)	Horizontal Deviation (2X sinθ − 44.4) (cm.)
5.0	−28.7*
10.0	−13.0
14.1	0.0
20.0	18.4**
30.0	49.8

*negative numbers represent the amount of overlap (ie. no windows)

**positive numbers represent the width of the window in the cut-off wall

$$\left[+\begin{array}{l}\text{(10\% of beam web, 8.4cm.)} \\ \text{(length of fin, 36cm.)}\end{array}\right] = 44.4$$

FIG. 9—*Major field deviation ("window" size at "key" elevation).*

Conclusion

This paper has compared the relative merits and quality control parameters of the conventional slurry trenching method and the vibratory beam injection method. Wall continuity, low permeability and high compressibility (plasticity) are all important features of groundwater cutoff walls, particularly permanent installations.

Many of the quality control parameters in both methods center around ensuring cutoff wall continuity. The conventional slurry trenching method provides for more positive control in two ways: (1) excavation allows for inspection of all trench dimensions, and (2) removal of material from the trench allows for visual inspection of the suitability of the excavated aquaclude material into which a trench is being keyed. The VBI method must settle for above ground indications of conditions below. Beam alignment tolerances and beam deflections limit the effectiveness of this method, especially at penetration depths greater than 14.1 m.

References

[1] D'Appolonia, D. J., *Journal of the Geotechnical Engineering Division*, American Society of Civil Engineers, Vol. 106, No. GT4, April 1980, pp. 399–419.

[2] Nash, J. K. T. L. and Jones, G. K., "The Support of Trenches Using Fluid Mud," in *Grouts and Drilling Muds in Engineering Practice*, Butterworth's, London, 1963.

[3] LaRusso, R. S., "Wanapum Development Slurry Trench and Grouted Cut-Off," in *Grouts and Drilling Muds in Engineering Practice*, Butterworth's, London, 1963.

[4] Ryan, C. R., "Technical Specifications, Soil-Bentonite Slurry Trench Cut-Off Wall," GEO-CON Inc., Pittsburgh, PA 15235.

[5] Jefferis, S. A., "Bentonite-Cement Slurries for Hydraulic Cut-Offs," in *Proceedings*, Tenth International Conference on Soil Mechanics and Foundation Engineering, Vol. I, Rotterdam, June 1981.

[6] Burgin, C. R., "Investigation of the Physical Properties of Cement-Bentonite Grouts for Improvement of Dam Foundations," thesis presented to the University of Florida in 1979 in partial fulfillment of the requirements for the degree of Master of Science.

[7] Ryan, C. R., "Slurry Cut-Off Walls: Methods and Applications," presented at GEO-TEC '80, Chicago, 18 March 1980, GEO-CON Inc., Pittsburgh, PA 15235.

[8] "Specifications. Slurry Cut-Off Wall Vibrating Beam Injection Method," Slurry Systems, a division of Thatcher Engineering Corp., Gary, IN.

[9] Hulsey, Dr. J., "Prediction of Beam Deflection in a Vibrated Slurry Wall Installation," presented at the North Carolina section of the American Society of Civil Engineers annual meeting, Charlotte, NC, 22 Sept. 1983.

[10] Schmednecht, Zlamal, Nelson abstract, "Vibrated Beam Technique for Thin Wall Barriers—Applications in Construction and Environmental Applications," Slurry Systems, a division of Thatcher Engineering Corp., Gary, IN.

[11] Gunter, G. C., Colonel, Corps of Engineers, Department of the Army, Contracting Officers' Report "Protest of Slurry Systems," division of Thatcher Engineering Corporation, Gary, IN, Under IFB No. DACW41-83-B-0183, GAO File B-212033, 29 June 1983.

[12] Mobile, Alabama District Army Corps of Engineers "Lock E" Test Program, Tennessee-Tomhigbee Waterway, Mobile District, Dec. 1978.

[13] Mobile, Alabama District Army Corps of Engineers "Design Memorandum No. 31," Lock "E," Lock, Slurry Wall and Earthwork, Tennessee-Tomhigbee Waterway, Mobile District, July 1979.

[14] Place, M., "Evaluation of Vibratory Beam Injection Grouting as a Construction Method for Vertical Cut-off Walls at Waste Containment Sites," March 1982.

DISCUSSION

F. Schmednecht[1] *(written discussion)*—What is Christopher Jepsen's field knowledge of vibrated beam technology? Does he know the phenomenon of the path of least resistance and passive earth strength? Also, why have there been so few field pumping tests of soil-bentonite walls?

C. Jepsen and M. Place (authors' closure)—The authors' paper represents primarily a literature search of the two cutoff wall construction methods. However, the authors also drew upon their joint experience and knowledge in the areas of soil sealing, cutoff wall design, and slurry properties testing to write the paper.

The phenomenon of a path of least resistance was recognized by the authors as being valid in rather ideal homogeneous conditions; however, numerous factors can readily nullify "the least resistance effect," such as (1) sudden changes in geology, (2) beam deflection caused by rocks, (3) successive nonplumb beam penetrations, (4) a necking in or collapse of previous beam imprints, and (5) insufficient beam overlap.

In short, a path of least resistance for the beam may not be always necessarily along the imprint of the previous beam penetration given all the internal and external factors that come into the picture.

The excellent performance of soil-bentonite cutoff walls in various dewatering projects throughout the country in the past 25 years has more or less obviated the need to "test" the soil-bentonite cutoff wall method with field pumping tests.

Y. B. Acar[2] *(written discussion)*—The author implies that slurry walls could be used in the containment of toxic wastes. Slurry walls are constructed of highly active clays. I would like to pose this question: Active clays are more susceptible to structural changes due to variations in pore-fluid chemistry. Does the author have an activity criteria in the use of such slurries for the containment of toxic wastes?

C. P. Jepsen and M. Place (authors' closure)—It is true that sodium bentonite is a very active clay and is susceptible to structural changes; however, as pointed out in this paper, the conventional soil-bentonite slurry wall should have at least 15 to 20% fines in it to assist the sodium bentonite in achieving the desired coefficient of permeability of the cutoff wall. Thus, any activity

[1]Slurry Systems, Inc., Gary, IN 46406.
[2]Assistant Professor, Department of Civil Engineering, Louisiana State University, Baton Rouge, LA 70803.

criteria for a soil-bentonite cutoff wall should be based on the specific backfill mixture proposed for a specific project.

For the VBI method in which specific predesigned self-hardening slurries are used, general activity criteria could obviously be determined to serve as guidelines for specific types of wastes.

The authors have not established an activity criteria for the use of slurries for waste containment; however, a survey of the performance of existing cutoff walls in use would be the most practical way of determining the long-term stability of cutoff walls against various types of wastes and leachates.

J. Evans[3] *(written discussion)*—The necking of thin cutoff walls was noted for U.S. Corps of Engineers test sections. Have the results of these test sections been published? Where? What are the conclusions from these Corps studies?

C. P. Jepsen and M. Place (authors' closure)—The authors are aware of at least two test installation sites. One test was done quite recently and, as of the date of presentation of this paper, the report was not published. The earlier test site was at "Lock E" of the Tennessee-Tomhigbee Waterway. The two reports generated from the earlier test are as follows:

1. Mobile, Alabama District Army Corps of Engineers "Lock E" Test Program, Tennessee-Tomhigbee Waterway, Mobile District, Dec. 1978.

2. Mobile, Alabama District Army Corps of Engineers "Design Memorandum No. 31, "Lock "E," Lock, Slurry Wall and Earthwork, Tennessee-Tomhigbee Waterway, Mobile District, July 1979.

To the best of the authors' knowledge, the results of the Corps studies are interpreted solely in the context of the specific conditions (geological or otherwise) under which the tests were conducted.

G. A. Leonards[4] *(additional closure)*—Necking of a vibrated beam wall was noted only at the Sunny Point test cell. It was caused by vibrations at 90° across a wall that had not yet set to its final consistency. In this case, a new bentonite-fly ash slurry was being tested for salt water resistance. A report on the Corps' study of the vibrated beam was due out in August 1984.

B. S. Beattie[5] *(written discussion)*—Why are polymer-treated clays more resistant to chemical leachate degradation than untreated clays, since the polymers used (sodium polyacrylates) are biodegradable, water soluble in nature, and have a shelf life of four to five years at best?

C. Jepsen and M. Place (authors' closure)—To the authors' knowledge, there is only one technical paper that discusses this question![6] In that paper, the results of only one permeameter column are presented. Thus, sweeping

---

[3]Project engineer, Woodward-Clyde Consultants, Plymouth Meeting, PA 19462.
[4]School of Civil Engineering, Purdue University, West Lafayette, IN 47907.
[5]Federal Bentonite, Montgomery, IL 60538.
[6]Alther, G. B., "The Role of Bentonite in Soil Sealing Applications," International Minerals & Chemicals Corp., Des Plaines, IL 60016.

conclusions cannot be made regarding the effectiveness of polymer-treated bentonites.

Data generated by the American Colloid Co. indicate that, in fact, polymer-treated bentonites are not more resistant to chemical leachate degradation. However, the author's company has determined that bentonite treated with inorganic dispersants do significantly improve contaminant resistance. Such specially treated products enjoy patent protection and have been available for the past seven to ten years.

Anonymous (written discussion)—The question of vibrated beam verticality and windows was raised pertaining to the Corps of Engineers test cells.

F. Schmednecht (closure)—Concerning vibrated beams versus open trench slurry wall methods, the only full-scale comparison known to me is at Wheatfield, Indiana, where both methods were used under the same conditions.

In 1975, five and one-half miles of vibratory beam wall was installed from on top of a dike with depths ranging from 11.6 to 13.7 m (38 to 45 ft). The slurry material was bentonite and cement. An in-place test cell yielded an average permeability of 8×10^{-8} cm/s. There has been no noticeable leakage to date.

In 1982, less than two miles of open-trench soil bentonite was installed from on top of the dike with depths averaging about 12.2 m (40 ft). It was installed with a backhoe and clamshell. To date, 214 lineal m of wall has been repaired because of piping and boils, and the problem still exists.

The preceding direct comparison under field conditions shows the relative quality of the two techniques applied to the same conditions.

George Alther,[1] *Jeffrey C. Evans,*[2] *Hsai-Yang Fang,*[3] *and Kevin Witmer*[4]

Influence of Inorganic Permeants upon the Permeability of Bentonite

REFERENCE: Alther, G., Evans, J. C., Fang, H.-Y., and Witmer, K., **"Influence of Inorganic Permeants upon the Permeability of Bentonite,"** *Hydraulic Barriers in Soil and Rock, ASTM STP 874,* A. I. Johnson, R. K. Frobel, N. J. Cavalli, and C. B. Pettersson, Eds., American Society for Testing and Materials, Philadelphia, 1985, pp. 64–73.

ABSTRACT: A total of 16 inorganic aqueous solutions were utilized as permeants to determine their effects upon the permeability of both contaminant resistant (polymerized) bentonite and untreated bentonite. It was found that, of the aqueous solutions tested, those with potassium (K^+) cations or chloride (Cl^-) anions or both induced the largest permeability increases with increasing electrolyte concentration. Conversely, solutions with sodium (Na^+) cations or carbonate ($CO_3^=$) anions had the least impact upon the permeability of bentonite. It was also observed that doubly charged cations ($+2$) have a greater initial effect on the permeability than do singly charged cations ($+1$). Furthermore, a "saturation limit" was in evidence for $+2$ cations, indicating that beyond certain concentrations the further addition of the soluble salts had only limited additional impact upon the permeability.

For selected bentonite-contaminant combinations, slurry cracking pattern tests were conducted. A correlation between the permeability changes and the results of the cracking pattern tests was demonstrated.

Based upon the findings of this study, it is concluded that the character of the solute anions, as well as the primary cations, affect the permeability of bentonite clays. Further, the Gouy-Chapman model of diffuse double layer was found to be generally consistent with the test data.

KEY WORDS: bentonite, permeability, slurries

Disposal of hazardous and toxic wastes in the subsurface environment has resulted in the widespread application of geotechnology to the design and construction of waste containment systems. Specifically, the use of soil-bentonite slurry trench cutoff walls and soil-bentonite liners requires an under-

[1]Geologist, International Minerals and Chemical Corp., Detroit, MI 48220.
[2]Project engineer, Woodward-Clyde Consultants, Plymouth Meeting, PA 19462.
[3]Professor and director, Geotechnical Engineering Div., Lehigh University, Bethlehem, PA.
[4]Research director, D. W. Landfill, Inc., Dewart, PA 17730.

standing of the effects of contaminants upon such barriers [1]. In an effort to better understand the interaction between waste fluids and bentonite, a study was undertaken examining the effects of aqueous inorganic solutions upon certain of the bentonite properties, particularly the permeability.

A relatively rapid method to determine a permeant's effect upon bentonite has been developed by adapting a standard water loss chamber (filter press) as used for American Petroleum Institute (API) water loss (filter loss) tests [2]. This modified test technique permits evaluation of the relative permeability of bentonite slurries in response to exposure to varying concentrations of the aqueous inorganic permeants. Tests were conducted with treated (polymerized) and untreated bentonites to permit comparisons in the magnitude and also direction of the permeant effects. This served to check for test errors as both treated and untreated bentonites should show the same trends. It should be noted that since the test is "rapid," the resulting data may not precisely model long-term effects.

To supplement the permeability tests, standard cracking pattern tests were conducted with a selected permeant to investigate any correlation between permeability changes and cracking pattern differences. Standard cracking pattern test results are produced by mixing a fixed amount of stabilized slurry with an equal volume of alternate fluid (one of the aqueous inorganic permeants that were previously tested). The resulting slurry is then poured onto a level glass plate and allowed to air dry [3]. The resulting cracking pattern is then analyzed with respect to crack density.

Ultimately, it is desirable not only to observe the effects of differing pore fluids on bentonite, but also to model such effects to permit an a priori prediction of the effects. Specifically, the Gouy-Chapman model of diffuse double layer may be applicable in describing the observed trends in both the permeability and cracking pattern tests.

Procedure

A rapid test to study the compatibility of a waste with a bentonite can be performed with a water loss chamber (Fig. 1), as used for API water loss or filter loss tests. In this test, a fully hydrated slurry of 6.25% was mixed in a Hamilton Beach mixer for 20 min. After 20 min, the salt was added and the slurry was mixed for another 10 min. The mixture was then left to age for 16 h. Next, the slurry was introduced into the chamber and subjected to a 6.9×10^5 Pa (100 psi) air pressure. Upon the formation of "filter cake," the steady-state permeability was calculated by Darcy's law, utilizing the flow rate, area and thickness of filter cake, and the hydraulic gradient. Concentrations of the aqueous salts used in the slurry solution were varied from 0 to 8% by weight.

Standard cracking patterns were produced utilizing a procedure standardized at Lehigh University for conducting cracking pattern tests on clay slurries. A clay-water slurry is prepared at 5% by weight clay concentration, and

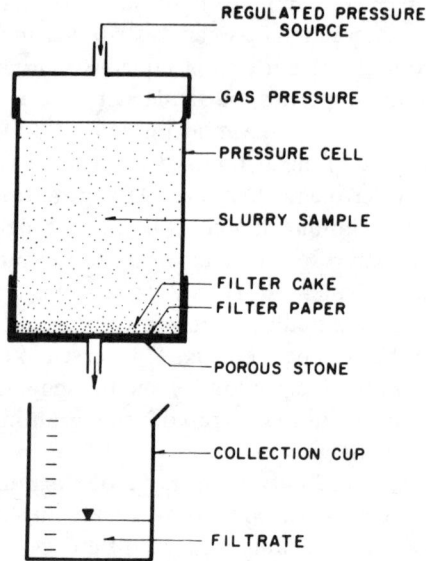

FIG. 1—*Schematic of filter-press test apparatus.*

the clay mineral is allowed to fully hydrate. Next, a fixed amount of slurry (25 mL) is extracted from the bulk sample. Next, an equal volume of aqueous solution (25 mL) is added and mixed with the clay-water slurry. The resulting slurry is then poured onto a glass plate and allowed to seek equilibrium with the gravitational stresses and subsequently to air dry to develop the characteristic cracking pattern.

Experimental Test Results

Sixteen aqueous salts were tested for their effects upon bentonite by means of a standard API water loss chamber. With two exceptions, all solutions were used with both treated and untreated bentonite slurries. The results of these experiments are presented in Table 1 along with the valence state. In order to initiate better interpretation, the data from this table were grouped by chemical association and plotted. These graphical results are presented as Figs. 2 through 5.

Standard cracking pattern tests were conducted using a combination of a bentonite-water slurry and a solution containing potassium chloride (KCl) varying from 0 to 10% by weight. The photographs of these specimens are presented as Fig. 6.

TABLE 1—Summary of permeability test results.

Contaminant	Bentonite Type (a) Untreated (b) Treated	0%	0.25%	0.5%	1%	2%	3%	3.2%	4%	5%	8%
NaCl[a]	(a)	4.14×10^{-8}	...	...	5.14×10^{-8}	5.71×10^{-8}	...	8.5×10^{-8}	...	1.31×10^{-7}	...
	(b)	3.85×10^{-8}	...	...	4.57×10^{-8}	5.71×10^{-8}	...	7.14×10^{-8}	...	1.06×10^{-7}	...
CaCl$_2$	(a)	3.85×10^{-8}	1.1×10^{-7}	1.63×10^{-7}	...	1.82×10^{-7}	...	...	...	1.83×10^{-7}	...
	(b)	3.85×10^{-8}	8.57×10^{-8}	1.43×10^{-7}	1.44	1.63×10^{-7}	...	...	...	1.66×10^{-7}	...
MgCl$_2$	(a)	3.85×10^{-8}	8.29×10^{-8}	1.4×10^{-7}	1.66×10^{-7}	1.6×10^{-7}	...	...	...	1.6×10^{-7}	...
	(b)	3.85×10^{-8}	6.29×10^{-8}	9.43×10^{-8}	1.14×10^{-7}	1.26×10^{-7}	...	...	...	1.23×10^{-7}	...
KCl[a]	(a)	3.85×10^{-8}	3.43×10^{-8}	4.57×10^{-8}	5.0×10^{-8}	...	1.57×10^{-7}	...	...	2.0×10^{-7}	...
	(b)	3.85×10^{-8}	4.91×10^{-8}	5.3×10^{-8}	7.72×10^{-8}	...	1.54×10^{-7}	...	...	2.78×10^{-7}	...
NH$_4$Cl[a]	(a)	3.85×10^{-8}	3.91×10^{-8}	5.17×10^{-8}	1.06×10^{-7}	1.66×10^{-7}	2.86×10^{-7}	...	...	...	...
	(b)	3.85×10^{-8}	3.62×10^{-8}	4.88×10^{-8}	1.0×10^{-7}	1.73×10^{-7}	2.69×10^{-7}	...	...	...	...
CaCO$_3$	(a)	3.85×10^{-8}	...	...	...	3.77×10^{-8}	4.17×10^{-8}	...	4.23×10^{-8}	4.29×10^{-8}	3.71×10^{-8}
	(b)	3.43×10^{-8}	...	...	...	3.71×10^{-8}	3.71×10^{-8}	...	3.71×10^{-8}	3.71×10^{-8}	4.43×10^{-8}
MgCO$_3$	(a)	3.43×10^{-8}	...	...	...	4.57×10^{-8}	4.97×10^{-8}	...	5.01×10^{-8}	5.89×10^{-8}	6.22×10^{-8}
	(b)	3.85×10^{-8}	...	...	...	4.11×10^{-8}	4.57×10^{-8}	...	4.86×10^{-8}	5.14×10^{-8}	6.0×10^{-8}
K$_2$CO$_3$[a]	(a)	3.85×10^{-8}	...	...	4.57×10^{-8}	6.29×10^{-8}	1.11×10^{-7}	...	1.66×10^{-7}	1.94×10^{-7}	...
	(b)	3.43×10^{-8}	...	...	5.72×10^{-8}	6.17×10^{-8}	1.11×10^{-7}	...	1.69×10^{-7}	2.06×10^{-7}	...
CaSO$_4$	(a)	3.43×10^{-8}	6.85×10^{-8}	1.06×10^{-7}	1.09×10^{-7}	...	9.42×10^{-8}	...	...	9.71×10^{-8}	...
	(b)	3.43×10^{-8}	5.7×10^{-8}	9.54×10^{-8}	8.86×10^{-8}	...	9.66×10^{-8}	...	...	9.77×10^{-8}	...
MgSO$_4$	(a)	3.43×10^{-8}	5.09×10^{-8}	9.2×10^{-8}	1.14×10^{-7}	...	1.26×10^{-7}	...	...	1.21×10^{-7}	...
	(b)	3.43×10^{-8}	4.0×10^{-8}	6.0×10^{-8}	1.11×10^{-7}	...	1.31×10^{-7}	...	...	1.37×10^{-7}	...
K$_2$SO$_4$[a]	(a)	3.43×10^{-8}	...	...	...	5.42×10^{-8}	8.86×10^{-8}	...	1.17×10^{-7}	1.46×10^{-7}	...
	(b)	3.43×10^{-8}	...	...	...	5.2×10^{-8}	7.94×10^{-8}	...	1.14×10^{-7}	1.4×10^{-7}	...
Na$_2$SO$_4$[a]	(a)	3.43×10^{-8}	...	...	...	4.29×10^{-8}	4.0×10^{-8}	...	4.0×10^{-8}	4.29×10^{-8}	...
	(b)	4.0×10^{-8}	...	...	...	3.94×10^{-8}	4.0×10^{-8}	...	4.0×10^{-8}	4.17×10^{-8}	...
Na$_2$SO$_3$	(a)	4.0×10^{-8}	...	...	...	3.9×10^{-8}	3.7×10^{-8}	...	3.66×10^{-8}	5.09×10^{-8}	...
	(b)	4.0×10^{-8}	...	...	...	4.0×10^{-8}	4.3×10^{-8}	...	4.57×10^{-8}	5.26×10^{-8}	...
NaNO$_3$[a]	(a)	4.0×10^{-8}	...	...	...	4.5×10^{-8}	5.8×10^{-8}	...	6.0×10^{-8}	8.0×10^{-8}	...
	(b)	4.0×10^{-8}	...	...	...	4.05×10^{-8}	4.05×10^{-8}	...	4.86×10^{-8}	6.57×10^{-8}	...
FeSO$_4$	(a)	4.7×10^{-8}	...	5.72×10^{-8}	9.72×10^{-8}	1.34×10^{-7}	...	...	1.66×10^{-8}	...	...
FeO	(a)	4.7×10^{-8}	...	8.86×10^{-8}	1.17×10^{-7}	1.22×10^{-7}	...	...	1.13×10^{-8}	...	...

[a]Monovalent cationic solutions.

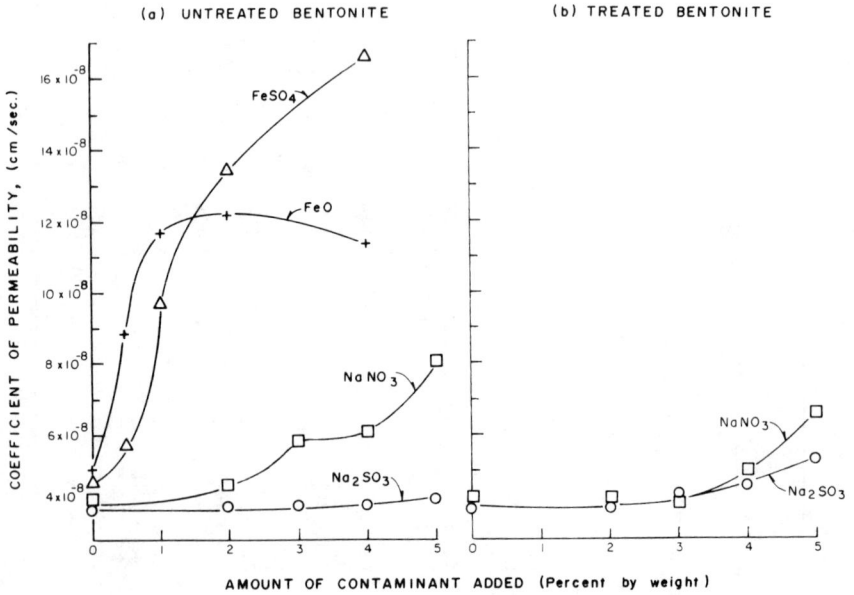

FIG. 2—Coefficient of permeability versus amount of contamination.

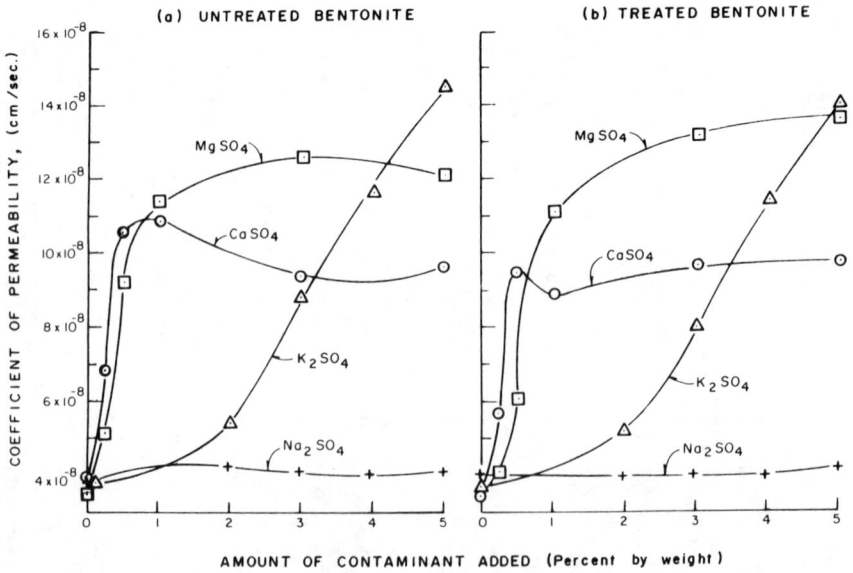

FIG. 3—Coefficient of permeability versus amount of contamination.

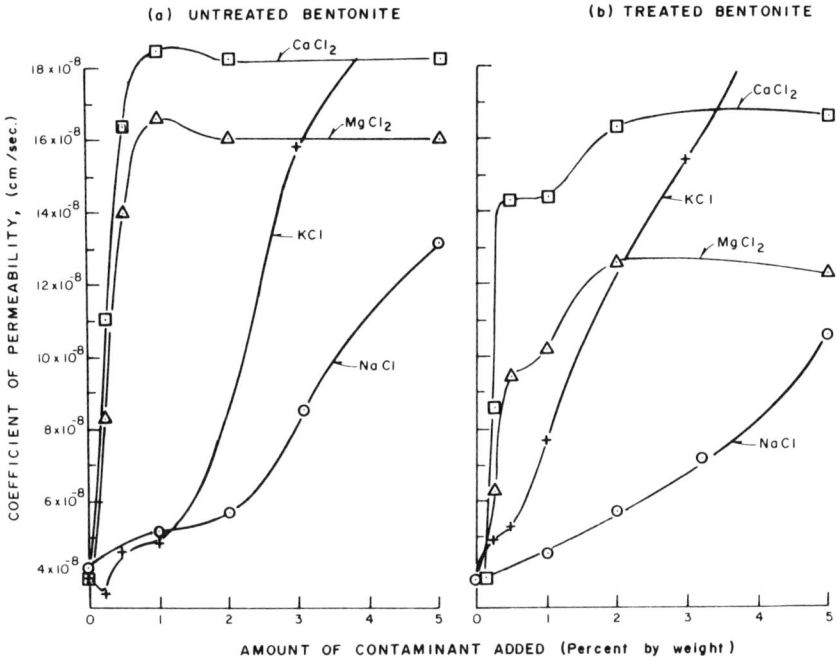

FIG. 4—*Coefficient of permeability versus amount of contamination.*

Discussion of Experimental Results

The permeability of bentonite generally increased in response to permeation with the aqueous salts tested; however, there are obvious differences in the permeability increases which are somewhat unique to each chemical. It appears (Figs. 2 through 4) that the presence of monovalent potassium cations has the most pronounced and most linear (relative to concentration) effect upon bentonite permeability. Monovalent chloride anions are also associated with extensive permeability degradation, and the rate of permeability increase (relative to concentration) appears to be controlled by this anion (see Figs. 2, 4, and 5).

Solutions containing divalent cations display a two-stage permeability change that is characterized by a rapid initial degradation followed by essentially a steady-state response. This indicates a "saturation limit" where the continued addition of salts does not result in further degradation of the permeability. This response is exhibited for the iron, magnesium, and calcium cations as shown in Figs. 2 through 4. Perhaps such a limit also exists for single valance substances at concentrations above those tested is this study.

The aqueous salts that least affect permeability are those containing sodium monovalent cations or carbonated monovalent anions, although trends

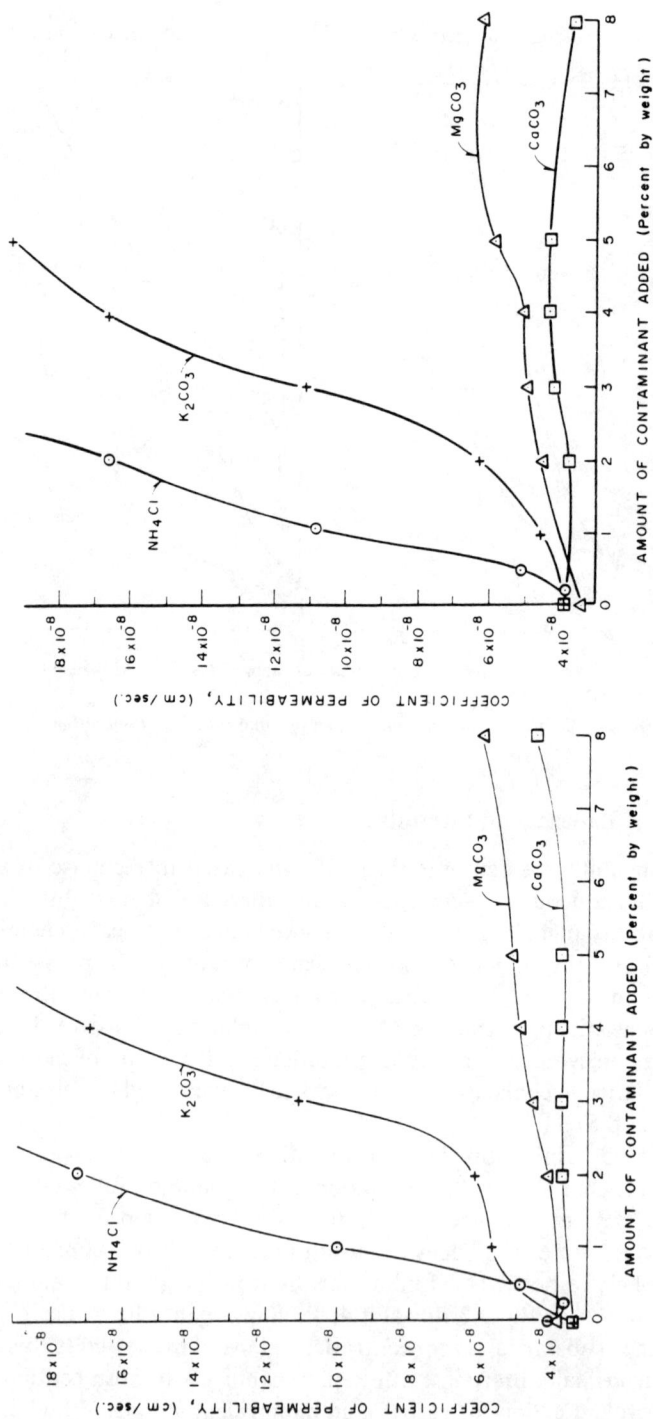

FIG. 5—Coefficient of permeability versus amount of contamination.

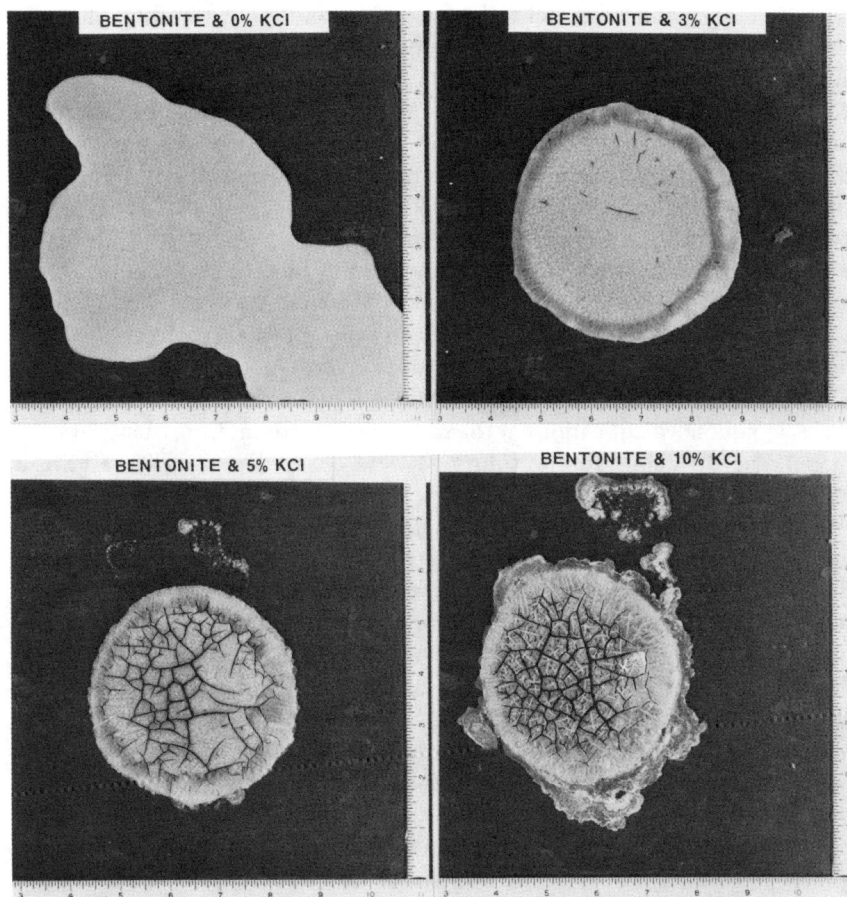

FIG. 6—*Cracking patterns observed in response to various pore fluid conditions.*

are not as well established as others. These data are shown on Figs. 2, 3, and 5.

As noted, tests were conducted utilizing both treated (that is, polymerized) and untreated (that is, nonpolymerized) bentonite. The polymerization would be expected to reduce the effect of the aqueous salt solutions upon the degradation of permeability of the bentonite filter cake. The data presented in Figs. 2 through 5 consistently present this trend. The permeability increases are consistently greater for the untreated bentonite than for the treated bentonite. The trend of the permeability increases, however, was consistent for both the treated and untreated bentonite. This further establishes the reproducibility of the data.

The findings just presented and discussed were examined for correlation with a physiochemical model to better understand the controlling phenomena. The observed trends were therefore examined for compatibility with results that would be expected from the Gouy-Chapman theory [4] and the findings of this comparison follow. The model predicts that as the electrolyte concentration in the pore fluid increases the thickness of the double layer tends to decrease. A tendency towards a more flocculated structure is caused by the decrease in the double layer thickness. Hence, an increasingly more flocculated and more permeable structure would be expected with increasing aqueous salt concentration. The permeability generally increased with increasing salt concentration as predicted by the model.

When all else is held constant, increasing the ion valance will also cause a decrease in the thickness of the double layer according to the Gouy-Chapman model. Thus, the salts with divalent cations would be more flocculated than those with monovalent cations at the same concentration. Consistent with the model, the divalent cationic solutions were observed generally to have a greater influence on the permeability than the monovalent cationic solutions.

Also, the model predicts that the smaller the hydrated ion, the closer it can approach the colloidal surface of the clay particle. Thus, all else being equal, the smaller the hydrated ion, the smaller the double layer and the greater the tendency for a more flocculated clay structure. Although this is the process predicted by the Gouy-Chapman theory, the data at hand are inconclusive. The data exhibit magnesium and calcium in compound form with the same anion [magnesium sulfate ($MgSO_4$) and calcium sulfate ($CaSO_4$), also, magnesium chloride ($MgCl_2$) and calcium chloride ($CaCl_2$)]. It is observed that $MgSO_4$ has a higher permeability than $CaSO_4$, while $CaCl_2$ has a higher permeability than $MgCl_2$. These compounds differ in the size of the hydrated cation, but it appears that the anion may have an influence.

Cracking patterns observed in response to various pore fluid conditions can be also explained by an extension of the Gouy-Chapman model. Specifically, the diffuse double layer theory may be applicable in quantifying the observed cracking mechanism. Generally speaking, in an increasingly flocculated soil structure, the volume change tendency would be evenly distributed in all directions. Hence, double layer reductions would tend to draw the particles closer together in all planes as drying progresses. Considering this, it is logical that higher tensile stresses would develop and that the release of these tensile stresses through cracking would occur more frequently with more flocculated structures. Therefore, the cracking pattern of increasingly more flocculated specimens would result in more cracks and a more densely populated cracking pattern. These phenomena were observed in the experiments as shown in Fig. 6.

Conclusions

From the experimental results just described, it is concluded that the permeability of the bentonite filter cake increases, over that measured with water, in response to permeation with the salt solutions tested. Further, the greater the concentration of the electrolyte in solution, the greater the permeability increase over the water permeability. In general, it appears that there is a saturation limit for the solutions containing divalent cations in which only limited further degradation of permeability occurs beyond some electrolyte concentration. It is further concluded that the Gouy-Chapman model can be utilized to explain permeability changes in response to permeation with electrolytic solutions. Individual parameters examined in this study were discussed in light of the Gouy-Chapman model, and this diffuse double layer model was found to be generally consistent with the test data. Note that decreasing double layer thickness consistently resulted in a more flocculated structure and a correspondingly more permeable structure. Standard cracking pattern tests were conducted with one of the inorganic permeants, and it was found that increasing the electrolyte concentration resulted in a more densely populated cracking pattern. Certain trends were found to exist that are more subtle and that were not satisfactorily explained solely by the Gouy-Chapman model. It is, therefore, concluded that additional testing is required in order to better understand the parameters that control clay pore fluid interactions.

Acknowledgments

Financial support for this work was provided by International Minerals and Chemical Corp., Detroit, Michigan and Woodward-Clyde Consultants, Plymouth Meeting, Pennsylvania.

References

[1] Evans, J. C. and Fang, H.-Y., "Geotechnical Aspects of the Design and Construction of Waste Containment Systems" in *Proceedings*, 3rd National Conference on the Management of Uncontrolled Hazardous Waste Sites, Washington, DC, Nov. 1982.
[2] American Petroleum Institute, "Recommended Practice for Standard Procedure for Testing Drilling Fluids," Specification RP 13B, 8th ed., Dallas, Apr. 1980.
[3] Andrews, R. E., Gawarkiewicz, J. J., and Winterkorn, H. F., "Comparison of Three Clay Minerals with Water, Dimethyl Sulfoxide, and Dimethyl Formalmide," Highway Research Record No. 209, 1967.
[4] Van Olphen, H., *An Introduction to Clay Colloid Chemistry*, Wiley, New York, 1977.

DISCUSSION

B. S. Beattie[1] *(written discussion)*—(1) Explain why an oil well grade feedstock bentonite material is superior to a foundry grade or taconite grade clay for toxic leachate containment? (2) Are there clays that accept chemical polymer treatment better than others and why?

G. Alther, J. C. Evans, H.-Y. Fang, and K. Witmer (authors' closure)—Oil-well grades have a lower water loss, that is, lower permeability. They have a better cation balance on the surface than foundry grades. They have a higher swelling potential than foundry grades.

W. Adaska[2] *(written discussion)*—You showed that contaminants such as calcium and potassium greatly increase the permeability of soil-bentonite slurry trenches. With cement-bentonite (C-B) slurries, cement which is high in calcium would reduce the swelling of the bentonite-water slurry immediately during mixing operations. Would not this chemical reaction tend to stabilize the permeability of a C-B slurry so that a C-B mix could be designed that would have a long-term stable permeability when exposed to contaminants such as calcium and potassium? Has any testing been done on permeability of C-B slurries exposed to contaminants?

G. Alther, J. C. Evans, H.-Y. Fang, and K. Witmer (authors' closure)—Such testing has been done, and your assumptions are correct. Please check the literature of such authors as Christopher Ryan and D. D'Appolonia for more data.

---

[1]Federal Bentonite, Montgomery, IL 60538.
[2]Senior energy and water resources engineer, Energy and Water Resources Dept., Portland Cement Association, Skokie, IL 60077.

Safdar A. Gill[1] and Barry R. Christopher[1]

Laboratory Testing of Cement-Bentonite Mix for Proposed Plastic Diaphragm Wall for Complexe LaGrande Reservoir Caniapiscau, James Bay, Canada

REFERENCE: Gill, S. A. and Christopher, B. R., **"Laboratory Testing of Cement-Bentonite Mix for Proposed Plastic Diaphragm Wall for Complexe LaGrande Reservoir Caniapiscau, James Bay, Canada,"** *Hydraulic Barriers in Soil and Rock, ASTM 874,* A. I. Johnson, R. K. Frobel, N. J. Cavalli, and C. B. Pettersson, Eds., American Society for Testing and Materials, Philadelphia, 1985, pp. 75–92.

ABSTRACT: A comprehensive laboratory investigation was carried out to determine a suitable mix of cement-bentonite for a slurry trench cutoff wall for Caniapiscau Reservoir project in James Bay, Canada in 1980. Specifications required the mix to have a permeability of less than 0.01 μm/s and to be able to sustain a plastic deformation of 10% without fissuring as measured at 90 days in a triaxial compression test with a lateral pressure of 196 kPa. Laboratory samples were prepared with different mixes of cement, bentonite, sand, and water and kept in a curing chamber of 100% humidity. Permeability determination and triaxial compression tests were performed after curing periods ranging from 10 to 120 days. Over 84 specimens were tested. The compression tests included unconfined tests, unconsolidated-undrained tests at a constant rate of strain, unconsolidated-drained tests with varying strain rates and at a controlled rate of stress, and a consolidated-drained test. Permeability tests were performed before and, in some cases, after loading. The specimens were examined for fissures and cracks after a strain of 10%. It was concluded that a mixture of 6% (by weight) Ultra Gel 180 bentonite and 160 kg of cement in each cubic metre of the mixture would meet the required specifications. Recommendations for standardized test procedures are made.

KEY WORDS: bentonite, cement, slurry trenches, permeability, test procedures, triaxial test, cutoff wall, strain, strain rate, deformation, plasticity, stress-strain curve

Slurry trench walls are increasingly being used as nonstructural cutoffs for seepage barriers in foundations of water-retaining structures [1–4]. Various

[1]Chief engineer and [1]principal engineer, respectively, STS Consultants, Ltd., Northbrook, IL 60062.

types of slurry trench cutoffs are summarized in a paper by Haug and Kozicki [5]. In the earliest construction, bentonite slurries were used primarily for stability of the trench, and the fluid was displaced by concrete or modified mixtures of soil and bentonite. The use of cement-bentonite slurry to support the trench during excavation and later as a self-hardening material to form the membrane is a relatively new technique [1,6]. The first slurry trench with soil backfill was used in North America in 1945 [7]. A detailed description of laboratory tests on soil and bentonite backfill used at Wanapum Dam in 1963 is given by LaRusso [8]. Typical mixtures of soil-bentonite for trench backfill, performance criteria of such slurry trench cutoffs, and a description of projects utilizing soil-bentonite backfill are given in several publications [3,6,8–10].

The use of cement-bentonite slurries was still in the experimental stage in 1968 [11], and the first cement-bentonite cutoff for a tailings disposal basin was used in the United States in 1976 [12] and for a dam in 1978 [13]. Published data on the testing of cement-bentonite mixtures as a self-hardening material in a trench have remained scarce. Cement-bentonite grouts have been in use much before the advent of slurry trench cutoffs. Jones [14] has studied cement-bentonite slurries for grouting and has presented data on workability, flow-ability, and the strength characteristics of mixtures with various proportions of cement, bentonite, and water. While that data can be used for the preliminary proportioning of cement-bentonite for slurry trenches from considerations of strength and self-hardening properties, no conclusion can be drawn regarding the plastic deformation behavior since no such tests were conducted. It is only recently that limited data on the strength and strain behavior of cement-bentonite for slurry trenches have been given by Millet and Perez [6] and by Jefferis [15].

This paper presents the results of a comprehensive program of laboratory tests on cement-bentonite mixtures. Testing was directed towards obtaining a suitable mixture of cement and bentonite which would meet the requirements specified for a plastic cutoff for the Caniapiscau Reservoir at James Bay, Canada. However, specific test requirements were not specified with respect to the type of triaxial test (that is, unconsolidated-undrained, consolidated-drained, or consolidated-undrained), the rate of loading, or the method and condition (before or after deformation) under which the permeability test should be performed. Also, the stress condition (unconsolidated or consolidated) for which the specimen deformation should be measured was not defined. The plastic deformation response of various mixtures under several stress and drainage conditions was evaluated. Hydraulic conductivity (permeability) was also examined before and after deformation. A need exists to establish standard procedures for evaluating plastic diaphragm walls to obtain a predictable design for field use. The test procedures and results of the tests utilized in this study should be compared with similar studies to evaluate and establish appropriate test procedures.

Requirements for the Plastic Diaphragm Cutoff

There are three important criteria for satisfactory behavior of a plastic cut-off wall: durability, deformability, and permanence. The eventual measure of performance is the hydraulic conductivity. Factors which influence the eventual hydraulic conductivity of the slurry trench cutoff are continuity and integrity, thickness of the wall, cutoff backfill properties, and connection details with surface structures. Except for the backfill properties, the other items are related to design and construction and those are beyond the scope of laboratory testing. Since in the case of a cement-bentonite cutoff the slurry used during the excavation becomes the permanent backfill, continuity and integrity are assured by controlling the slurry properties. The properties of slurry required for maintaining the stability of the sides of the trench during the excavation are the same as for a soil-bentonite cutoff or for a concrete cutoff placed in a slurry trench.

The behavior of bentonite and cement could be affected by the permeant and the nature of the surrounding soil. Thus, for permanent applications it is necessary to investigate the permeability characteristics of the proposed mixtures with in situ soils and the prevailing permeant. A deformable cutoff is desired to sustain movements from the superimposed loads and from possible ground movements during earthquakes. In the present project, only two requirements were specified for laboratory testing, namely:

1. A coefficient of permeability of less than 0.01 μm/s.
2. Ten percent plastic deformation without fissuring as measured at 90 days in a triaxial compression test at a lateral pressure of 196 kPa.

No testing for durability was required, perhaps due to the absence of detrimental salts in the groundwater and the subsoils.

The specifications did not indicate the required testing method, such as unconfined tests, consolidated-drained tests, or unconsolidated-undrained triaxial tests and as such did not indicate the point for which the 10% deformation was to be measured. Also not included were the procedures and conditions (before and after deformation) for performance of the permeability tests.

Test Samples

The proportioning of trial mixes was selected based on experience and considering the range of materials commonly used in cement-bentonite slurries. Bentonite used in the mixes was Ultra Gel 180 as manufactured by American Colloid Co. of Skokie, Illinois. The cement conformed to ASTM Specification for Portland Cement (C 150-832), Type 2, which is the same as Canadian Standard Type 20. Recognizing the possible influence of the permeant and the natural materials on the permeability and performance of the final product, sand and water were obtained from the site. To compare laboratory and field conditions, the mixes were prepared in two batches, each having the same cement-bentonite proportion. One mix was prepared with distilled water and the other mix with sand and water from the site. The proportions of cement, bentonite, sand, and water for each of the selected samples are summarized in Table 1. In the preparation of each of the samples, the proportion of the ingredients was calculated for a volume of approximately 1500 cm³. This would provide sufficient material to mold the specimens from that batch.

The measured quantity of bentonite was slowly added to the water in a glass cylinder while continuously stirring the water-bentonite mixture with a glass rod. The bentonite was added in small portions at a time, and the mixture was stirred with a glass rod for a period of approximately 1 min or until the entire bentonite was dissolved. Next, the entire mixture was stirred and shaken vigorously by moving the cylinder rapidly up and down while keeping the top covered with the palm of the hand. The mix was then left overnight for hydration. The next day the entire bentonite was found to be thoroughly suspended with no sedimentation noted at the bottom of the cylinder. The solution was thoroughly mixed again, and cement was slowly added to the bentonite solution in a manner similar to that for the bentonite. When the cement had been thoroughly mixed, sand was added and mixed in the solution for those samples which had a sand proportion. It was noted that the sand and cement were thoroughly suspended in the bentonite solution.

The specimens were prepared by placing the mixture into 51-mm (2-in.)-diameter, approximately 102-mm (4-in.)-long plastic molds which were sealed at the lower end by a plastic sheet. The placing method of the mixture consisted of pouring the fluid into the mold in three approximately equal layers. The first layer was tapped on the palm of the hand to remove trapped air at the bottom of the specimen. Each layer was rodded with a small glass rod until no air pockets were visible around the sides of the specimen. The top of the specimen was leveled with a straight edge. The mold was then placed in a humid curing chamber that complies with ASTM Specification for Moist Cabinets, Moist Rooms and Water Storage Tanks Used in the Testing of Hydraulic Cements and Concretes (C 511-80) and was maintained at a relative humidity of at least 98% at a temperature of 23°C ± 2°C.

In the first batch of six samples, a thin plastic sheet was also placed on top of each mold. It was observed that in several specimens the material was held up by this plastic sheet as it solidified, and horizontal cracks occurred in the top portion of the specimen during the curing. This was apparently due to a vacuum that developed as the specimen hydrated. To avoid such an occurrence, the upper plastic sheet was removed and was not utilized in the later batch of specimens. A total of 84 specimens representing the 14 mixes was prepared.

TABLE 1—*Proportion of trial mixes of cement-bentonite.*

Mix No.	Cement, kg/m³	Bentonite, %	Sand, %	Water	Weight Per Cubic Meter				
					Cement, kg	Bentonite, kg	Sand, kg	Water, kg	Total Weight, kg
1	163	5	5	site	163	59.1	59.1	901.6	1182.8
2	163	5	0	distilled	163	57.2	0	924.5	1144.7
3	178	5	10	site	177.8	59.7	59.7	896.5	1193.7
4	178	5	0	distilled	177.8	57.8	0	919.6	1155.2
5	193	5	5	site	192.7	60.2	60.2	891.4	1204.5
6	193	5	0	distilled	192.7	58.3	0	914.7	1165.7
7	207	5	10	site	207.5	62.8	125.7	860.8	1256.8
8	148	6	5	site	148.2	70.8	59.0	901.7	1179.7
9	148	6	0	distilled	148.2	68.5	0	925.5	1142.2
10	163	6	5	site	163	71.4	59.5	896.5	1190.4
11	163	6	0	distilled	163	69.1	0	919.8	1151.9
12	119	6	5	site	118.6	69.4	57.9	912.0	1157.9
13	148	4	5	site	148.2	46.6	58.2	911.7	1164.7
14	163	4	5	site	163.0	47.0	58.8	906.6	1175.4

Testing Procedures

Both the permeability and compression tests were performed in a triaxial chamber. Testing was performed at different curing periods. The specimens were taken out of the curing chamber just before the testing time and extruded from the molds. Each specimen was trimmed to the required size, encased in a rubber membrane, and set up in the triaxial chamber. The chamber was then filled with water. A pressure of 59 kPa was applied to the chamber water, and simultaneously a pressure of 49 kPa was applied to the pore fluid in the specimen (back pressure) by applying pressure to the drain lines connected to the porous stones located at the top and bottom of the specimen. The 10-kPa pressure differential was necessary to prevent the flow of water between the membrane and the specimen. The cell pressure and the back pressure were increased simultaneously in increments of 49 kPa to maximum values of 402 and 392 kPa, respectively. Each 49-kPa increment was maintained for a minimum of 15 min to allow equilibrium of pressure. A water head differential of approximately 300 mm was applied between the top and bottom of the specimen to aid in flushing out air from the specimen. The higher head was applied to the base of the specimen. The cell pressure and back pressure were maintained until all air was flushed out of the specimen.

Complete saturation of the specimen was determined by obtaining Skempton's B coefficient. "B-check" consists of applying a 49-kPa cell pressure to the sample with drain lines closed and monitoring the subsequent pore pressure. The B-value is then equal to the pore pressure increase divided by the cell pressure increase. When this value approaches 1 (0.95 or greater), the specimen is assumed saturated. The saturation level is necessary to obtain reliable permeability and pore pressure results during testing.

Permeability Test

After saturation of the specimen was obtained, a permeability test was performed in the triaxial chamber by maintaining a head difference between the top and bottom of the specimen. For the permeability test, the head difference was increased to approximately 900 mm, which resulted in an effective gradient ranging from 8 to 10 across the specimen.

Again, the higher head was applied to the base of the specimen. The amount of water flowing through the specimen and the subsequent change in head difference were monitored. The permeability was then calculated from Darcy's law. The testing method is regarded as a static rising head test.

In some of the specimens, the permeability test was repeated following the triaxial shear test to evaluate changes in permeability from consolidation and deformation during the shearing.

Triaxial Test

After performing the permeability test, the drain lines were closed and the chamber pressure was increased to 588 kPa to obtain an effective confining pressure of 196 kPa on the specimen. For the initial phase of tests, the drain lines were closed and the specimen was axially loaded until a strain of 10% was achieved. By closing the drain lines, an unconsolidated-undrained shear test was performed on the specimen. Readings were taken for load and deformation during the test. From these results, the principal stress difference versus strain was calculated and plotted. The specimen was then removed from the chamber and observed for fissures and cracks. In these unconsolidated-undrained tests, strain was applied at a constant rate, and the density and water content of the specimens were obtained both before and after the tests.

After a review of the initial results, the triaxial test procedures were changed to allow the specimen to drain, thus simulating the condition during initial construction of the trench and embankment. The testing was changed to an unconsolidated-drained condition by opening the drain lines during the application of the load. Initially, a low rate of strain (0.06 mm/min) was applied. It was observed during the testing that when the confining pressure was applied the specimen was consolidating by the lateral pressure at a much faster rate than the vertical load on the specimen. The vertical load resulted from the controlled rate of strain during the testing. To overcome this problem, several modifications were made in the testing. These modifications consisted of initially reducing the cell confining pressure to values as low as 49 kPa and finally increasing the strain rate for the unconsolidated drained test to 1.52 mm/min, which is the limit of our loading equipment.

Some tests were performed at a controlled rate of stress instead of at a controlled rate of strain. The stress-controlled test was performed by applying uniform constant load increments for a fixed period of time. This time period was set to allow for complete dissipation of pore pressure resulting from the axial load. Loading was continued to obtain a strain of 10% or more.

One stress-controlled, consolidated-drained test was performed on a sample of Mix 1, cured for 42 days to simulate long-term conditions after construction of the embankment over the trench. The specimen was consolidated under an effective consolidation pressure of 49 kPa for this test.

The type of tests performed, the confining pressures, and the strain rates are summarized in Table 2. All of the 120-day tests were unconfined with a strain rate of 1.52 mm/min.

After performing the tests, the specimens were taken out of the triaxial chamber and examined for the presence of fissures, cracks, and deformations.

Test Results

Permeability test results are summarized in Table 3. Pertinent data about the deformation behavior of the various specimens—including peak stress,

TABLE 2—*Summary of triaxial test conditions.*

Mix No.	10 Days			30 Days			60 Days			90 Days		
	Test Type	σ_3, kPa	Strain Rate, mm/min	Test Type	σ_3, kPa	Strain Rate, mm/min	Test Type	σ_3, kPa	Strain Rate, mm/min	Test Type	σ_3, kPa	Strain Rate, mm/min
1	UU[a]	196	0.006	UD[a]	49	0.006	UD	196	stress control	UD	196	1.52
2	...	..	..	UD	98	0.05	UD	196	stress control	UD	196	1.52
3	UU	196	0.76	UD	196	0.76	UD	196	stress control	UD	196	1.52
4	UU	196	0.06	UD	196	0.50	UD	196	stress control	UD	196	1.52
5	UU	196	0.06	UD	196	1.51	UD	196	stress control	UD	196	1.52
6	UU	196	0.06	UD	196	1.52	UD	196	stress control	UD	196	1.52
7	UU	196	0.06	UD	196	1.52	UD	196	stress control	UD	196	1.52
8	UU	196	0.06	UD	196	1.52	UD	196	1.52	UD	196	1.52
9	UU	196	0.06	UD	196	1.52	UD	196	1.52	UD	196	1.52
10	UU	196	0.06	UD	196	1.52	UD	196	1.52	UD	196	1.52
11	UU	196	0.06	UD	196	1.52	UD	196	1.52	UD	196	1.52
12	UD	196	1.52	UD	196	1.52	UD	196	1.52	UD	196	1.52
13	UD	196	1.52	UD	196	1.0	UD	196	1.52	UD	196	1.52
14	UD	196	1.52	UD	196	1.52	UD	196	1.52	UD	196	1.52

[a]UU = unconsolidated-undrained. UD = unconsolidated-drained.

TABLE 3—*Summary of permeability test results.*

	Permeability (μm/s) at Curing Period of				
Mix No.	10 Days	30 Days	42 Days	60 Days	90 Days
1	0.2	0.04	0.06 (0.05)[a]	0.02 (0.02)	0.1
2	0.2	. . .	. . .	0.1 (0.03)	0.06
3	0.1	0.2	. . .	0.01 (0.01)	0.1
4	0.3	0.2	. . .	0.06 (0.02)	0.006
5	0.1	0.1	. . .	0.07 (0.003)	0.6 (0.4)
6	0.3	0.03	. . .	0.09 (0.02)	0.09
7	0.3	0.2	. . .	0.1 (0.05)	0.09
8	0.2	0.04	. . .	0.02 (0.02)	0.05
9	0.06	0.07	. . .	0.04 (0.02)	0.05
10	0.06	0.03	. . .	0.03 (0.02)	0.02
11	0.09	0.07	. . .	0.03	0.09
12	0.3	0.2 (0.02)	. . .	0.1 (0.03)	0.1
13	0.2 (0.01)	0.3 (0.04)	. . .	0.4	0.2
14	0.3 (0.02)	0.2 (0.02)	. . .	. . .	0.1

[a]Numbers in parentheses indicate values obtained after triaxial test.

strain at peak stress, and visual description of sample condition after peak strain—are summarized in Table 4. The 60 and 90-day tests showed no bulging or cracking.

Several representative stress-strain plots are shown in Figs. 1, 2, and 3.

Discussion of Results

The following variables were involved in this testing program: (a) proportion of ingredients; (b) permeant; (c) curing period; and (d) method of testing.

The three important variables in the mixtures were the proportions of cement, bentonite, and sand. The permeants used were distilled water and site water.

With regard to the first objective of the study, that is, the hydraulic conductivity, the results show the following:

1. The coefficient of permeability of all the mixtures was in the range of 0.1 to 0.01 μm/s. Within the range of material proportions tested, it appears difficult to obtain permeability less than 0.01 μm/s.

2. The addition of 5% sand to the cement-bentonite mixture did not affect the permeability of the mixture. The addition of 10% sand appeared to increase the permeability slightly.

3. An increase in the proportion of bentonite tended to decrease the permeability. At 6% bentonite, there was a one order of magnitude decrease in permeability, compared to 4% bentonite. An increase of bentonite from 5 to 6%, however, decreased the permeability by less than 100%.

4. An increase in the cement content tended to increase the permeability.

TABLE 4—Summary of test results.

Sample No.	Peak Stress for Tests[a], kPa				Strain at Peak Stress for Tests[a], %				Sample Condition after Testing	
	A	B	C	D	A	B	C	D	Bulging	Cracking
1	27.5	263	130	151	9.5	9.9	11.9	11.0	B (slight)	A (minor), B
2	...	44.1[b]	109	155	...	11.0	13.7	11.3	A (large slump)	
3	57.9	26.5[b]	469	36.3	9.7	12.3	4.0	11.4	B	A (severe), B
4	28.4	75.5[b]	203	375	10.2	12.5	14.2	10.1	A (little), B	A (minor), B (at top)
5	47.1	228	260	161	2.2	12.2	13.6	10.0	B (slight)	A (severe), B (minor)
6	29.4	110	68.6	97.1	9.9	11.3	10.0	10.2	A, B	A (some), B (minor)
7	31.4	139	219	142	9.4	11.2	15.2	10.8	A	A (slight), B (slight)
8	53	92.2[b]	249	170	9.5	11.0	10.3	10.1	A (some)	A (minor)
9	110	106[b]	94.1	137	0.9	11.0	10.6	10.3	A (little)	A (minor)
10	74.5	129[b]	146	211	2.1	10.8	10.0	10.1	A (little), B	A (minor), B (slight)
11	53	151[b]	216	133	1.1	11.0	12.3	11.1	A (some)	A (minor)
12	54[b]	94.1[b]	84.3[b]	...	12.2	12.1	10.3	...	...	
13	53	86.3	54.9	178	11.5	10.8	11.3	11.1	...	
14	70.6	182	...	151	11.5	13.2	...	11.0	...	

[a] A = 10 days; B = 30 days; C = 60 days; D = 90 days.
[b] Stress value may not represent true stresses due to deformation resulting from consolidation under chamber pressure.

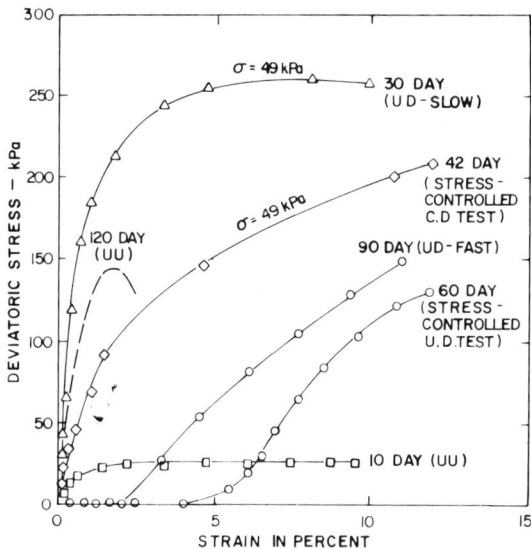

FIG. 1—*Mix 1: Variations in stress-strain behavior for different test conditions.*

5. Site water or distilled water as the permeant did not make any significant difference in the coefficient of permeability.

6. There is a trend towards a decrease in the permeability with the age of the mixture. However, the influence is less significant after a curing period of about 60 days.

7. Permeabilities tested after the consolidation and triaxial testing indicated a reduction in the value by one order of magnitude. This confirms the findings by Jefferis [15] that confining pressure reduces the permeability.

As can be seen from Fig. 1, the type of test has a significant influence on the plastic deformation characteristics as well as on the strength. An unconfined compression test is not appropriate to determine the strains which a cement-bentonite mixture can sustain without cracking or fissuring. In an unconfined compression test, strains on the order of 1 to 3% cause shear failure in all mixes commonly used for cement-bentonite cutoff. Similar strain values at peak stresses were observed in another series of tests performed by us in 1977 for a cement-bentonite cutoff proposed for Lake Chicot Pumping Station project in Arkansas. When placed in a triaxial chamber at a confining pressure and allowed to drain, the specimens were found to undergo lateral and axial deformations faster than the strain rate applied from the axial loading mechanism. This resulted in large deformations prior to the specimens actually supporting any load, as can be seen in the 60 and 90-day tests in Fig. 1 and which is even more apparent in Fig. 2. The most appropriate test for the

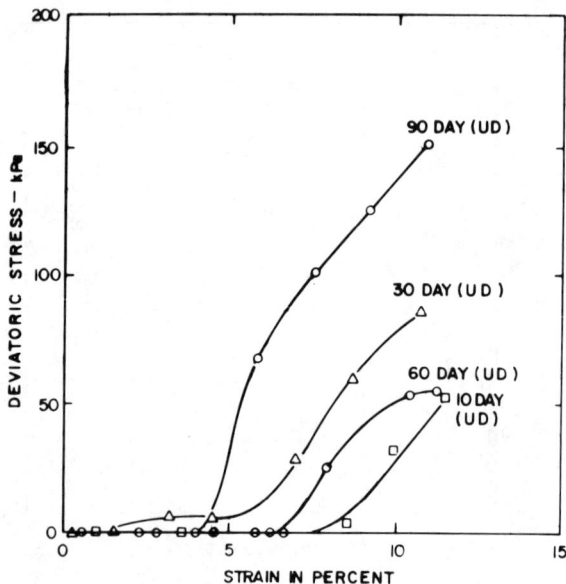

FIG. 2—*Mix 13: Large deformations due to rapid consolidation during loading of specimens in unconsolidated drained test.*

evaluation of field conditions appears to be the consolidated-drained test with variable confining pressures.

With regard to deformation response, virtually no cracking, fissuring, or bulging was observed in the specimens tested in the drained condition after 10% deformation. For the undrained condition, many of the unconsolidated-undrained specimens and all of the unconfined compression specimens reached peak stress at relatively low strains on the order of 1 to 3%. Most of these samples were observed to be severely cracked with fissures formed by shear failure planes. However, the more plastic samples that reached 10% deformation in the unconsolidated-undrained test generally showed minor or no cracking. Another indicator of the absence of fissures and cracks after 10% deformation in the drained condition is the decrease in permeability noted after the triaxial tests.

It was not the intention of this study to define the strength of the material but to observe the deformation characteristics at 10% strain. For this purpose, the specimens were removed from the test chamber at about 10 to 12% deformation for observation. From a review of the stress-strain curves, it was evident that many of the samples tested in the unconsolidated-drained test condition did not reach failure at the maximum applied strain as can be seen in Figs. 1, 2, and 3. The values of peak stress indicated in Table 4 are the maximum values attained at maximum applied strain and are not to be considered as the ultimate strength of the specimens.

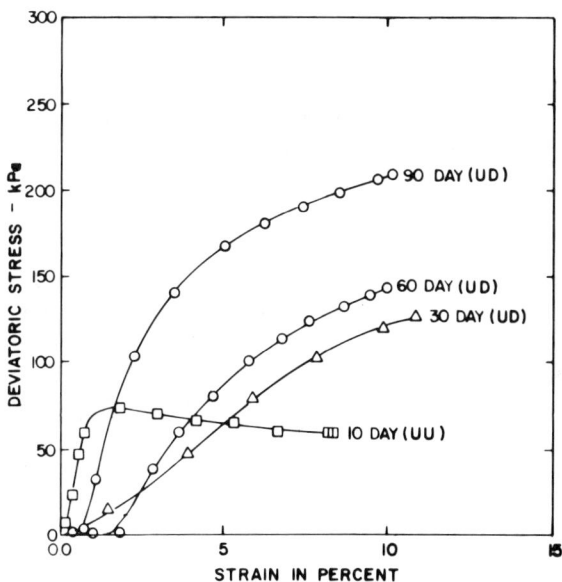

FIG. 3—*Mix 10: Stress-strain behavior of selected design mix.*

Regarding the strength behavior of cement-bentonite specimens, one of the variables previously studied by others is the cement/water ratio. The published results [6] indicate an increase in strength with the cement/water ratio. However, results obtained from this study did not establish the trend towards increase in the strength with an increase in the cement/water ratio. The tests in this study were for samples within a relatively narrow range of cement/water ratio (generally varying between 0.16 and 0.24 except for Sample 12, which had a cement/water ratio of 0.13). The lack of trend as observed in this testing program may be due to the fact that peak stresses were not attained in our tests and may also reflect the influence of testing methods. Even for the case of unconfined compression tests performed at the 120-day curing period, there was no clear trend towards increase in the strength with the cement/water ratio.

Another influencing factor in this erratic trend observed from the test results seems to be the variation in the moisture content and density of the materials tested. In certain mixes, the moisture content and density were noted to vary between the specimens with no apparent trend. This occurred even though considerable effort was made to prepare consistent specimens. Thus, these variations are attributed to the curing process and side friction in the mold and, in addition, the problem with plastic covers on the mold may have resulted in incomplete consolidation during curing. Erratic strength results with respect to curing time have also been attributed to this problem.

There appears to be a relation between increase in strength with decrease in water content and increase in density.

The overall increase in strength with time appears to be related to the apparent frictional component developed in the unconsolidated-drained test. The unconsolidated-undrained tests initially performed on the 10-day specimens compared with the unconfined compression test performed on the 120-day specimens indicated that the cohesion component of the strength either increased slightly or remained relatively constant.

The stress-strain curves reflect sample consolidation with the application of the confining pressure. In many of the samples, initial portions of the curves show large strains with very little deviatoric stress. In such cases, the axial deformation of the sample was greater than the applied strain in axial direction. The samples with higher permeability had such larger initial strain. In each case the 10-day tests showed a more plastic behavior due to the sample being not fully hardened.

Conclusions

Based on the tests performed for this project, it was concluded that cement-bentonite behaves as a deformable plastic material. It attains a stable and semisolid state in less than ten days with cement proportions as low as 119 kg/m^3. It can sustain a strain of more than 10% without cracking with cement content as high as 207 kg/m^3. A mixture of 6% bentonite with 160 kg/m^3 of cement would be adequate to meet the requirements specified for the cutoff wall for the James Bay project. For this mixture, a strength of about 50 kPa would be obtained at a strain of 10% when tested after 30 days of curing as shown in Fig. 3. The ultimate strength of this mix under a consolidated-drained condition would be likely twice that value.

The most significant conclusion from this study is the need to standardize the type of testing for cement-bentonite mixes. Standard laboratory procedures must be developed for sample preparation, specimen molding, and specimen-curing conditions. Methods used in this program are suggested as a guideline. Test methods should be established to properly evaluate the deformation and hydraulic conductivity properties of the mixture. For strength and deformation evaluation, two tests are recommended. Both unconsolidated-drained tests under low confining pressure (50 kPa) to evaluate deformation response during construction and consolidated-drained tests under design stress conditions to evaluate shear strength should be performed. Shear tests should be performed to maximum strain level (for example, 10%). A permeability test should be performed in the triaxial chamber before and after all shear tests. Deformation response should be based on permeability test results.

In preparing specimens for testing, some consideration should be given to curing under load to simulate effective stresses in situ.

One final observation that should be noted is that permeability values for cement-bentonite diaphragm wall materials should not be expected to be less than 0.01 μm/s. Additional consideration should be given to the permeability of the filter cake (anticipated to be on the order of 0.0001 μm/s for 5% or higher bentonite mixtures). Hopefully, the methods and conclusions of this test program will provide us with a basis for developing standardized procedures.

Acknowledgments

The testing described herein was performed for Bencor Corporation of America, Dallas and Petrifond Foundation Co. of Montreal. Their permission to publish the results is gratefully acknowledged.

References

[1] Boyes, R. in *Civil Engineering*, Morgan-Grampion Construction Press, London, Aug. 1980, pp. 47–51.
[2] Gill, S. A., *Journal of the Construction Division*, American Society of Civil Engineers, Vol. 106, No. C02, June 1980, pp. 155–167.
[3] Wilson, S. D. and Marsal, R. J., *Journal of the Construction Division*, American Society of Civil Engineers, 1979, pp. 46–48.
[4] Wilson, S. D. and Squier, R., *Proceedings*, Seventh International Conference on Soil Mechanics and Foundation Engineering, state of the art volume, 1969, p. 165.
[5] Haug, M. D. and Kozicki, P., *Canadian Journal of Civil Engineering*, Vol. 10, 1983, pp. 527–537.
[6] Millet, R. A., and Perez, J.-Y., *Journal of the Geotechnical Engineering Division*, Proceedings of the American Society of Civil Engineers, Vol. 107, No. GT8, Aug. 1981, pp. 1041–1056.
[7] Kramer, H., *Engineering News-Record*, Vol. 76, 27 June 1946, pp. 76–80.
[8] LaRusso, R. S. in *Grouts and Drilling Muds in Engineering Practice*, Butterworths, London, 1963, pp. 196–201.
[9] Boyes, R. G. H., *Structural and Cut-off Diaphragm Walls*, Wiley, New York, 1975.
[10] Ryan, C. R., "Slurry Cut-Off Walls, Design and Construction," technical course on slurry wall construction, design, techniques, and procedures, Chicago, 19–20 April 1976.
[11] Bush, E. and Bares, F., *Journal of the Soil Mechanics and Foundation Division*, American Society of Civil Engineers, No. SM2, March 1968, pp. 578–581.
[12] "Cement-Bentonite Slurry Wall, Saves Time, Money as Tailings Dam Cutoff," *Engineering News-Record*, 2 Dec. 1976, p. 20.
[13] "First Cement Bentonite Dam Seepage Cutoff Walls," *Geotechnical Engineering and Construction*, Dec./Jan. 1979, p. 7.
[14] Jones, G. K. in *Grouts and Drilling Muds in Engineering Practice*, Butterworths, London, 1963, pp. 22–28.
[15] Jefferis, S. A., *Proceedings*, Tenth International Conference on Soil Mechanics and Foundation Engineering, Oslo, Sweden, Vol. I, 1981, pp. 435–440.

DISCUSSION

D. Rath[1] *(written discussion)*—What kind of cement is recommended for slurry walls and what consideration is given to possible reactions which may occur between the bentonite and cement type used?

S. A. Gill and B. R. Christopher (authors' closure)—The type of cement will depend on the quality of water and does not appear to be directly related to the type of bentonite. However, no studies were made for this project. To our knowledge, there is no published data on the effect of cement on bentonite. Typically ASTM Type I cement (ASTM C 150) is used in slurry walls with regular bentonite. Type III cement (ASTM C 150) is recommended in aggressive environments where typically "Saline Seal" bentonite is used.

J. Evans[2] *(written discussion)*—No permeability data were presented; did any of the mix designs achieve the desired permeability of 1×10^{-7} cm/s?

S. A. Gill and B. R. Christopher (authors' closure)—Coefficients of permeability of all mixtures was in the range of 0.1 to 0.01 μm/s. Within the range of material proportions tested, it appears difficult to obtain permeabilities less than 0.01 μm/s. It may be unrealistic to expect a 0.01 μm/s permeability from cement-bentonite mixtures within the trench. However, the possibility of a bentonite "cake" of a much lower permeability forming along the walls of the trench should be evaluated with respect to the permeability of the entire wall. Thus the effective permeability of the wall including the cake may provide an overall effective permeability of less than 0.01 μm/s.

P. M. Jarrett[3] *(written discussion)*—The unconsolidated-drained test is a most unconventional form of test to use! One would expect results to be completely dependent on sample size. Could the authors comment on the reasons for the use of this test, the analysis of results, and the procedure used? Was pore water pressure monitored?

S. A. Gill and B. R. Christopher (authors' closure)—No doubt the unconsolidated-drained test is inappropriate for analysis of strength. However, the test is valid to simulate as-placed conditions, to evaluate hydraulic fracture potential, and to evaluate the influence of rapid pore water pressure decrease on permeability. In the field, the slurry wall will be subjected to consolidation after it has been completed as the dam is being constructed over it. The concern in such construction is whether the wall can tolerate loads during construction of the dam without fracturing, thus under an unconsolidated-

---

[1] IMC, Belle Fourche, SD.
[2] Woodward-Clyde Consultants, Plymouth Meeting, PA 19462.
[3] Royal Military College of Canada, Kingston, Ont., Canada.

drained condition. Since the ultimate criteria is to achieve a desired permeability after such conditions have been imposed, it would also appear appropriate to perform permeability tests after the specimen has been exposed to such conditions.

An additional reason for performing this type of test was to demonstrate the inadequacy of the specification in defining the test procedures. Since no procedure was defined, we felt it necessary to demonstrate that the required strains could easily be obtained in an unconsolidated-drained test even though they could not be obtained in a more conventional consolidated-drained test. Pore water pressure was not monitored and in actuality the test was an unconsolidated, partially drained test.

V. P. Drnevich[4] *(written discussion)*—Was any attempt made to study the effects of chamber fluid on test results? For long-term tests (more than a few days long), will not osmotic pressures through the membranes affect results?

S. A. Gill and B. R. Christopher (authors' closure)—As relatively short-term (three to four days) tests were performed in this study, no attempt was made to study the effects of chamber fluid on test results. The possibility of osmotic pressures through the membrane affecting test results is an interesting question and merits further research for long-term tests in flexible-wall permeameters.

Y. Acar (additional closure)—This author would like to reemphasize the importance of this specific testing consideration. Figure 4 of our paper indicates that effluent concentration reached only 25 to 30% of the influent concentration even after 10 to 14 pore volumes of permeation which took about 3 to 4 months of testing. We have found that a significant portion of the contaminant migrated into the cell water through the latex membrane. If provisions are not taken to avoid such an occurrence, it is expected that tests in flexible-wall permeameters using low concentration of contaminants as permeant are not expected to demonstrate the full effect of the contaminant on the soil.

B. S. Beattie[5] *(written discussion)*—Since cement poisons the active bentonite during curing, why was a high yield, more expensive clay preferred over a standard American Petroleum Institute (API) 13*a* drilling mud-type clay?

S. A. Gill and B. R. Christopher (authors' closure)—We are not aware of harmful effects of cement on bentonite. The use of bentonite panels as a waterproofing membrane along concrete walls is widespread with no reported cases (to our knowledge) of long-term deterioration. Cement and normal bentonite are compatible. Bentonite is also used in cement-water grouts to increase fluidity and to minimize shrinkage on setting. In this particular case, it was considered more economical to use Ultra Gel 180 bentonite as compared with a lower yield bentonite. The bentonite used did meet the requirements of API Standards 13*a*.

[4] University of Kentucky, Lexington, KY 40506.
[5] Federal Bentonite, Montgomery, IL 60538.

Anonymous (written discussion)—Did you conduct any freeze-thaw tests for this project? What were the results? Are they public?

S. A. Gill and B. R. Christopher (authors' closure)—No freeze-thaw tests were performed for this project. In our opinion, such tests are irrelevant as the cutoff was to be covered by the dam and would not be subjected to ambient freeze-thaw conditions.

David C. Anderson, [1] *Wayne Crawley,* [1] *and J. David Zabcik* [1]

Effects of Various Liquids on Clay Soil: Bentonite Slurry Mixtures

REFERENCE: Anderson, D. C., Crawley, W., and Zabcik, J. D., **"Effects of Various Liquids on Clay Soil: Bentonite Slurry Mixtures,"** *Hydraulic Barriers in Soil and Rock, ASTM STP 874*, A. I. Johnson, R. K. Frobel, N. J. Cavalli, and C. B. Pettersson, Eds., American Society for Testing and Materials, Philadelphia, 1985, pp. 93–103.

ABSTRACT: Slurry-wall backfill materials were placed in double-ring permeameters to evaluate their permeability to water, xylene, and methanol. Permeability determinations were made using double-ring permeameters because these devices separate flow that occurs near the sidewalls from flow occurring through the central portion of the specimens. The use of double-ring devices appears to overcome some of the limitations of fixed-wall and flexible-wall permeameters. Fixed-wall devices may have sidewall flow, and flexible-wall devices may be inappropriate for testing soft specimens such as a slurry mixture.

Permeability tests on materials that are to contain waste leachates are typically done using water [0.01 N calcium sulfate ($CaSO_4$)] as the permeant liquid. The results of this study show that permeability values determined with water may be misleading. Low permeabilities were obtained with water, but both xylene and methanol caused large permeability increases in the soil-slurry mixtures.

KEY WORDS: permeability, organic liquids, double-ring permeameters, sidewall flow, fixed-wall permeameter, flexible-wall permeameter, slurry wall, bentonite, materials tests, clays

Previous studies to evaluate the permeability of clays have shown large permeability increases when the tests were conducted using concentrated organic liquids [1–8]. As a result, questions have been raised as to whether similar permeability increases would take place in slurry-wall backfill materials prepared from clay soils mixed with a bentonite slurry. This study was undertaken to evaluate the permeability of a clay soil-bentonite slurry mixture when the material was permeated by organic liquids. Double-ring permeameters were used to conduct the tests.

Previous permeability tests have been conducted using either fixed-wall or flexible-wall permeameters. There are, however, disadvantages to the use of these devices to determine the permeability of soil-slurry mixtures. Sidewall

[1]Senior associate, associate, and senior associate, respectively, K. W. Brown & Associates, Inc., College Station, TX 77840.

leakage may be substantial in tests on loosely packed slurry-wall backfill tested in fixed-wall permeameters. Confining pressure may reduce the permeability of soft slurry-wall backfill tested in flexible-wall permeameters.

Double-ring permeameters appear to have the potential to overcome the problems of sidewall flow and confining pressure. Confining pressures are not used with these devices, and flow near the sidewall is separated from flow through the central portion of the specimen. Double-ring permeameters were first suggested by McNeal and Reeve as a method of eliminating boundary flow errors [9]. More recently, it has been suggested that divided outflow permeameters (such as double-ring devices) may give more reliable permeability results [10]. Consequently, double-ring permeameters may have significant advantages for permeability testing of slurry-wall backfill.

Materials and Methods

Permeameters used were similar to fixed-wall permeameters, described earlier by Brown and Anderson [1], except for the specially modified base plate that separates flow near the permeameter sidewalls from flow through the central portion of the soil core (Figs. 1 and 2). Preparation of specimens

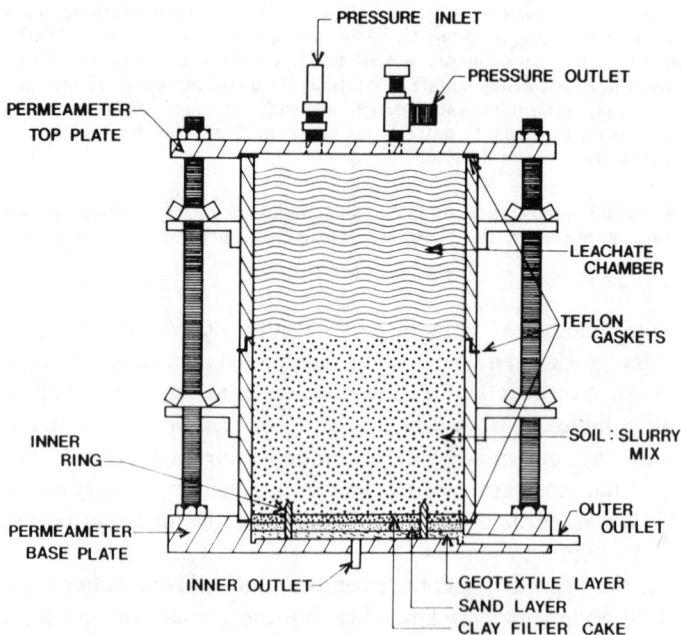

FIG. 1—*Schematic of a double-ring permeameter for use with soil-slurry mixture.*

[2]Saline Seal-100, manufactured by American Colloid Co., Skokie, IL 60077.
[3]Brinkman, R., technical representative of American Colloid Co., private communication, 1982.

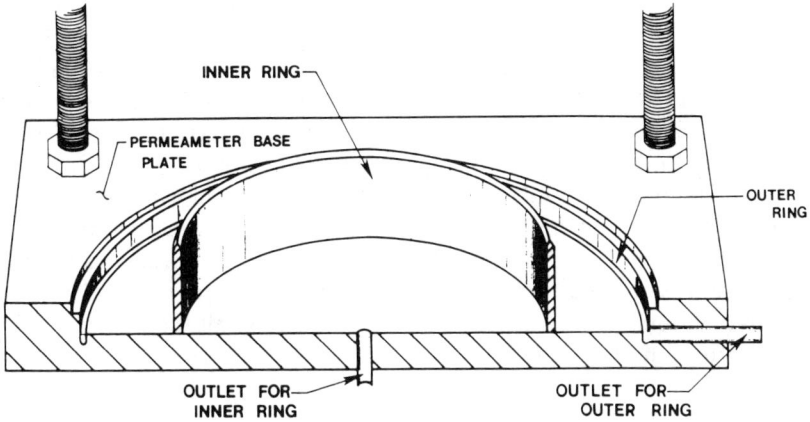

FIG. 2—*Details of the modified base plate to the double-ring permeameter.*

for testing included placement of a filter cake followed by deposition of a clay soil-slurry mixture through a 9% solution of "contaminant resistant" bentonite.[2] A baseline permeability for each specimen was determined using water (0.01 N CaSO$_4$). This was followed by permeability determinations using either xylene or methanol.

A slurry of contaminant-resistant bentonite was prepared with distilled water as suggested by the clay manufacturer using 90 kg bentonite/m^3 water.[3] While the distilled water was mixed in a blender (on low speed), the clay was slowly added until the 9% solution was obtained. The slurry was then mixed for 2 min at low speed and an additional minute at high speed. Finally, the slurry was poured through a screen to assure that all clay aggregates were reduced to less than 2 mm in diameter. The 9% bentonite slurry was poured into the 15-cm double-ring permeameters. To simulate the filter cake that forms on the sidewall of a slurry trench, the slurry was left undisturbed to allow the clays to settle out and form a 1-cm-thick layer over an underlying sand layer. A cross section of the permeameters used shows these layers (Fig. 1).

Soil-slurry mixtures were then prepared using the same calcareous smectitic clay soil described in earlier studies [1,3]. Air-dried soil was mixed with the 9% bentonite slurry until a homogeneous paste was obtained. Soil and slurry were mixed to obtain a slump of 125 to 150 mm (as measured with a standard concrete slump cone). Moisture content and slump of the soil-slurry mixtures averaged 69% (dry weight) and 140 mm, respectively.

Soil-slurry mixtures were then poured into the permeameters displacing the bentonite slurry. Soil-slurry mixtures were worked with a jabbing action using putty knives until nearly all pockets of slurry were displaced by the soil-slurry mixture. Care was taken not to disturb the continuity of the underlying filter cake. At this point the specimens were ready to be tested.

The permeability tests began by applying water (0.01 N CaSO$_4$) as the permeant liquid. After one month of permeation by water, an additional 5-cm-thick layer of soil-slurry mixture was added to each 15-cm-diameter permeameter. After an additional week of permeation by water, one specimen was extruded for tests to determine water content, pore volume, and other measurements needed to determine permeability (Table 1).

Consolidation of materials in an actual field slurry wall will increase with depth in the trench. Although no overburden pressure was used to hasten consolidation of the specimens in the present study, the specimens did undergo several centimeters of consolidation within the first month following soil-slurry placement. Permeability was calculated on all samples using the following form of Darcy's law

$$k = \frac{V}{ATi} \qquad (1)$$

where

k = permeability constant, cm s^{-1},
V = volume of liquid discharged, cm^3,
A = cross-sectional area of flow, cm^2,
T = time over which discharge is measured, s, and

$$i = \text{hydraulic gradient} = \frac{\Delta h}{L} \qquad (2)$$

TABLE 1—*Properties of soil-slurry mixture after placement in the permeameters.*

Soil Property	15-cm Diameter Double-ring Permeameter	
	Inner Compartment	Outer Compartment
Total volume, cm^3	1823	1807
Water content, % dry weight	69	69
Estimated pore volume, cm^3	1258	1247
Cross-sectional area of flow, cm^2	91.6	90.8
Soil core length, cm	19.9	19.9
Hydraulic head, cm H$_2$O	700	700
Hydraulic gradient	36.2	36.2

where

Δh = difference in pressure between the top and bottom of the specimen, cm of H_2O, and

L = length of soil-slurry mixture, cm.

Permeability tests on soil-slurry mixtures should be conducted at hydraulic gradients below the critical value above which permeability increases will occur [11]. To assure that tests with simulated waste leachates were conducted below the critical gradient, stable baseline permeabilities were obtained with water (0.01 N CaSO$_4$). Water was then replaced with either the nonpolar organic liquid (xylene) or the polar organic liquid (methanol). Physical and chemical properties of the liquids are given in Table 2. Permeability, leachate volume, and time were recorded for each pore volume of liquid passed during the tests so that permeability could be plotted against pore volume of liquid passed.

Results and Discussion

After allowing the soil-slurry mixture to consolidate during four weeks of passing water (0.01 N CaSO$_4$) and after placing an additional 5-cm soil-slurry layer in the permeameter, stable baseline permeabilities were obtained. Next, the samples were permeated by either xylene or methanol; the results of these permeability tests follow.

Permeabilities and breakthrough curves for the inner and outer compartments of the soil-slurry mixture treated with xylene are given in Fig. 3. Breakthrough of xylene in the effluent was determined by recording the volume of each liquid layer (water and xylene) in the effluent collection bottles. Both the inner and outer chambers exhibited breakthrough of xylene at very low pore volume values (<0.2 pore volume). Permeabilities of the inner and outer chambers increased concurrently with xylene breakthrough.

The soil-slurry mixture in the outer chamber underwent a permeability increase greater than two orders of magnitude by the time two full pore volumes

TABLE 2—*Physical and chemical properties of the liquids studied.*

Property	Water	Xylene	Methanol
Density at 20°C, g cm^{-3}	0.98	0.87	0.79
Viscosity at 20°C, cP	1.00	0.81	0.54
Dielectric constant at 20°C	80.4	2.4	31.2
Dipole moment, debyes	1.83	0.40	1.66
Water solubility at 20°C, g L^{-1}	...	0.20	miscible
Molecular weight	18	106	32
Melting point, °C	0	−47	−98
Boiling point, °C	100	139	65

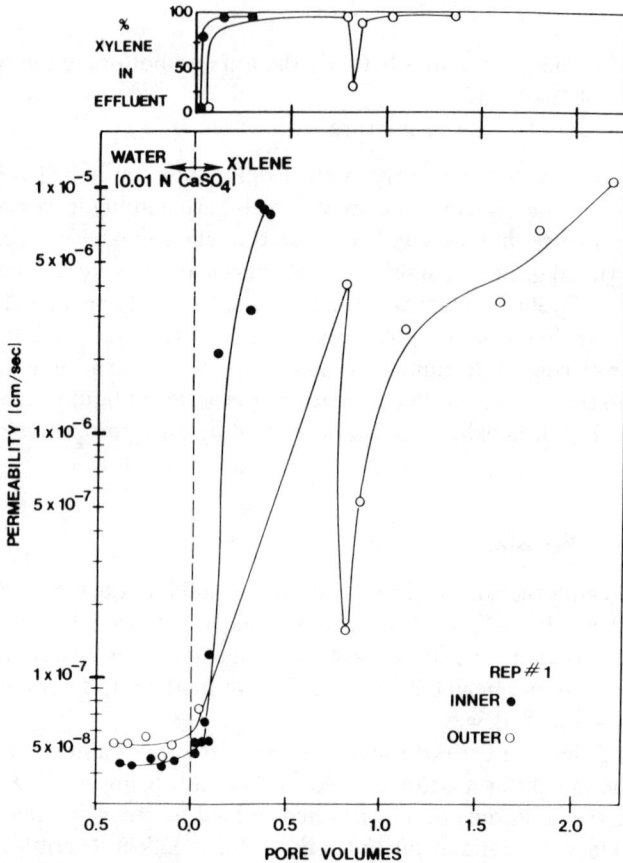

FIG. 3—*Permeability to water (0.01 N CaSO₄) and xylene of soil-slurry mixture composed of calcareous smectitic clay soil mixed with a 9% solution of "contaminant resistant" bentonite placed in 15-cm-diameter double-ring permeameters (Replicate 1).*

of xylene had passed. Permeability of the mixture in the inner chamber, however, had increased two orders of magnitude after passage of only one third of a pore volume.

One interesting aspect of the xylene breakthrough in the outer chamber of Replicate 1 was the increase in water in the effluent at 0.80 pore volume (Fig. 3). This coincided with a greater than one order of magnitude drop in permeability. Apparently the xylene was initially moving down a channel along the sidewall. This sidewall channel appeared to temporarily self-heal, forcing the xylene to percolate through the soil matrix and displace additional water. The apparent self-healing, however, was not sufficient to prevent shrinkage within the soil matrix, and the permeability subsequently increased approximately two orders of magnitude.

Permeability values for Replicate 2 are given in Fig. 4. This specimen exhibited a higher initial permeability to water in the outer chamber than in the inner chamber. The result was a breakthrough of xylene and a permeability increase in the outer chamber before xylene breakthrough in the inner chamber. These results indicate that specimens selected for testing should have nearly equal inner and outer compartment baseline permeabilities.

Specimens permeated with methanol had nearly equal initial permeabilities to water in both the inner and outer chambers, and both chambers underwent large permeability increases (Fig. 5) upon addition of the organic liquid. These increases occurred before passage of 0.75 pore volume of methanol. Both inner chambers and one of the outer chambers underwent greater than two orders of magnitude increases in permeability. The outer chamber on the

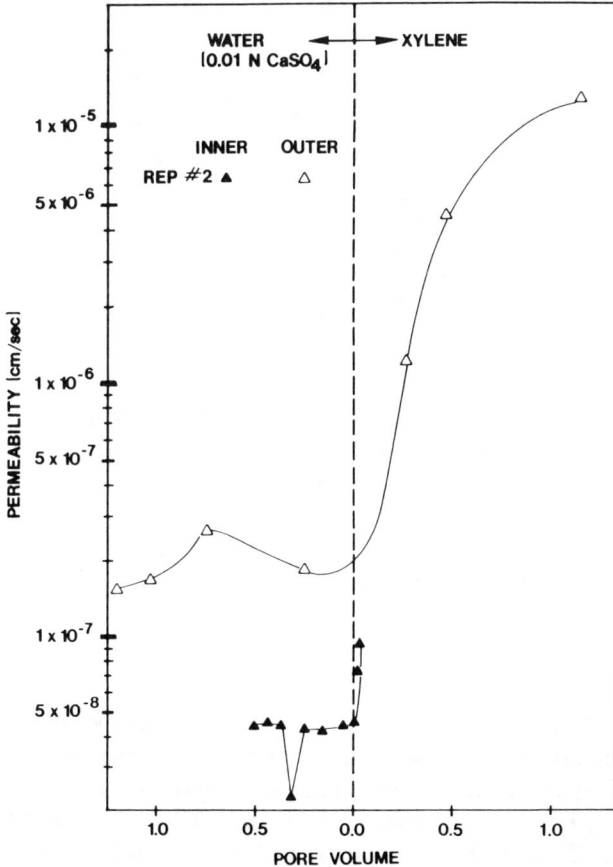

FIG. 4—Permeability to water (0.01 N CaSO₄) and xylene of soil-slurry mixture composed of calcareous smectitic clay soil mixed with a 9% solution of "contaminant resistant" bentonite placed in 15-cm-diameter double-ring permeameters (Replicate 2).

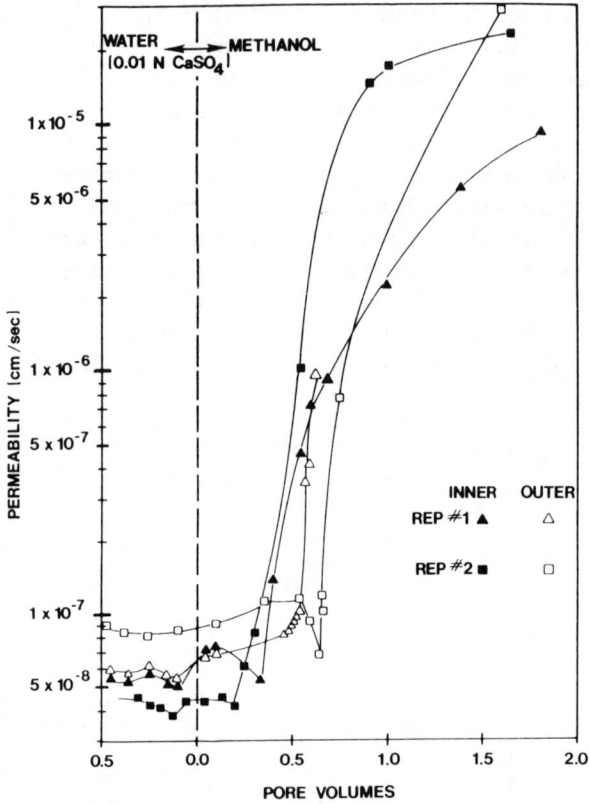

FIG. 5—*Permeability to water (0.01 N CaSO₄) and methanol of soil-slurry mixture composed of calcareous smectitic clay soil mixed with a 9% solution of "contaminant resistant" bentonite placed in 15-cm-diameter double-ring permeameters (Replicates 1 and 2).*

second replicate specimen appeared to be exhibiting a similar permeability increase when the tests were discontinued.

Conclusions

Stable baseline permeabilities were obtained in the double-ring permeameters when the clay soil-bentonite slurry mixtures were permeated by water. Large permeability increases were obtained, however, when the mixtures were subsequently permeated by organic liquids. This occurred with both the nonpolar liquid (xylene) and the polar liquid (methanol). Permeability increases occurred in both the inner and outer chambers of the permeameters.

By using double-ring permeameters, it was possible to differentiate flow near the sidewalls from flow through the central portion of the specimens, thus giving a more accurate value for the permeability than could otherwise be obtained.

Acknowledgments

We wish to thank Walter E. Grube, Jr. of the U.S. Environmental Protection Agency Municipal Environmental Research Laboratory for his technical assistance and encouragement. Work described in this report was partially supported by the U.S. Environmental Protection Agency. Mention of commercial products does not constitute endorsement or recommendation for use.

References

[1] Brown, K. W. and Anderson, D. C., "Effects of Organic Solvents on the Permeability of Clay Soils," EPA Report for Contract No. R 806825010, Municipal Environmental Research Laboratory, Cincinnati, OH, 1983.

[2] Schram, M., "Permeability of Soils to Four Organic Solvents and Water," M.S. thesis, University of Arizona, Tucson, AZ, 1983.

[3] Anderson, D. C., "Organic Leachate Effects on the Permeability of Clay Soils," M.S. thesis, Texas A & M University, College Station, TX, 1981.

[4] White, R., "Remolded Soil Samples for Proposed Waste Landfill Site North of Three Rivers, Texas," Report No. 76791, Trinity Engineering Testing Corp., Corpus Christi, TX, 1976.

[5] Buchanan, P. N., "Effect of Temperature and Adsorbed Water on Permeability and Consolidation Characteristics of Sodium and Calcium Montmorillonite," Ph.D. dissertation, Texas A & M University, College Station, TX, 1964.

[6] Van Schaik, J. C., Canadian Journal of Soil Science, Vol. 54, 1974, pp. 331–332.

[7] Macey, H. H., Transactions, British Ceramic Society, Vol. 41, 1942, pp. 73–121.

[8] Brown, K. W., Green, J., and Thomas, J. in Land Disposal, Incineration and Treatment of Hazardous Waste, D. W. Shultz, Ed., Proceedings of the Ninth Annual Research Symposium, EPA-600/9-83-002, U.S. Government, Washington, DC, 1983.

[9] McNeal, B. L. and Reeve, R. C., Proceedings, Soil Science Society of America, Vol. 28, 1964, pp. 713–714.

[10] McIntyre, D. S., Cunningham, R. B., Vatanakul, V., and Stewart, G. A., Soil Science, Vol. 128, No. 3, 1979, pp. 171–183.

[11] Xanthakos, P. P., Slurry Walls, McGraw-Hill, New York, 1979.

DISCUSSION

T. B. Edil[1] *(written discussion)*—The author concludes that the double ring separates the side flow. Does he have a test where he has side flow but no cracks in the center to prove this?

D. C. Anderson, W. Crawley, and D. Zabcik (authors' closure)—A test of this type is built into every permeability study. After compaction and before permeation by the test liquid, all samples are permeated by water. This yields a baseline permeability for both the inner and outer chambers. If there is significant sidewall flow, it is immediately apparent as a higher permeability in the outer chamber. Generally, 20 to 30% of the samples are discarded because of unacceptably high rates of sidewall flow. Routine inspection of these discarded samples has not revealed any obvious imperfection such as the cracks or aggregations seen in samples permeated by organic liquids.

T. B. Edil (written discussion)—How does he apply Darcy's law to the double-ring permeameter?

D. Anderson, W. Crawley, and D. Zabcik (authors' closure)—Darcy's law was applied as shown in the materials and methods section using values given in Table 1. Values for the cross-sectional area were determined as follows (see Fig. 2):

1. For the inner chamber, the area used was that inside the inner ring.
2. For the outer chamber, the area used was that inside the outer ring minus the area inside the inner ring.

C. M. Jones[2] *(written discussion)*—For the cracks shown in clay liners, what was the degree of compaction?

D. C. Anderson, W. Crawley, and D. Zabcik (authors' closure)—The soil-slurry mixtures discussed in the paper were not compacted. During the presentation, however, examples of laboratory and field clay liners with cracks were shown. The field clay liners had been compacted but the degree of compaction is unknown. The laboratory clay liner had been compacted to 95% of standard Proctor.

C. M. Jones (written discussion)—If the clay is compacted to a high density, such as 90% of standard Proctor, how much cracking occurs?

D. C. Anderson, W. Crawley, and D. Zabcik (authors' closure)—Many factors affect the extent, if any, that compacted clays will crack upon either

---

[1]Professor of civil and environmental engineering, University of Wisconsin, Madison, WI 53706.
[2]Soils engineering technical specialist, U.S. Bureau of Reclamation, Denver, CO 80225.

desiccation or exposure to concentrated organic liquids. For instance, Seed and Chan (1959)[3] found that both water content and the method of compaction would affect the shrinkage potential.

G. W. Gee[4] *(written discussion)*—Is there any justification to running your fixed-wall permeability tests without any vertical confinement?

D. C. Anderson, W. Crawley, and D. Zabcik (authors' closure)—The double-ring test was developed to evaluate the compatibility of barrier materials for various permeant liquids rather than to exactly duplicate barrier trench conditions. Little or no vertical confining pressure would, however, be representative of conditions near the surface in a barrier trench. In addition, many barrier materials become stiff after consolidating (as with bentonite-soil materials) or setting (as with soil cement or asphalt-cement emulsions). These barrier materials can have localized areas of low vertical confining pressure due to arching.

The first series of tests evaluated two types of barrier material (asphalt-cement emulsion and "contaminant-resistant" bentonite clay) exposed to concentrated organic liquids (miscible and immiscible). These tests were all conducted using no vertical confining pressure. Additional tests are planned that will evaluate these barrier materials permeated by several dilutions of miscible organic liquids and under a range of overburden pressures.

G. W. Gee (written discussion)—Can't vertical confinement, including overburden pressures, etc., drastically affect your permeability test results?

D. Anderson, W. Crawley, and D. Zabcik (authors' closure)—Overburden pressure generally results in a decreased permeability for most barrier materials. The question that remains to be answered is whether larger overburden pressures would reduce the permeability increases observed in bentonite materials exposed to concentrated organic liquids at low overburden pressures.

G. W. Gee (written discussion)—Wouldn't there tend to be considerably less swelling and shrinkage of "active" soils, such as smectites, if the sample is confined vertically or put under some vertical stress?

D. Anderson, W. Crawley, and D. Zabcik (authors' closure)—Increasing vertical confining pressure would result in decreased vertical swelling. Such a vertical load, however, would have little effect on lateral shrinkage and would tend to increase vertical shrinkage.

G. W. Gee (written discussion)—What would be the effect of very low concentration (500 ppm) xylene on smectite clay permeability?

D. Anderson, W. Crawley, and D. Zabcik (authors' closure)—Several studies have found that xylene at its solubility limit (less than 200 ppm in water at 25°C) has no discernable effect on the permeability of clay liners. If concentrations greater than the solubility limit are found, a separate, concentrated, immiscible layer of xylene may well be floating on the sampled water.

[3]Seed, H. B. and Chan, C. K., "Structure and Strength Characteristics of Compacted Clays," *Journal of Soil Mechanics and Foundation Division*, Proceedings, ASCE, 85, SM5, Oct. 1959.
[4]Battelle, Pacific Northwest Laboratory, Richmond, WA 99352.

Clay and Soil-Admix Liners

David E. Daniel,[1] David C. Anderson,[2] and
Stephen S. Boynton[3]

Fixed-Wall Versus Flexible-Wall Permeameters

REFERENCE: Daniel, D. E., Anderson, D. C., and Boynton, S. S., **"Fixed-Wall Versus Flexible-Wall Permeameters,"** *Hydraulic Barriers in Soil and Rock, ASTM STP 874,* A. I. Johnson, R. K. Frobel, N. J. Cavalli, and C. B. Pettersson, Eds., American Society for Testing and Materials, Philadelphia, 1985, pp. 107–126.

ABSTRACT: Permeameters are of two general types: fixed-wall and flexible-wall cells. A controversy has developed over which type of cell is best suited for measuring the hydraulic conductivity of relatively impermeable, fine-grained soils. The various types of permeameters are discussed and their relative advantages and disadvantages are listed. Differences in applied stress, boundary leakages, and degree of saturation are the major differences between cells. It is concluded that no one type of cell is best suited to all applications. Data show that the type of permeameter used has little effect for laboratory-compacted clay permeated with water but can have a major effect for clays permeated with concentrated organic chemicals. Fixed-wall cells are perhaps best suited to testing laboratory-compacted clays that will be subjected to little or no effective overburden pressure in the field. Flexible-wall cells are better suited to testing undisturbed samples of soil (to minimize boundary leakages) and testing soils that will be subjected to significant effective stress.

KEY WORDS: permeability, hydraulic conductivity, permeameter, clay, compacted clay, triaxial, triaxial cell, flexible-wall cell, fixed-wall cell, consolidation cell, compaction mold

In the past several years a controversy has developed over the use of fixed-wall versus flexible-wall permeameters for measuring the hydraulic conductivity (k), or coefficient of permeability, of clayey soils. Fixed-wall devices have the advantages of low cost and convenience for testing compacted soils. However, there may be imperfect contact between the soil and the inside of the fixed-wall cell, which can lead to so-called "sidewall" leakage and to erroneously large measurements of k. Sidewall leakage may be particularly im-

[1]Assistant professor of civil engineering, University of Texas, Austin, TX 78712.
[2]Associate, K. W. Brown and Associates, College Station, TX 77840.
[3]Geotechnical engineer, Geotechnical Engineers, Inc., Winchester, MA 01890.

portant when the soil is permeated with a concentrated organic chemical because the chemical may cause the soil to shrink and to pull away from the walls of the permeameter. Flexible-wall devices not only tend to minimize sidewall leakage but also are convenient for testing with back pressure, for measuring volume change within the soil specimen, and for controlling both the horizontal and the vertical effective stress. However, if the effective stress applied to the soil in a flexible-wall cell exceeds the effective stress in the field, the measured k may be too low.

The advantages and disadvantages just mentioned provide background for the issue of whether fixed-wall or flexible-wall permeameters should be used for measuring the hydraulic conductivity of fine-grained soil. The primary purposes of this paper are (1) to identify as completely as possible the advantages and disadvantages of each type of permeameter, (2) to present comparative measurements of hydraulic conductivity obtained with fixed- and flexible-wall permeameters, and (3) to make specific recommendations for the types of permeameters that should be considered for various applications.

Fixed-Wall Permeameters

Compaction-Mold Permeameters

Compaction-mold permeameters are used extensively for measuring the hydraulic conductivity of compacted clays and sand-bentonite mixtures. The permeameter is assembled by clamping a compaction mold between two end plates. A typical design involves a standard 10.14-cm-diameter compaction mold [ASTM Test Methods for Moisture-Density Relations of Soils and Soil-Aggregate Mixtures, Using 5.5-lb (2.49-kg) Rammer and 12-in. (304.8-mm) Drop (D 698-78), Method A], although other sizes are also used. The soil is compacted into the mold, the ends of the soil specimen are trimmed flush with the mold, the permeameter is assembled, and permeation is initiated. A collar that contains a reservoir of permeant liquid may be included in the permeameter (Fig. 1*a*) or may be omitted (Fig. 1*b*). If the permeant liquid is not stored in the collar above the compaction mold, it is stored in a separate reservoir that is connected to the permeameter with suitable tubing. The permeant liquid is often pressurized to achieve a suitably large hydraulic gradient to reduce testing times. Atmospheric pressure is normally maintained in the effluent line. The rate of flow is measured either by collecting the effluent liquid or by monitoring the rate of inflow. No attempt is made to restrict swelling of the soil on the assumption that the soil will be free to swell in the field. Other types of compaction-mold cells permit application of a vertical stress. Back pressure is normally not used, but there is no reason why it could not be employed. Further details on the method of testing may be found in Refs *1–3*.

This type of permeameter is the simplest and most economical for testing

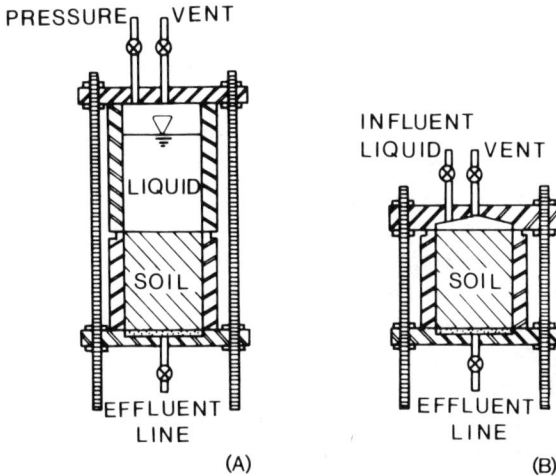

FIG. 1—*Compaction-mold permeameters with* (a) *a reservoir of permeant liquid contained within a collar located directly above the soil specimen and* (b) *a separate reservoir of permeant liquid.*

compacted clays. The disadvantages of the compaction-mold device are the possibility of incomplete saturation of the soil if back pressure is not used, lack of capability for measuring the amount of shrinkage or swelling, lack of complete control over the stresses that act on the soil, and the potential for sidewall leakage.

Sidewall leakage is of particular concern when clays are permeated with organic chemicals. Anderson and Brown [1,2] permeated several compacted clays with water and then switched the permeant liquid to various concentrated organic chemicals. Hydraulic conductivities to water were 1×10^{-8} cm/s or less. The organic chemicals generally caused two to three orders of magnitude increase in k compared to the values obtained with water. The large increase in hydraulic conductivity was apparently caused primarily by shrinkage of the soil or a type of chemical desiccation. The shrinkage led to formation of microcracks, which in turn led to high hydraulic conductivity. The possibility exists that some portion of the observed increase in hydraulic conductivity was caused by sidewall leakage. More recently, Brown, Green, and Thomas [4] tagged the organic solvents with brightly colored tracers and have examined the soil specimens after the tests were completed. Internal microcracks and fractures were stained by the tracer but there was no visible discoloration along the exterior of the cylindrical specimens. These observations suggest that sidewall leakage was nil.

Independent tests on kaolinite were recently performed at The University of Texas to evaluate the importance of sidewall leakage. The liquid and plastic limits of the kaolinite were 58 and 35%, respectively. Kaolinite was compacted into a compaction-mold permeameter at optimum moisture content. The hydraulic conductivity of the specimen to water was 6.0×10^{-8} cm/s. The permeant liquid was then changed to heptane (a concentrated organic solvent with a low dielectric constant) and k quickly increased to 1.3×10^{-5} cm/s and remained there for several pore volumes of flow. The top of the permeameter was removed, at which time the condition depicted in Fig. 2a existed. A visible gap was observed between the soil and the wall of the per-meameter at the top of the cell (the end exposed directly to heptane) but not at the other end of the cell. The opportunity for sidewall leakage clearly existed at the top of the specimen, but the depth over which sidewall leakage might have occurred was unknown. To investigate the magnitude of sidewall leak-age, approximately 2 mm of soil was first trimmed from the upper surface of the specimen as indicated in Fig. 2b. The permeameter was reassembled, the soil was repermeated with heptane, and k was found to be 1.9×10^{-5} cm/s, which is almost identical to the value measured before the surface of the clay was trimmed. Next, the permeameter was dismantled, and a bead of high-vacuum silicone grease was applied to the contact between the soil and the wall of the permeameter as indicated in Fig. 2c. The purpose of the grease was to seal off the soil-wall interface from direct contact with the reservoir of permeant liquid. The permeameter was reassembled, the soil was reper-meated with heptane, and the value of k dropped by one order of magnitude to 1.2×10^{-6} cm/s. In this case, it was clear that part of the three-order-of-magnitude increase in k that was observed when the permeant liquid was changed from water to heptane was the result of sidewall leakage, but much of the observed increase was not the result of open gaps existing along the sidewall. The bulk of the increase in hydraulic conductivity was probably the result of internal cracks that formed when the clay shrank as it was exposed to heptane [1,4].

The key question is whether concentrated organic chemicals would also cause open cracks to develop in the field and, if so, whether the cracking pattern in the laboratory would match the pattern in the field. Field tests on

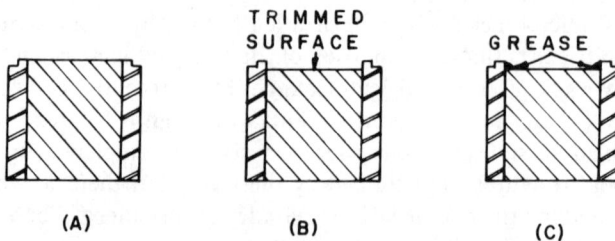

FIG. 2—*Compaction mold containing a soil specimen that was permeated with heptane. Condition* (a) *represents the initial test situation. Condition* (b) *represents the trimming of 2 mm of soil from the surface of the soil specimen. Condition* (c) *represents an attempt to seal off any sidewall leakage with a bead of high-vacuum silicone grease.*

prototype clay liners indicate that liners in the field do indeed undergo large increases in k when exposed to concentrated organic solvents and that the values measured in compaction-mold permeameters are similar to values measured in the field [4]. The prototype liners were subjected to little overburden pressure, and, in this sense, the laboratory and field conditions matched well. If the prototype liners had been subjected to significant overburden pressure, perhaps the cracks in the liner would have remained closed and k might have been much less.

Double-Ring Permeameter

Anderson et al [5] have developed a modified compaction-mold permeameter known as a double-ring permeameter. As indicated in Fig. 3, a 15-cm-diameter mold contains a specimen of soil while a ring built into the base plate separates the outflow that occurs through the central portion of the soil from the outflow that occurs near or along the sidewall. If there is significant sidewall leakage, the rate of flow into the outer collection ring will be much greater than the rate of flow into the inner ring. If the rates of flow into the two rings do not agree, one could either assume that the rate of flow into the inner ring is more nearly correct or, even better, reject the test and set up a new one. Because of uncertainties in hydraulic gradient and patterns of flow in cases with substantial sidewall leakage, the latter option is a better choice.

The double-ring permeameter is assembled much like the compaction-mold device shown in Fig. 1, except that two drainage lines rather than one extend from the base plate.

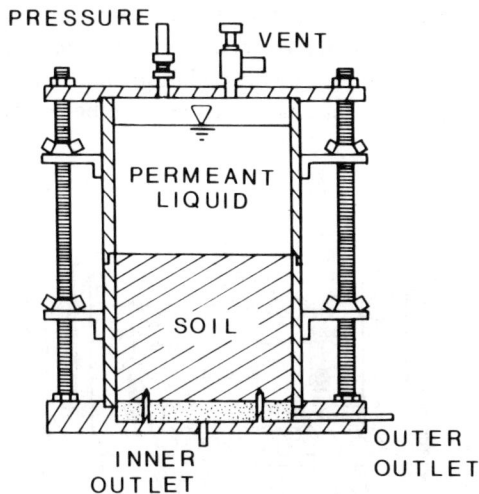

FIG. 3—*Double-ring compaction-mold permeameter developed by Anderson et al* [5].

There is not a great deal of experience with this device. The permeameter has been used more extensively on slurry-wall backfill materials than on compacted clay. The cell has worked well so far [5] and shows excellent promise. The disadvantages are the same as those described previously for the compaction-mold permeameter, except that one has some indication of the relative magnitude of sidewall leakage.

Consolidation-Cell Permeameter

A diagram of a consolidation-cell permeameter is shown in Fig. 4. The soil is trimmed into a sharpened steel ring with a diameter of 5 to 8 cm and a thickness of 1 to 2 cm, and the ring is clamped into position. Next, the reservoir that surrounds the ring is filled with water (or some other liquid), and the soil specimen is consolidated by applying vertical pressures of the desired magnitude. Vertical deformation of the soil specimen can be monitored with a dial gage or displacement transducer. Permeation is initiated by raising the hydraulic head at the base of the specimen and flowing permeant liquid up through the soil. Additional details on the apparatus and methods of testing may be found in Refs 3 and 6. Consolidation cells which have back pressure capability are also available [7–9].

The advantages of consolidation-cell permeameters are the relative ease of trimming undisturbed samples into the fixed ring, the ability to apply a vertical pressure (which not only simulates conditions in the field but also helps to squeeze the soil back against the wall of the consolidation ring and thus minimize sidewall leakage), and the ability to measure vertical deformations of the soil. The primary disadvantages are the potential for sidewall leakage that exists for all fixed-wall permeameters, the difficulty of performing tests at low effective stresses, the difficulty associated with trimming some soils (such as sandy or gravelly clays) into the consolidation ring, and the thinness of the specimens. Thin specimens are not considered to be desirable because larger

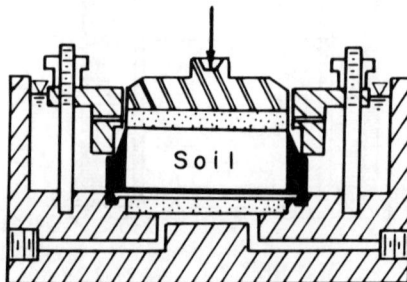

FIG. 4—*Consolidation-cell permeameter.*

specimens generally have a better chance of being representative of conditions in the field [3].

Fixed-Cylinder Permeameter

Perhaps the simplest fixed-wall permeameter consists of a tube with two end plates clamped to either end. A sharpened steel ring can be attached to the tube to permit trimming of undisturbed soil samples into the fixed ring, or a section of the thin-walled sampling tube can be used as the ring of the permeameter. A separate reservoir is used to store the permeant liquid.

Several variations exist. For example, to minimize sidewall flow, end plates may be outfitted with undersized porous disks as recommended by Hawley and Northey [10] and as sketched in Fig. 5. The calculation of hydraulic conductivity is discussed in Ref 10. Experience with this method is very limited, but Hawley and Northey's work points the direction toward an analytical assessment of sidewall leakage.

The problem of trimming the soil into the containing ring without creating any gaps between the soil and ring is clearly the most important disadvantage with fixed-cylinder cells. With specimens that are tested directly in the sampling tube, sidewall leakage is still a possibility. The inability to apply a vertical stress to the specimen aggrevates the problem. Attempts to trim undisturbed soil into clear plastic tubes have often led to visible gaps between the ring and soil.

FIG. 5—*Fixed-cylinder permeameter with undersized porous disks embedded in the end plates* [10].

Flexible-Wall Permeameters

Flexible-wall permeability tests are performed in triaxial cells or modified triaxial cells. Several designs are described in the literature [3,6,11–13]. A schematic diagram of the cell used at The University of Texas is shown in Fig. 6. Interchangeable base pedestals and top caps permit the testing of specimens with diameters from 4 to 15 cm. Double drainage lines to both the top and bottom of the specimen facilitate the flushing of air bubbles from hydraulic lines and direct measurement of the pressure drop across the soil specimen using a differentially acting electrical pressure transducer. Separate pressure controls maintain the cell pressure and the two pressures acting on the ends of the soil specimen. The specimen is normally back pressured prior to permeation to ensure full saturation.

The flexible-wall cell has several advantages. Undisturbed samples are easily tested because minimal trimming is required and irregular surfaces on the specimen are easily accommodated. Back pressure is normally used which helps to saturate the soil fully. Saturation can be confirmed during application of back pressure [6,11–13]. One can also measure the vertical and volumetric deformations of the soil and control the vertical and horizontal stresses.

There are several disadvantages of flexible-wall cells. The membranes used to confine the soil are normally made of latex, butyl, or neoprene rubber, which can be attacked and destroyed by certain chemicals. However, the specimens may be wrapped with a sheet of Teflon and then placed over the Teflon to protect the membrane [14]. In order to maintain contact between the membrane and soil specimen, the pressure in the cell liquid must be higher than the pore pressure in the specimen. In order to test with an ele-

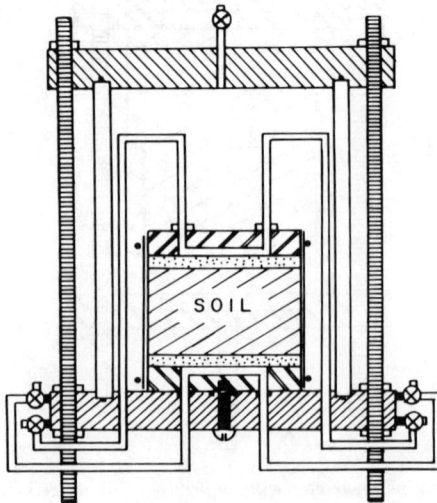

FIG. 6—*Flexible-wall permeameter.*

vated hydraulic gradient, the effective stress at one end of the sample must be fairly large. For example, suppose that a soil specimen has a length of 7.6 cm (3 in.) and is to be tested with a hydraulic gradient of 100 and a minimum back pressure of 207 kPa (30 psi). The pressure drop across the soil sample must be 74 kPa (10.8 psi) to achieve a hydraulic gradient of 100. If the pressure at one end of the soil specimen is 207 kPa (30 psi), the pressure at the other end must be 281 kPa (40.8 psi), and the cell pressure must be at least 281 kPa (40.8 psi) to maintain contact between the membrane and soil. Therefore, the effective confining pressure must be at least 74 kPa (10.8 psi) at one end of the specimen. It would not be possible to test this sample at a maximum effective confining pressure less than 74 kPa (10.8 psi) unless a lower hydraulic gradient was used. It is always true that (1) the maximum effective confining pressure cannot be less than the pressure drop across the sample; (2) the only way to test samples from shallow depth without exceeding the confining pressure in the field is to use low hydraulic gradients; and (3) if a permeability test is to be continued until one or more pore volumes of flow is achieved, it will be impractical to test most clays at low hydraulic gradients because the testing times will be too long.

The influence that effective confining pressure can have on k is illustrated by tests performed by Boynton [6]. Boynton desiccated slabs of compacted clay and then trimmed cylindrical specimens from the cracked clay. The specimens were set up in a flexible-wall cell, subjected to a 14-kPa (2-psi) effective confining pressure, and then permeated with water. The hydraulic conductivity was measured, and the effective confining pressure was gradually increased incrementally to a maximum of 103 kPa (15 psi). The gradient pressure was 7 kPa (1 psi). The measured relationship between k and effective confining pressure is plotted in Fig. 7 for one of the specimens. Also shown in Fig. 7 are the results obtained with an uncracked, control specimen. For the desiccated specimen, there does not seem to be much difference in use of an effective confining pressure of 14 versus 28 kPa (2 versus 4 psi), but there is an order-of-magnitude drop in k when the effective confining pressure is increased from 28 to 55 kPa (4 to 8 psi). These data demonstrate that high effective confining pressure tends to close cracks that may be present in the specimen.

A third disadvantage of flexible-wall permeameters is that the cell and appurtenant hydraulic system tend to be more expensive than for other types of permeameters.

Comparative Tests

Tests on Compacted Clay Permeated with Water

Permeability tests were performed on two soils using compaction-mold, consolidation-cell, and flexible-wall permeameters. The two soils were ka-

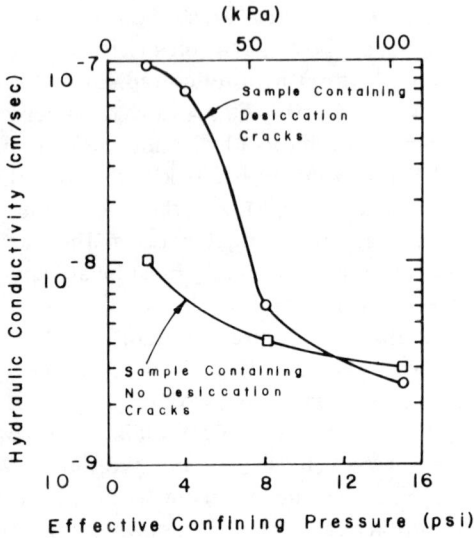

FIG. 7—*Hydraulic conductivity plotted as a function of effective-confining pressure for two specimens of compacted clay tested in a flexible-wall permeameter at a hydraulic gradient of less than 10* [6].

olinite and fire clay purchased from Georgia Kaolin Co., Elizabeth, New Jersey, and Elgin Brick Co., Austin, Texas, respectively. Liquid and plastic limits of kaolinite were 58 and 35% while those of fire clay were 53 and 18%. More than 95% of both soils passed the No. 200 sieve.

The soils were prepared by first mixing the air-dried materials with water to the desired water content and then compacting the moist soil into 10.16-cm-diameter compaction molds following ASTM D 698-78, Method A. If a compaction-mold permeameter was to be used, the soil was left in the mold. Material to be tested in the consolidation cell permeameter was extruded from the mold, and the central portion of the compacted specimen was trimmed into the consolidation ring. The specimens for the flexible-wall tests were obtained by extruding the soil sample from the compaction mold and trimming approximately 1 cm from each end to remove any soil that might have been smeared across the surface of the compacted specimen. An average effective stress of approximately 103 kPa (15 psi) was used in the consolidation cell and flexible-wall device. A back pressure of 413 kPa (60 psi) was employed in flexible-wall tests. Hydraulic gradients ranged from 20 to 100. The tests were generally continued for one to three weeks to ensure that steady-state conditions had been reached.

The measured hydraulic conductivities are plotted as a function of molding water content in Figs. 8 and 9. Although there are distinct variations in the conductivities measured from one permeameter to another, the differences are small and form no particular pattern. No one type of permeameter consis-

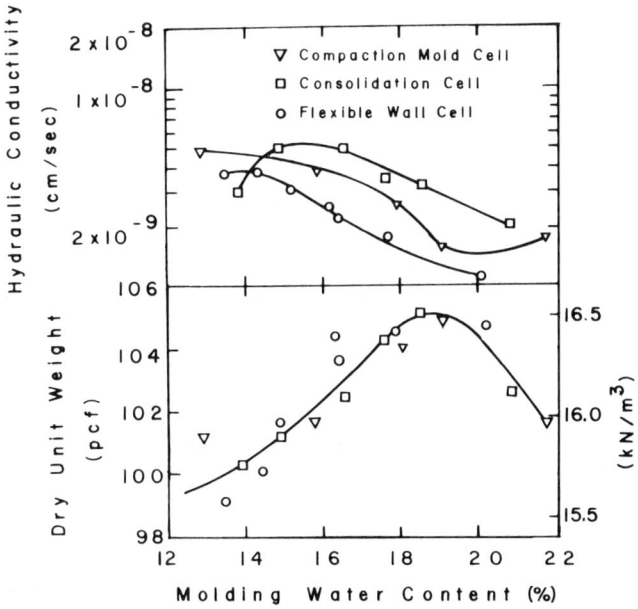

FIG. 8—*Hydraulic conductivity and dry unit weight versus molding water content for fire clay.*

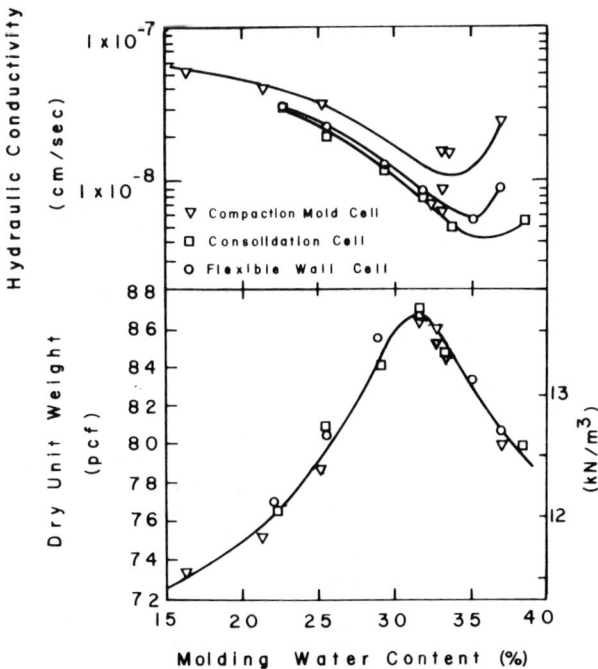

FIG. 9—*Hydraulic conductivity and dry unit weight versus molding water content for kaolinite.*

tently produces larger k's than other types. For practical purposes, the measured hydraulic conductivities are about the same for all three permeameters.

The subtle differences in hydraulic conductivities measured with different permeameters may be attributed to a variety of differences in the equipment, test procedures, and applied stresses in the various tests. Some of these differences are listed in Table 1. It would appear that either the parameters listed in Table 1 were not very significant for these soils or the various differences tended to offset one another.

It should be noted that great care was taken in mixing the soil uniformly, in breaking down large clods of clay during mixing, and in compacting reasonably homogeneous specimens. When testing a natural soil from the field, one often finds large clods of clay which are not readily broken down during the

TABLE 1—*Differences in various test parameters depending on type of permeameter.*

	Type of Permeameter		
Test Parameter	Compaction Mold	Consolidation Cell	Flexible Wall
Side-wall leakage	leakage is possible	applied vertical stress makes leakage unlikely	leakage is unlikely
Void ratio, e	relatively high e because applied vertical stress is zero	relatively low e because a vertical stress is applied	relatively low e because an all-round confining pressure is applied
Degree of saturation	specimen may be unsaturated	specimen may be unsaturated	application of back-pressure is likely to cause essentially full saturation
Voids formed during trimming	impossible; soil is tested in the compaction mold and is not trimmed	voids may have formed, but application of a vertical stress should help in closing any voids	voids are not relevant; the flexible membrane tracks the irregular surface of the soil specimen
Portion of sample tested	all of the compacted specimen is tested, including the relatively dense lower portion and the relatively loose upper portion; the dense lower portion may lead to measurement of relatively low k	only the central portion of the specimen is tested; the upper and lower third of the specimen are trimmed away	1 cm of soil is trimmed off both ends of the compacted sample

compaction process. There may be more of an opportunity for sidewall leakage to develop when testing soils from the field compared to testing soils that were prepared under carefully controlled conditions.

Comparative Tests for Compacted Kaolinite Permeated with Methanol

Comparative tests were performed with compaction-mold permeameters and flexible-wall permeameters on kaolinite compacted to optimum moisture content (34%) and then permeated with methanol. The measured hydraulic conductivities are plotted as a function of the number of pore volumes of flow in Fig. 10. Hydraulic conductivities were not corrected for the effects of density and viscosity of methanol; correction for these effects would have reduced k to methanol by 30%, which is too small an adjustment to be of practical interest. With both types of cells, the hydraulic conductivity initially decreased and then increased to a value that was greater than the initial value. The increase was probably caused by collapse of the diffuse double layer coupled with development of shrinkage cracks. The important point is that the increase in hydraulic conductivity is greater for the fixed-wall cell than for the flexible-wall cell. It is possible that sidewall leakage may have caused the

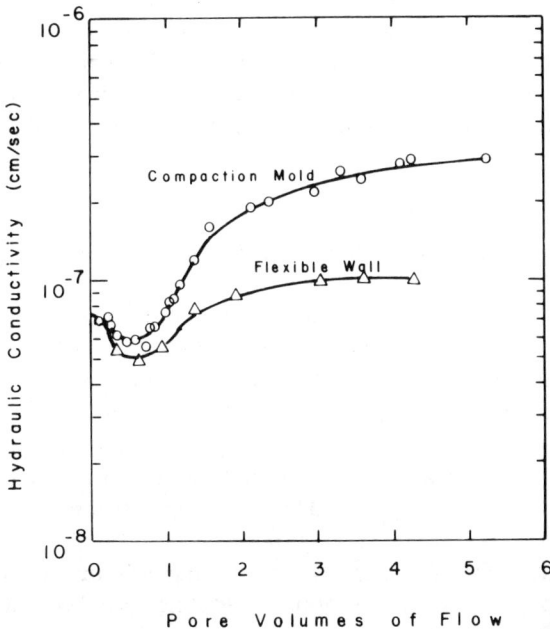

FIG. 10—Hydraulic conductivity versus number of pore volumes of flow for compacted kaolinite permeated with methanol. The hydraulic conductivity to water was measured independently on replicate samples and was between 6×10^{-8} and 8×10^{-8} cm/s.

hydraulic conductivity measured in the fixed-wall cell to be too high. It is also possible that confinement provided in the flexible-wall cell may have prevented the opening of cracks in the soil and therefore led to measurement of k that was too low.

Discussion

Relative Advantages and Disadvantages

With rigid-wall permeameters, the major advantages are low cost, simplicity, applicability to testing compacted soils, compatibility with a wide range of permeant liquids, and lack of a need to apply high confining pressure. The major disadvantages are incomplete control over stresses, inability to measure deformations in most fixed-wall cells, difficulty in trimming soil samples into the containing ring, and potential for sidewall leakage. Use of double-ring compaction-mold permeameters shows promise for minimizing uncertainties associated with sidewall leakage.

Flexible-wall permeameters have the advantage of minimizing sidewall leakage, permitting control over vertical and horizontal stresses, enabling one to measure vertical and volumetric deformations, permitting convenient back pressuring of the soil, enabling one to verify saturation, and providing the possibility of testing specimens with a range of diameters in the same cell. The disadvantages are higher cost, compatibility problems between the flexible membrane and some permeant liquids, and, perhaps most importantly, a need to apply significant confining pressures at high hydraulic gradient in order to maintain contact between the membrane and the soil.

Selection of a Permeameter

The following cases are considered ideal for the various types of permeameters:

1. *Compaction-Mold Permeameter*—This permeameter is ideally suited for testing laboratory-compacted specimens of clay that will be subjected to low overburden stress in the field. Any permeant liquid may be used. Some sidewall leakage may occur, particularly when the soil is permeated with a concentrated organic solvent. However, comparisons between laboratory and field measurements of k for clays permeated with organic solvents have been quite good for several prototype field liners subjected to low overburden pressure [4].

2. *Double-Ring Permeameter.* This permeameter is ideally suited to laboratory-compacted clay or laboratory-prepared slurry-wall backfill that will be subjected to low overburden stress in the field. The device is especially useful for tests in which sidewall leakage might be a major problem [5]. Such conditions would include testing at high hydraulic gradients, testing with concen-

trated organic chemicals or other caustic chemicals, or testing soils which by their very nature might have very irregular sidewalls, for example, compacted clays containing gravel or large clods of clay.

3. *Consolidation-Cell Permeameter.* This device is ideally suited to testing undisturbed samples of relatively compressible soil (such as soft to medium clay and slurry-wall backfill material) that will be subjected to significant overburden pressure. The applied vertical stress should tend to close any gaps along the sidewall and help to ensure proper results.

4. *Fixed-Cylinder Permeameter.* When compacted soils are used, this device is virtually identical to the compaction-mold permeameter. For undisturbed samples, this type of device is not recommended because sidewall leakage can be a major problem. However, if the undisturbed sample is tested directly in the sampling tube, problems with sidewall leakage are sometimes reduced.

5. *Flexible-Wall Permeameter.* This device is ideally suited to soil samples with irregular sidewalls, for example, undisturbed samples and compacted samples that contain gravel or large clods of clay. The device may be difficult to use with certain caustic chemicals due to incompatibility between the membrane and chemical. The flexible-wall cell is also ideal for soils that will be subjected to substantial overburden pressure, for example, natural samples obtained at depth, slurry-wall backfill materials that will be placed in the lower portions of a relatively deep trench, or compacted clay that will be overlain by significant thicknesses of soil or solid waste.

Conclusions

All permeameters have important advantages and serious limitations. One should take these into account when selecting the type of cell to use for a particular application. No one type of cell should be required or recommended because the best cell to use will vary from project to project.

More is known about how the type of permeameter affects the hydraulic conductivity of compacted clay than any other type of soil. Comparative data on two compacted clays demonstrate that with careful attention to sample preparation and compaction, the various types of permeameters yield virtually the same hydraulic conductivity to water.

When concentrated organic chemicals are used to permeate compacted clay, compaction-mold permeameters tend to yield higher hydraulic conductivities than flexible-wall cells. The available data do not support the belief that sidewall leakage alone causes the differences in measured k. The major difference between the compaction-mold cell and flexible-wall cell is the difference in the applied stresses. In a compaction-mold device, the applied effective stress is usually zero, and the only source of effective stress is residual stress from the compaction stage and seepage pressure. As the soil tends to shrink from reaction to the organic chemical, the hydraulic conductivity increases. At least three independent sets of tests have shown that this increase is not caused primarily by sidewall leakage. In contrast, the confining pressure used in flexible-wall tests tends to prevent the opening of cracks and therefore leads to a lower k than obtained with a fixed-wall cell. With concentrated organic chemicals, the critical question is which, if either, test correctly models conditions in the field. Clearly, neither test quite duplicates field conditions, but the compaction-mold device seems more appropriate for compacted clays subjected to low overburden pressure while the flexible-wall device may be more appropriate for clays subjected to high overburden pressure.

Comparative data are not available for slurry-wall backfill materials or natural deposits of clay, nor are data available for inorganic acids and bases or other classes of chemical wastes.

Acknowledgments

All three authors have benefited from discussions with individuals who are too numerous to mention. Special appreciation is extended to Walter E. Grube, Jr. of the U.S. Environmental Protection Agency (EPA) Municipal and Environmental Research Laboratory for his support and technical assistance. The data reported in Fig. 10 for the flexible-wall permeameter were collected by David E. Foreman. The work described in this paper was partially supported by grants from the National Science Foundation (Grant No. CEE-8204967) and the U.S. EPA (Cooperative Agreement CR-810165). Although the research described in this article has been funded in part by the U.S. EPA, it has not been subjected to the agency's peer and administrative review and, therefore, does not necessarily reflect the view of the agency and no official endorsement should be inferred. Mention of trade names or commercial products does not constitute an endorsement or recommendation for use.

References

[1] Brown, K. W. and Anderson, D. C., in *Proceedings*, Sixth Annual Research Symposium on Land Disposal of Hazardous Waste, Report No. EPA-600/9-80-010, U.S. Environmental Protection Agency, Cincinnati, OH, Mar. 1980, pp. 123–134.
[2] Anderson, D. C., "Organic Leachate Effects on the Permeability of Clay Soils," M.S. thesis, Texas A & M University, Soil and Crop Sciences Dept., College Station, TX, 1981.
[3] Olson, R. E. and Daniel, D. E., in *Permeability and Groundwater Contaminant Transport, ASTM STP 746*, American Society for Testing and Materials, Philadelphia, 1981, pp. 18–64.
[4] Brown, K. W., Green, J. W., and Thomas, J. C., in *Proceedings*, Ninth Annual Research Symposium on Land Disposal of Hazardous Waste, Report No. EPA-600/9-83-018, U.S. Environmental Protection Agency, Cincinnati, OH, Sept. 1983, pp. 114–125.
[5] Anderson, D. C., Crawley, W., and Zabcik, D., this publication, pp. 93–103.
[6] Boynton, S. S., "An Investigation of Selected Factors Affecting the Hydraulic Conductivity of Compacted Clay," M.S. thesis, University of Texas, Austin, TX, 1983.

[7] Lowe, J., Zaccheo, P. F., and Feldman, H. S., *Journal of the Soil Mechanics and Foundations Division*, American Society of Civil Engineers, Vol. 90, No. SM5, Sept. 1964, pp. 69-86.

[8] Smith, R. E. and Wahls, H. E., *Journal of the Soil Mechanics and Foundations Division*, American Society of Civil Engineers, Vol. 96, No. SM2, Mar. 1969, pp. 519-539.

[9] Wissa, A. E. Z., Christian, J. T., Davis, E. H., and Heiberg, S., *Journal of the Soil Mechanics and Foundations Division*, American Society of Civil Engineers, Vol. 97, No. SM10, Oct. 1971, pp. 1393-1413.

[10] Hawley, J. G. and Northey, R. D., in *Proceedings*, Tenth International Conference on Soil Mechanics and Foundation Engineering, Stockholm, Vol. 1, June 1981, pp. 617-620.

[11] Carpenter, G. W., "Assessment of the Triaxial Falling Head Permeability Testing Technique," Ph.D. dissertation, University of Missouri, Rolla, MO, 1982.

[12] Berre, T., *Geotechnical Testing Journal*, American Society for Testing and Materials, Vol. 5, No. 1/2, March/June 1983, pp. 3-17.

[13] Dunn, R. J., "Hydraulic Conductivity of Soils in Relation to the Surface Movement of Hazardous Wastes," Ph.D. dissertation, University of California, Berkeley, CA, 1983.

[14] Daniel, D. E., Trautwein, S. J., Boynton, S. S., and Foreman, D. E., *Geotechnical Testing Journal*, American Society for Testing and Materials, Vol. 7, No. 3, Sept. 1984, pp. 113-122.

DISCUSSION

Tuncer B. Edil[1] (*written discussion*)—(1) The need to apply high gradients results in high confining pressure regardless of the type of permeameter, that is, flexible or rigid wall. Similar to flexible-wall permeameter, adequate hydraulic gradient cannot be applied concurrently with low confining or consolidation stress in rigid-wall permeameters. The difference is, in the former consolidation it is three-dimensional, in the latter it is one-dimensional. Therefore, I disagree with the authors in listing this aspect as a disadvantage for flexible-wall permeameters only. (2) including side flow and saying that this is conservative is not meaningful because the difference in performance is dramatic when there is side flow, which practically rules out using a certain liner. (3) I agree with the conclusion that flexible- and rigid-wall permeameters give similar results with water; however, our research shows that the exception to this is the case when one attempts to apply back pressure with a rigid-wall permeameter—then channelling results.

D. E. Daniel, D. C. Anderson, and S. S. Boynton (authors' closure)—
With regard to Tuncer Edil's first point, the authors agree that the use of high hydraulic gradient leads to high seepage forces and, therefore, to high effec-

[1]Professor of civil and environmental engineering, University of Wisconsin—Madison, Madison, WI 53706.

tive stress. However, externally applied stress seems to have more impact on hydraulic conductivity than effective stresses associated with seepage pressure. With flexible-wall permeameters, one is forced to use high externally applied effective stress with high gradients to maintain contact between the membrane and soil sample; such is not the case with most rigid-wall devices. With regard to the discusser's second and third points, data were presented in the paper which suggested that sidewall leakage may not be the major cause of high hydraulic conductivity of clays permeated with organic chemicals. Furthermore, the few data that are available on field performance of clays permeated with organic chemicals show good agreement between field hydraulic conductivities and values measured in rigid-wall devices. The authors do not contend that laboratory tests in which significant sidewall leakage occurs necessarily model field conditions. However, it is the authors' belief that if rigid-wall permeameters show high hydraulic conductivity, there is cause for concern and perhaps a need for more extensive testing using devices such as the double-ring permeameter, the flexible-wall permeameter, or field tests.

J. F. Christiansen[2] (*written discussion*)—The authors should be congratulated for a well written paper on a topic which has largely been evaded in the past. The controversy, if this is the correct term, transcends the past several years and goes beyond the question of laboratory apparatus and procedure to include the broader aspect of field versus laboratory testing. With the advent of the environmental awakening, the need for hydraulic conductivity data increased dramatically, resulting in a revival of the long-standing debate on how the data can best be obtained.

Permeability tests conducted in the field are often touted as a panacea to our problems. Granted, the test section can be made large enough to be statistically representative of the various macrofeatures. It should not be overlooked, however, that in the field, test and boundary conditions are often poorly defined, test area preparation methods will influence the result, and frequently the sensitivity of the measurement apparatus is inadequate to assure definitive results.

It is widely acknowledged that hydraulic conductivity is one of the soil properties most difficult to measure accurately. Volumes have been written about the multitude of factors which influence the measurement of hydraulic conductivity in the field as well as in the laboratory. Over the years, it has been distressing to see purportedly serious research conducted with apparatus priding itself in simplicity and ease of operation and then being applied to one of the most complex and potentially serious problems we have, that is, containment of hazardous waste.

Our primary objective must be to achieve increased accuracy in predicting seepage quantities or rates of waste plume advancement. This can be accomplished only by systematically categorizing the numerous causes of variance between actual and expected performance. We must resist the temptation to

[2]Director of geotechnical testing, Empire Soils Investigations, Inc., Groton, NY 13073-0037.

encumber the laboratory permeability test with those "sources of error" that are clearly beyond its capacity to control. As an example, take a permeability test on a fissured specimen. In this case, the assumption of a uniformly distributed pressure drop over the length of the specimen does not hold, and the actual gradient variation is indeterminate. The result can at best be termed "apparent hydraulic conductivity" and will not be representative of the permeability of the soil matrix, nor have anything but a coincidental relationship to the crack pattern that may prevail in the field. It may be possible to assess the effect of macrofeatures such as desiccation cracking, worm and root holes, and animal burrows by appropriate computer modeling. The important thing is that we strive in our design and construction procedures to minimize the effect of these factors. If contact with a particular waste stream results in shrinkage and cracking, that soil is clearly the wrong choice of material and the resulting "permeability" is barely of academic interest.

So, in developing apparatus and test procedures, let us concentrate on reducing the magnitude of the many errors which can be introduced in the laboratory. This can only be accomplished if the apparatus has the capability to control these factors or if observations can be made of phenomena occurring during the test, so that their influence on the test result can be assessed.

The major deficiency of conventional rigid-wall permeameters is that these controls and observational opportunities are ignored. Consequently, one can never be certain that the result is correct. The double-ring permeameter is an improvement in this respect, but only permits a detection of sidewall leakage.

Since hydraulic conductivity varies with degree of saturation, it is not possible to make a valid comparison between tests on different materials or apparatuses unless it is known that the degree of saturation is the same. Back pressure saturation provides a common basis for comparison. It is not stated in the paper that the two types of fixed-wall apparatuses used in the study were back pressure saturated. Neither is it mentioned if double-ring compaction permeameter was used. Consequently, the conclusions concerning the relationship between hydraulic conductivity and type of apparatus should be questioned. If all of the tests had been performed with a gradient superimposed on an elevated back pressure, such as in the flexible-wall permeameter, it is likely that the spread in results for different types of apparatus would have been greater.

The concern over the effect of confining pressure in the flexible-wall permeameter is somewhat overstated and misdirected. Just because there is no provision for measuring lateral pressure in a rigid-wall permeameter, there is no reason to believe that there isn't any, or that it remains constant throughout the test. Anybody who has attempted to extricate a compacted specimen from a mold will know that lateral resistance can be high. With the flexible-wall permeameter one has the option to select confining and gradient pressures in accordance with the demands of a particular project. At times it will be important to minimize the effects of confining and gradient pressures. Then a reasonable choice has to be made, in consultation with the client, between the expediency of obtaining test results and strict duplication of conditions applicable in the field.

The value of flexible-wall permeameters lies in the flexibility to vary and measure the effect of a variety of test conditions, and the ability to verify that the specimen has stabilized under the test conditions at the time the data for determination of the hydraulic conductivity are obtained. Furthermore, it is equally applicable to the testing of undisturbed as well as compacted specimens.

D. E. Daniel, D. C. Anderson, and S. S. Boynton (authors' closure)—Mr. Christiansen criticizes fixed-wall permeameters on the basis that one can never be certain that the measurements obtained from fixed-wall devices are correct. However, the fixed-wall device yields about the same result as the flexible-wall device for compacted clay that is permeated with water. Because the fixed-wall cell is much easier to use, the authors see no reason not to use the fixed-wall cell when it yields about the same results as a flexible-wall cell. Because no type of laboratory permeameter precisely models field conditions, the meaning of "correct" results from laboratory tests becomes clouded. The authors believe that the person performing laboratory permeability tests should think about the field conditions that the laboratory test is supposed to model, then review the limitations of each method of testing, and, finally, determine which type or types of permeameter to use.

None of the fixed-wall tests that were discussed in the paper involved the use of back pressure. The double-ring permeameter was not used to obtain any of the test results that were reported in the paper.

Rodney W. Lentz,[1] *William D. Horst,*[2] *and
Janardanan O. Uppot*[3]

The Permeability of Clay to Acidic and Caustic Permeants

REFERENCE: Lentz, R. W., Horst, W. D., and Uppot, J. O., **"The Permeability of Clay to Acidic and Caustic Permeants,"** *Hydraulic Barriers in Soil and Rock, ASTM STP 874,* A. I. Johnson, R. K. Frobel, N. J. Cavalli, and C. B. Pettersson, Eds., American Society for Testing and Materials, Philadelphia, 1985, pp. 127-139.

ABSTRACT: Results of a laboratory study of the effects of acidic and caustic permeants upon the permeability of various fine-grained soils are reported. Triaxial falling head permeability tests were performed on specimens of kaolinite, kaolinite-bentonite mixture, and magnesium montmorillonite. Permeants used were hydrochloric acid with pH values of 1, 3, and 5 and water and sodium hydroxide with pH values of 9, 11, and 13. In no case for any of the clays or permeant pH did the permeability increase during the passage of six pore volumes of permeant, which indicates that no significant dissolution of clay minerals occurred. The only permeant that caused significant change in permeability was sodium hydroxide at a pH of 13, which caused a reduction in permeability of the magnesium montmorillonite by a factor of 13. This was found to be due to precipitation of magnesium hydroxide in the pores. Results are compared with findings reported by other researchers.

It is concluded that because of the wide variety of reactions possible between clays and permeant, future reports of permeability changes should give as much detail as possible about the chemistry and mineralogy of the soil studied. The equipment used and testing technique also have significant effects on reported test results and should be described in detail. Also, it is suggested that research is needed into possible reactions which could be induced within the soil pores to create favorable changes in permeability.

KEY WORDS: permeability, permeameters laboratory testing, hazardous waste, contaminated permeant, clay soils, earth linings, soils testing, triaxial devices

The problem of managing hazardous waste and its ultimate disposal is a concern that has increased dramatically within the past decade. One of the alternatives available for the management of hazardous waste is to deposit it in a secure landfill. Clay linings are often used as barriers in these landfills to prevent the contamination of groundwater by the toxic leachate produced in

[1]Assistant professor, University of Missouri—Rolla, Rolla, MO 65401.
[2]Assistant engineer, Anderson Engineering, Inc., Springfield, MO 65802.
[3]Associate professor, University of New Haven, West Haven, CT 06516.

the landfill. The effectiveness of the clay lining in preventing movement of the leachate depends on its ability to maintain a very low permeability while in contact with these toxic permeants. However, the permeability of a clay soil is dependent upon numerous variables, and the concern is that leachates, in contact with the clay lining, may affect its permeability. An unexpected increase in permeability could allow much more rapid travel of the leachate through the clay lining and, potentially, into the environment.

Variables which influence the permeability of any given fine-grained soil include degree of saturation, soil structure, density, adsorbed cations, and the nature of the permeating fluid [1,2][3]. Changes in permeability of some soils when permeated with certain chemical permeants have been known to occur and have been attributed mainly to ion exchange [1,3,4] and mineral dissolution [3,5,6].

Ion exchange may, depending on the adsorbed cation in the clay and the chemistry of the permeant, cause either an increase or a decrease in the thickness of the adsorbed double layer surrounding clay particles. If the double layer increases in thickness, the permeability should reduce due to less void space available for flow. If the double layer thickness is reduced, the effective void space is increased and permeability would be expected to increase. In addition, the possibility exists that due to the reduced effective particle size some soil particles may move through the voids causing piping. This may increase permeability if particles are washed from the soil or decrease permeability if the particles move to clog the pores.

Dissolution of soil minerals [3,5-8] is a distinct possibility under adverse pH conditions, with caustics tending to degrade the silica tetrahedra and acidic permeants dissolutioning the octahedral layer.

Results of recently published research indicate that some organic permeants such as acetone, methanol, and xylene can cause increases in clay permeability by up to three orders of magnitude [5,6,9]. These cases represented compacted samples in rigid-wall permeameters. The large permeability increases reported probably are due to the permeants causing volume decrease, which causes cracking or leaves space near the rigid wall for preferential flow of permeant and subsequent piping of soil particles.

This paper presents the results of an investigation of the effect of inorganic acidic and caustic permeants on the permeability of three different clay soils.

Equipment

A comparison of results between various researchers is hindered by different testing equipment and techniques, which may influence the results reported.

For instance, as just noted, if a rigid-wall permeameter is used under conditions where volume decrease occurs, cracks will form in the specimen and the measured values of permeability will show a large increase. This large increase will not represent either the intact portion of the clay or the field condition. In the field the distribution of the cracks will be different than in the small laboratory sample.

For identical conditions of volume decrease, a flexible-wall triaxial permeameter with confining pressure applied would measure a much smaller increase in permeability because the flexible membrane would conform to the specimen sides, reducing the amount of permeant bypassing the specimen. Thus the measured permeability would be representative of the intact soil. This condition more nearly models the field condition if in situ lateral stresses are able to deform the soil to compensate for volume decrease due to dissolution or shrinkage. To the extent the confining pressure in the permeameter causes more compression than can occur in the field, the measured permeability will be too small. If the confining stresses are selected with care, it appears that the flexible wall triaxial permeameter is more suitable than the rigid-wall permeameter.

The permeameters used in this study were of the triaxial type, which allows the specimen to be encased in a flexible membrane and subjected to confining pressure and back pressure. The use of back pressure ensures saturation. The flexible membrane reduces problems of leakage between the specimen and the sides of the permeameter, which would give spuriously high values of permeability.

Drain lines connected to the bottom and top cap of the soil specimen allow the application of a gradient across the specimen. The lines are connected to calibrated standpipes, which are used to monitor the flow of permeant through the specimen. Pressure regulators are used to control the application of air pressure for confinement and back pressure and to apply pressure to the inflow and outflow standpipes for creating a gradient across the specimen. Pressures are measured with a pressure transducer.

Polypropylene fittings and polyvinyl chloride valves were used in the construction of the equipment in order to avoid corrosion by the harsh permeants. Valves were installed to allow recharging the standpipes with permeant without depressurizing the specimen. A schematic of the experimental apparatus is shown in Fig. 1.

Material Tested

Three soil types representing a range of clay mineralogy were used in the investigation. One was a kaolinite (Kentucky ball clay). Another was a magnesium montmorillonite, a mixed-order clay made up primarily of smectite with minor amounts of illite and kaolinite. The third clay was a mixture of 85% ball clay and 15% bentonite, by weight. Chemical analyses of the kaolinite and the magnesium montmorillonite are given in Table 1. The mineral composition of the magnesium montmorillonite is presented in Table 2.

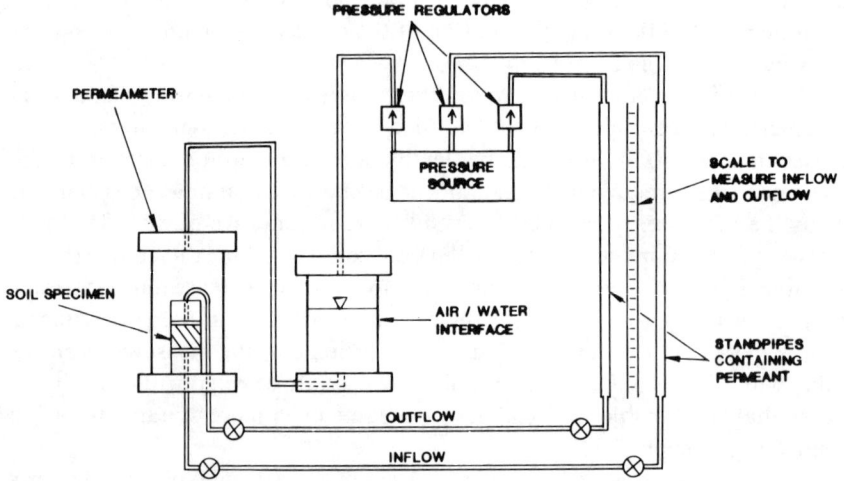

FIG. 1—*Schematic of experimental apparatus.*

TABLE 1—*Chemical analyses of kaolinite and magnesium montmorillonite (data furnished by the suppliers).*

	Kaolinite, %	Magnesium Montmorillonite, %
SiO_2	55.6	65.6
Al_2O_3	28.6	21.7
Fe_2O_3	1.0	5.8
TiO_2	1.8	0.8
CaO	0.1	0.1
MgO	0.4	0.3
Na_2O	0.1	0.1
K_2O	1.1	2.1
Loss on ignition (LOI)	11.4	3.8

TABLE 2—*Mineral composition of magnesium montmorillonite.[a]*

	Percent
Smectite	55 to 65%
Illite	10 to 20
Kaolinite	10 to 20
Quartz	5 to 15
Heavy minerals (FeS_2)	<0.001

[a]Data provided by the supplier, Southern Clay Co., Cape Girardeau, Missouri.

Permeants

Three different permeants were used in this investigation in order to cover the pH range from 1 to 13. De-aired tap water (pH = 7) was used as the permeant for determining the reference permeability of each soil. The acidic permeant was prepared by adding sufficient quantities of concentrated hydrochloric acid (HCl) to de-aired tap water in order to achieve the desired pH. Acidic permeants were prepared at pH values of 1, 3, and 5. The caustic permeants were prepared by adding sodium hydroxide (NaOH) to de-aired tap water to yield pH values of 9, 11, and 13. A pH meter was used to determine when the desired level was reached.

Sample Preparation

In order to achieve reproducibility of test results, the soil preparation method consisted of consolidating the soil from a slurry. This sample preparation technique had been previously evaluated by Carpenter [10] and found to produce specimens with excellent uniformity and reproducibility. The slurry was prepared by mixing the soil at a water content above its liquid limit and then allowing hydration for two days. The mixing water was distilled and contained 1 g/L of magnesium sulfate heptahydrate (epsom salts). All three soils tested were prepared in the same manner. This procedure retained magnesium as the primary ion on the montmorillonite and produced a highly flocculated structure. The concentration of magnesium ions was not high enough to flocculate the bentonite. After hydration the slurry was placed in consolidation cylinders, vibrated to remove air bubbles, and consolidated to 191.5 kPa (27.8 psi). This pressure was selected arbitrarily since the objective was only to obtain reproducibility of samples. All specimens were 35.6 mm (1.4 in.) in diameter and consolidated to lengths no greater than 100 mm (4 in.). After consolidation the soil was extruded, wrapped, and stored until tested.

Procedure

Specimens for testing were trimmed to a height of 13 to 25 mm (0.5 to 1 in.) and placed in the permeameter between de-aired Carborundum filter stones. Whatman No. 5 hardened filter paper was used to prevent intrusion of soil into the filter stones. After a specimen was mounted in the permeameter, a confining pressure of 34.5 kPa (5 psi) was applied, and the standpipe valves were opened. The cell pressure and the inflow and outflow pressures were then increased in increments of 34.5 kPa (5 psi) until a back pressure of 241.5 kPa (30 psi) was achieved. The soil was allowed to equilibrate under these pressures, after which a gradient was established across the specimen by simultaneously increasing the cell confining pressure and the inflow pressure.

The pressure was increased until a gradient of 400 to 500 was reached. A gradient of this magnitude was used so that a number of pore volumes of permeant could be passed through the specimen in a reasonable length of time.

   High hydraulic gradients have been reported to increase permeability in some cases and to decrease it in others [1]. Decreases in permeability have been attributed to particle migration, which causes clogging of pores and usually occurs only in specimens of low density. For specimens of magnesium montmorillonite identical to those used in the present study, Carpenter [10] found the measured permeability to decrease by a factor of two as the gradient was increased from 50 to about 250 and to remain approximately constant at higher gradients.

   Establishment of the hydraulic gradient resulted in an effective stress gradient across the specimen, causing it to undergo consolidation. Typically, 20 to 30 min were required to stabilize inflow and outflow, whereupon readings of the inflow head and outflow head were taken at time intervals. Depending upon the specimen thickness and permeability, these intervals varied from every two hours to once a day. Using these readings, the permeability was calculated by the falling head equation

$$k = \frac{a \cdot L}{A \cdot (t_i - t_f)} \cdot \ln(h_i/h_f) \qquad (1)$$

where

k = permeability, cm/s,
a = cross-sectional area of standpipe, cm^2,
L = specimen thickness, cm,
A = cross-sectional area of specimen, cm^2,
$t_i - t_f$ = elapsed time, s,
h_i = initial head difference, cm, and
h_f = final head difference, cm.

The values of head difference (h_i and h_f) were determined as the difference in permeant levels between inflow and outflow standpipes plus the difference in applied air pressure between inflow and outflow standpipes expressed in centimeters of water. Readings were continued until six pore volumes of permeant had passed through the soil. Tests lasted, depending upon the soil used and any permeant-induced changes in permeability, anywhere from 3.5 to 30 days.

   A fresh soil specimen was used for each level of pH. For the three soil types and the seven levels of pH, this required testing 21 specimens.

Results and Discussion

Neutral Permeant

Specimens of each soil were permeated with de-aired tap water as the neutral permeant until six pore volumes had passed through. This established reference values of permeability. The results of these tests are plotted in Fig. 2, which shows that the neutral permeant caused no change in permeability. If the use of high hydraulic gradient had resulted in particle migration, this figure should have shown permeability changing with an increasing number of pore volumes.

Acid Permeant

To investigate the possibility of acid causing alterations of the clay with an attendant change in permeability, specimens of each soil were permeated with HCl. The results indicated that for all three soil types HCl permeants with pH levels of 1, 3, and 5 caused no change in permeability during the passage of six pore volumes. The results from the tests using the strongest concentration, a pH of 1, are plotted in Fig. 3 for each soil type. These results indicate that it is unlikely that any dissolutioning of clay minerals occurred. Had dissolutioning taken place, the permeability should have increased as the volume of soil solids was reduced, creating larger voids. This might have been partially offset by the confining pressure, which would tend to compress the soil and close any voids created by dissolutioning. However, due to the flocculated soil structure, it is unlikely that such voids would be completely compen-

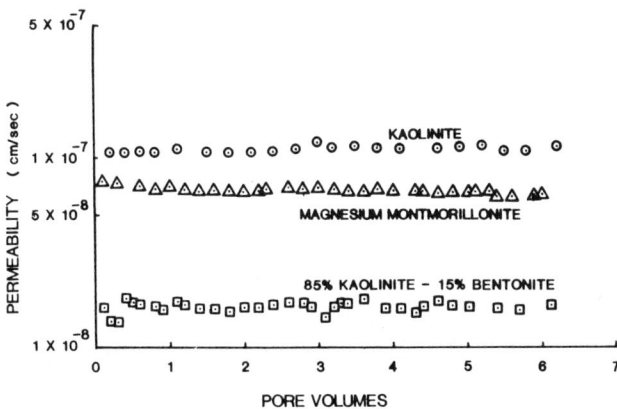

FIG. 2—*Permeability versus pore volumes using reference permeant, pH = 7.*

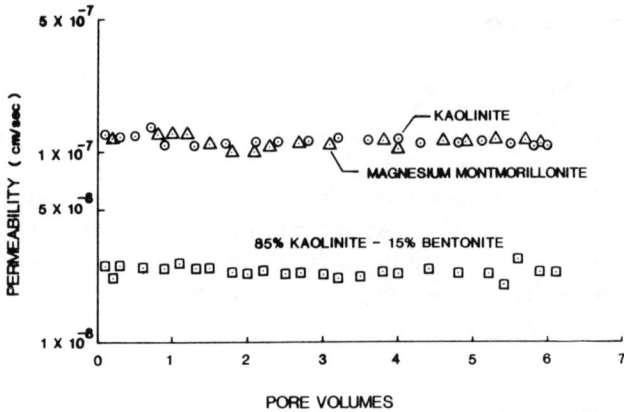

FIG. 3—*Permeability versus pore volumes using hydrochloric acid permeant, pH = 1.0.*

sated for by compression. Thus, if any significant amount of dissolution occurred, there should have been a noticeable increase in permeability.

Chemical reactions between the soil and the permeant are very time-dependent. The high gradient used means that the permeant was not in contact with the soil particles for as long a time as it would be at lower gradients typical of field conditions. The contact time ranged from 3.5 to 30 days. Thus any dissolutioning which occurs during the permeability tests would become less obvious as the duration of the test decreases.

The findings of this study differ somewhat with results reported by D'Appolonia [3], who found that soil-bentonite backfill and slurry-wall filter cake material permeated with a 5% solution of hydrochloric acid suffered apparent dissolutioning of clay minerals and an increase in permeability of up to one order of magnitude. The hydraulic gradients used were reported to be in the range of 100 to 200.

On the other hand, Gordon and Forrest [11] tested a compacted clay using sulfuric acid with pH of 1.5 and found no appreciable change in permeability throughout the test. They performed chemical analyses of the soil and permeant, which indicated that carbonates within the soil reacted with the sulfuric acid permeant to produce calcium sulfate precipitate in addition to carbon dioxide gas and water.

From these contrasting results it appears that the likelihood of clay mineral dissolution taking place in any given situation is difficult to predict and should be investigated by long-term tests of the specific soil being evaluated using the permeant it is expected to retain.

Caustic Permeant

Specimens of each soil were tested using sodium hydroxide permeants at pH levels of 9, 11, and 13. The results indicated little effect on permeability due to pH of 9 and 11. However, the permeant having pH of 13 caused permeability decreases varying from about a factor of $2\frac{1}{2}$ to a factor of about 13. In no case did the permeability increase. The results of the tests using pH of 13 are shown in Fig. 4. The results of all tests are summarized in Table 3.

The moderate decrease in permeability of the kaolinite and kaolinite-bentonite is probably the result of using tap water instead of distilled water to prepare the permeants. The water has a hardness approaching 300 mg/L as calcium carbonate. At a pH of 13, some calcium hydroxide precipitate in the permeant may have entered the clay specimens and blocked some of the pores.

The most significant change in permeability was in the magnesium montmorillonite. One possible explanation for a decrease could be ion exchange, whereby the monovalent sodium ion replaces the divalent magnesium ion. This would cause expansion of the adsorbed double layer surrounding the clay particles, thereby reducing the effective void area available for permeant flow. Because the divalent magnesium ion is more strongly attracted to the exchange sites on the clay surface than is the monovalent sodium ion, the exchange would not take place until the concentration of sodium ions increased enough at pH of 13 to replace the magnesium by mass action.

It can be seen in Fig. 4 that the permeability of the magnesium montmorillonite decreased to nearly the same low value as for the kaolinite-bentonite mix. Because the magnesium montmorillonite was observed to have a floccu-

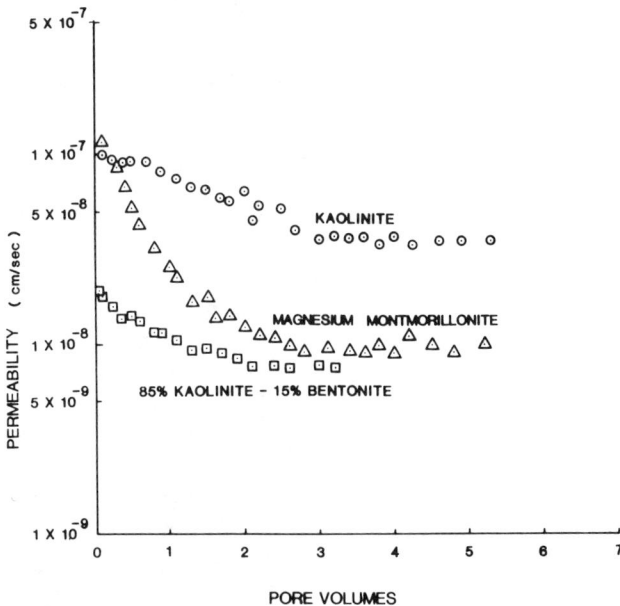

FIG. 4—*Permeability versus pore volumes using sodium hydroxide permeant, pH = 13.0.*

TABLE 3—*Summary of permeability test results.*

pH	Kaolinite			Kaolinite, 85% Bentonite, 15%			Magnesium Montmorillonite		
	k_i^a	k_f^b	k_i/k_f	k_i	k_f	k_i/k_f	k_i	k_f	k_i/k_f
1	1.25×10^{-7}	1.10×10^{-7}	1.09	2.50×10^{-8}	2.25×10^{-8}	1.11	1.20×10^{-7}	1.10×10^{-7}	1.09
3	1.25×10^{-7}	1.15×10^{-7}	1.09	3.00×10^{-8}	2.20×10^{-8}	1.36	8.00×10^{-8}	7.40×10^{-8}	1.08
5	1.10×10^{-7}	1.05×10^{-7}	1.05	2.50×10^{-8}	2.20×10^{-8}	1.14	6.50×10^{-8}	6.00×10^{-8}	1.08
7	1.15×10^{-7}	1.20×10^{-7}	0.96	1.65×10^{-8}	1.65×10^{-8}	1.00	7.40×10^{-8}	6.40×10^{-8}	1.16
9	1.15×10^{-7}	1.05×10^{-7}	1.10	2.60×10^{-8}	2.30×10^{-8}	1.13	5.00×10^{-8}	4.40×10^{-8}	1.14
11	1.20×10^{-7}	1.10×10^{-7}	1.09	2.65×10^{-8}	2.20×10^{-8}	1.20	1.20×10^{-7}	7.80×10^{-8}	1.54
13	1.00×10^{-7}	3.40×10^{-8}	2.94	1.92×10^{-8}	7.80×10^{-9}	2.46	1.15×10^{-7}	9.00×10^{-9}	12.78

[a]k_i = initial permeability, cm/s.
[b]k_f = final permeability after passage of six pore volumes of permeant, cm/s.

lated structure and the bentonite observed to be dispersed, it seemed unlikely that expansion of the double layer by ion exchange should be sufficient to reduce the permeability of the magnesium montmorillonite to the same low permeability as the dispersed bentonite. This suggested that some other reaction such as a precipitate forming in the voids of the magnesium montmorillonite was possibly contributing to the decreased permeability.

To test this hypothesis, specimens of magnesium montmorillonite were permeated with sodium chloride at concentrations of 0.1 N and 1 N, thus providing a high concentration of sodium ions which would also cause ion exchange with the magnesium. The results are shown in Fig. 5 and indicate no change in measured permeability, which would mean that the observed permeability decrease at pH of 13 was not due to ion exchange alone. A sample of leachate was obtained after passing 0.1 N sodium chloride through the magnesium clay. Some of the pH 13 NaOH solution was added to the leachate sample. A white fluffy precipitate was observed. This precipitate was analysed by X-ray diffraction and found to consist of magnesium hydroxide and calcium carbonate. This would indicate that after the magnesium ions were replaced by ion exchange with sodium ions they combined with the hydroxide ion, forming a precipitate in the clay voids resulting in a significant reduction in permeability.

Again the results obtained in this investigation contrast with the finding of D'Appolonia [3]. He found that when soil-bentonite backfill and slurry wall filter cake material was permeated with a 5% solution of NaOH the permeability increased by a factor of 5 to 10. Gordon and Forrest [11] tested consolidated tailings dam slimes using sodium carbonate at pH of 8 and 9. The pH of 9 resulted in permeability two orders of magnitude lower than when a pH of 8

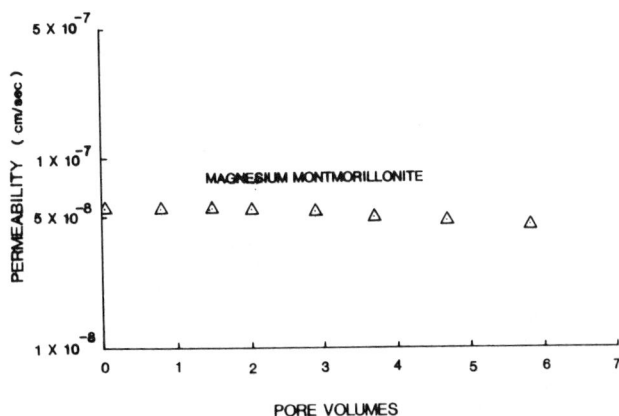

FIG. 5—*Permeability versus pore volumes using sodium chloride permeant (1 N sodium chloride).*

was used. This decrease was attributed to the effects of chemical reactions taking place within the soil.

Conclusions

It is concluded that for the conditions tested, pH, except for very high values, has little effect upon the permeability of the clays investigated. Any dissolution of clay minerals was not large enough to cause an increase in permeability at either very low or very high pH. Also, formation of precipitates within the pores of the clay is yet another result of reaction (apart from ion exchange and dissolution of minerals as postulated in the literature) between a permeant and a clay that can cause changes in permeability. Because of the wide variety of reactions possible between various clays and permeants, researchers reporting the results of permeability changes should report as much detail as possible about the chemistry and mineralogy of the soil being affected. Also, the equipment and testing technique used can have significant effect on reported results and should be described in detail. The results of this investigation also suggest the need for more research into possible reactions which could be induced within the pores to create favorable changes in permeability.

Acknowledgments

Financial support for this investigation was provided by the University of Missouri Weldon Spring Endowment Fund. This support is gratefully acknowledged. The authors also wish to express their appreciation for the advice and counsel so willingly provided by Professor John B. Heagler.

References

[1] Olson, R. E. and Daniel, D. E. in *Permeability and Groundwater Contaminant Transport, ASTM STP 746*, American Society for Testing and Materials, Philadelphia, 1981, pp. 18–64.
[2] Zimmie, T. F. in *Hazardous Solid Waste Testing: First Conference, ASTM STP 760*, American Society for Testing and Materials, Philadelphia, 1981, pp. 293–304.
[3] D'Appolonia, D. J., *Journal of the Geotechnical Engineering Division*, American Society of Civil Engineers, Vol. 106, No. GT4, 1980, pp. 399–417.
[4] Mesri, G. and Olson, R. E., *Clays and Clay Minerals*, Vol. 19, 1971, pp. 151–158.
[5] Brown, K. W. and Anderson, D. in *Proceedings*, Seventh Annual Research Symposium, Cincinnati, EPA-600/9-81-0026, U.S. Environmental Protection Agency, Cincinnati, 1981, pp. 119–130.
[6] Brown, K. W. and Anderson, D. in *Proceedings*, Sixth Annual Research Symposium, Cincinnati, EPA-600/9-80-010, U.S. Environmental Agency, Cincinnati, 1980, pp. 123–134.
[7] Grim, R. E., *Clay Mineralogy*, McGraw-Hill, New York, 1968, pp. 434–436.
[8] Pask, J. A. and Davis, B., Technical Paper 664, U.S. Bureau of Mines, Washington, DC, 1945, pp. 56–78.

[9] Green, W. J., Lee, G. F. and Jones, R. A., *Journal of the Water Pollution Control Federation*, Vol. 53, No. 8, 1981, pp. 1347–1354.

[10] Carpenter, G. W., "Assessment of the Triaxial Falling Head Permeability Testing Technique," Ph.D. dissertation, University of Missouri—Rolla, Rolla, MO, 1982.

[11] Gordon, B. B. and Forrest, M. in *Permeability and Groundwater Contaminant Transport, ASTM STP 746*, American Society for Testing and Materials, Philadelphia, 1981, pp. 101–120.

Allen H. Gipson, Jr.

Permeability Testing on Clayey Soil and Silty Sand-Bentonite Mixture Using Acid Liquor

REFERENCE: Gipson, A. H., Jr., **"Permeability Testing on Clayey Soil and Silty Sand-Bentonite Mixture Using Acid Liquor,"** *Hydraulic Barriers in Soil and Rock, ASTM STP 874*, A. I. Johnson, R. K. Frobel, N. J. Cavalli, and C. B. Pettersson, Eds., American Society for Testing and Materials, Philadelphia, 1985, pp. 140–154.

ABSTRACT: A laboratory permeability testing program was designed and performed to evaluate the effect of an acid liquor on two alternative materials planned for use in a compacted soil lining to minimize seepage loses from a phosphogypsum storage area. The alternative materials considered for use in the compacted soil lining included (1) on-site natural clayey soils and (2) on-site silty sands mixed with commercial bentonite (Volclay saline seal, No. 100). The acid liquor has a pH of 2.2 and is high in calcium, calcium oxides, sodium, chloride, and sulphate. The test results indicated that the acid liquor permeant resulted in lower permeabilities of the clayey soils than with tap water permeant. The tests on the commercial bentonite-silty sand mixtures with acid liquor permeant indicated that the permeability increased with time.

KEY WORDS: permeability, acid liquor, phosphogypsum, impermeable barriers, clay linings, soil linings, bentonite

A phosphogypsum disposal field covering about 1 600 000 m² (400 acres) with a height of 60 m (200 ft) is planned. The phosphogypsum results from the production of phosphoric acid for use in fertilizer manufacturing. The phosphogypsum is transported from the processing plant to the gypsum field in a slurry form and is then deposited in the disposal area. Within the disposal area, the slurry fluid will flow from the deposition points around the periphery to a water pond located in the center of the field. The fine sand and silt-size phosphogypsum particles will settle out of the fluid and be deposited in the field. The fluid will be decanted for reuse. This is commonly referred to as the upstream method of deposition in tailing dam terminology. The field bottom will be lined to minimize seepage loses. The laboratory testing pro-

[1]Vice president and associate, Woodward-Clyde Consultants, Englewood, CO 80111.

gram used to evaluate the permeability properties of soils available for lining construction at the site is outlined in the following paragraphs.

The site soils vary from silty sands and clayey sands to very plastic clays. Five soils within this range were selected for laboratory testing. Inasmuch as the chemical constituents of permeant liquids might have an effect on soil permeability—and there is a paucity of information available in the literature on the effect of acid liquor associated with phosphogypsum on clayey soils and silty sand-bentonite mixtures—tests were conducted with a permeant liquid considered to be similar to the liquid environment that will eventually exist in the gypsum field. Tests also were made with Denver tap water of low total dissolved solids (TDS) as the permeant for comparison.

Several types of laboratory tests were made and the mineralogical and chemical constituents of the soils were determined. Laboratory compaction tests were conducted to determine the densities and moistures at which the permeability specimens were molded to represent probable field lining conditions. Since the principal purpose of the testing was to evaluate the effects of the expected permeant liquid on soils classed as impervious, the permeability testing was conducted over an extended time period. The tests performed, using U.S. Bureau of Reclamation (USBR) standard 8-in. (20.3-cm) diameter permeameters (Test Designation E-13) [1], were conducted for about 60 days. The 5.08-cm (2-in.)-diameter permeability tests were performed over periods of six months to one year using small, thin soil specimens as an alternative to accelerate the permeant flow.

Details of the laboratory testing program, the results, and the conclusions follow.

Laboratory Tests and Testing Procedures

Soil Materials Tested

Four clayey soil samples were selected to represent the range of soils available at the site for use as compacted lining materials. These soil samples ranged from very high plasticity clays to low plasticity clayey sands. A fifth sample was prepared to represent silty sands available at the site for use in compacted silty sand-bentonite mixtures. The samples selected are described in Table 1.

Classification Tests

Laboratory testing was performed to provide information for use in classifying the soils according to the Unified Soil Classification System. Gradation tests, including sieve and hydrometer analysis, were made on each of the five samples in accordance with ASTM Standard for Particle-Size Analysis of Soils [D 422-63 (1972)]. Plasticity tests including liquid limit, plastic limit,

TABLE 1—*On-site soil sample description.*

Soil No.	Description
1	very high plasticity clays (CH)[a] with very high percentage of fines (minus No. 200 sieve size)
2	high plasticity clays (CH)[a] with high percentage of fines
3	medium plasticity clays (CL)[a] with high percentage of fines
4	low plasticity clayey sands (SC)[a] with medium percentage of fines
5	nonplastic, slightly silty sands (SP-SM)[a], low percentage of fines, for mixing with 5, 7.5, 10, 15, and 29% bentonite by weight

[a]Unified Soil Classification according to ASTM Classification of Soils for Engineering Purposes [D 2487-69 (1975)].

plasticity index, and shrinkage limits tests were conducted on each of the plastic soil samples. These tests were conducted in accordance with the following ASTM standards: Test Method for Liquid Limit of Soils [D 423-66 (1972)], Test Method for Plastic Limit and Plasticity Index of Soils [D 424-59 (1971)], and Test Method for Shrinkage Factors of Soils [D 427-61 (1974)], respectively. The tests were performed on samples before and after permeation with the acid liquor as discussed in following paragraphs. For the liquid limit and plastic limit tests, the wet preparation method was used. Changes in moisture content were effected mainly through evaporation; however, when the samples were drier than desired, distilled water was added to return the sample to the desired moisture content. Generally, testing was conducted to minimize the amount of moisture that needed to be added.

Compaction Tests

Standard "Proctor" compaction tests were conducted in accordance with ASTM Test Methods for Moisture-Density Relations of Soils and Soil-Aggregate Mixtures, Using 5.5-lb (2.49-kg) Rammer and 12-in. (304.8-mm) Drop (D 698-78) on each of the four plastic soils, Nos. 1–4. Modified compaction tests were made on each silty sand (Soil 5)-bentonite mixture. These tests were conducted in accordance with ASTM Test Methods for Moisture-Density Relations of Soils and Soil-Aggregate Mixtures Using 10-lb (4.54-kg) Rammer and 18-in. (457-mm) Drop (D 1557-78).

Where the compaction and other laboratory tests required the addition of moisture, the moisture was added and the samples stored in airtight containers and allowed to season for a period of 48 h before testing. This procedure was followed so that the small clay aggregations would become plastic and thus provide uniform specimens after molding that did not contain clay lumps. These samples were not oven-dried before testing, except for the gradation tests, so that possible drying changes on the mineralogical or physical properties of the clay fractions would not take place.

Permeability Tests

Permeability tests were conducted on each of the four plastic soils, Nos. 1–4, in 8-in. (20.3-cm)-diameter permeameters in accordance with the USBR Test Designation E-13 except that the specimen heights were 3.81 cm (1.5 in.) instead of the 7.6-cm (3-in.) height normally used and the soil was compacted in two rather than three layers. This change was made in an effort to accelerate the flow of permeant liquids through the specimens. The soil specimens were compacted to 98% of the maximum dry density at optimum moisture content as determined from the standard "Proctor" compaction tests [ASTM Test Methods for Moisture-Density Relations of Soils and Soil-Aggregate Mixtures, Using 5.5 lb (2.49-kg) Rammer and 12-in. (304.8-mm) Drop (D 698-78)] for each soil. Care was taken during molding to assure that the existence of clay lumps was minimized in an effort to enhance uniformity of the soil mass. Permeability tests (20.3-cm diameter) were also conducted on the sand-bentonite mixture containing 29% bentonite. The mixture was compacted to 85% of the maximum dry density at optimum moisture content determined from the modified compaction test [ASTM Test Methods for Moisture-Density Relations of Soils and Soil-Aggregate Mixtures Using 10-lb (4.54-kg) Rammer and 18-in. (457-mm) Drop (D 1557-78)]. The permeant liquid for these tests was Denver tap water having a low TDS content. In these tests, the flow of the water from constant head tanks was from the bottom to the top of the permeameters. A single load of 4.9 kg/cm^2 (10 000 psf) was applied to each specimen, representing the expected average load of the phosphogypsum on the lining, equivalent to about 30 m (100 ft) of phosphogypsum.

Permeability tests utilizing each of the four plastic soils and the sand-bentonite mixture with the 29% bentonite were also conducted in the 8-in. (20.3-cm)-diameter permeameter cylinder generally in accordance with the USBR Test Designation E-13. For each of these tests a 2.54-cm (1-in.)-thick layer of gypsum was placed on the bottom porous stone. The phosphogypsum classifies as a silt (ML) according to the Unified Soil Classification. This was followed by a layer of compacted plastic soils or the soil-bentonite mixture. The specimens were prepared and compacted in the same manner as just described. The permeant liquid was the acid liquor representative of the expected field liquor.

Permeability tests were also performed utilizing soil specimens 5.08 cm (2 in.) in diameter. These specimens were designed and fabricated of materials nonreactive with the acid liquor. The testing equipment was such that the height of the specimens could be varied to suit the testing requirements. The permeant liquid was introduced at the bottom, and the flow was to the top of the permeameter specimens. For each of the four plastic soils and four sand-bentonite mixtures tested in this equipment, a 1.59-cm (^{5}/$_8$-in.)-thick layer of soil was compacted in the testing cylinder. The four plastic soils were compacted to 98% of the maximum dry density determined by the standard "Proctor" tests (ASTM D 698-78). In addition, four sand-bentonite mixtures were similarly tested. These mixtures contained 5, 7.5, 10, and 15% bentonite by dry weight of the silty sand (Soil 5). These mixtures were compacted to 85% of the maximum dry density determined by the modified "Proctor" tests (ASTM D 1557-78). These soils and sand-bentonite mixtures were placed at moisture contents above optimum conditions to more nearly represent the natural soils at the site which are expected to have high moisture contents. The clayey soils and bentonite-silty sand mixes were placed and compacted in two equal lifts. Each lift was compacted to the desired height (density) utilizing a 4.9-cm-diameter flat-faced tamper utilizing energy input through hammer blows to compact the individual layers to the desired height. The surface of each layer was scarified after compaction. Following placement and compaction of the soil layer, a 0.95-cm (³/₈-in.)-thick layer of gypsum was placed in a firm condition on the bottom side of the soil layer and the remainder of the apparatus assembled as shown on Fig. 1. The soil samples were presaturated by applying the permeant liquor to one end of the specimens and vacuum to the other end for a period of several hours. A vacuum pressure of about 55 cm (22 in.) of mercury was used. The sand-bentonite specimens were presaturated in the same manner except that tap water was used for the permeant liquid for the initial 72 h. Single loads of 4.9 kg/cm^2 (10 000 psf) were applied to the specimens. The permeant liquid was the acid liquor.

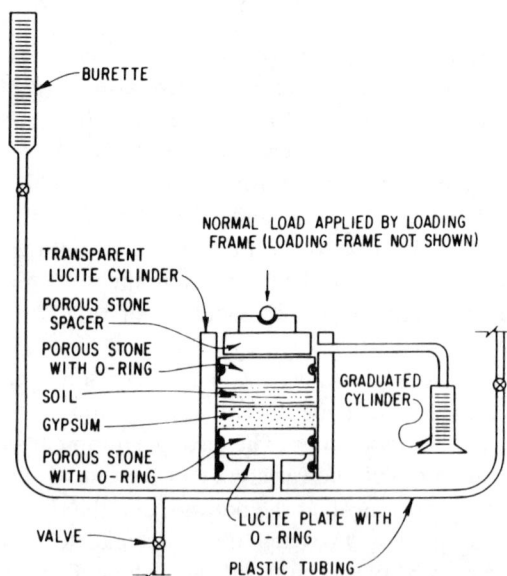

FIG. 1—A 5.08-cm (2-in.)-diameter permeameter.

Chemical Analyses

The following chemical analyses were performed: (1) analysis of the four plastic soils; (2) analysis of standard 10:1 liquid to solid leachate ratio of the four plastic soils in their natural state; and (3) analysis of the acid liquor used in the permeability tests.

Mineralogical Analyses

The four natural plastic soils were analyzed for principal mineral constituents. After completion of the 20.3-cm-diameter permeability tests in which these soils had been permeated with the acid liquor, the analyses were repeated to see if any change had taken place. The samples tested for the after acid liquor permeation were obtained by carefully scraping off the contact surface between the gypsum layer and the soil and then taking the minimum amount of soil that would be required to perform the classification and index property tests.

Testing Program Results

The gradation test results for the five soils are shown in Fig. 2. Liquid limit, plasticity index, and shrinkage limit test results on the soils are summarized in Table 2. The after-test results were obtained by testing samples taken from

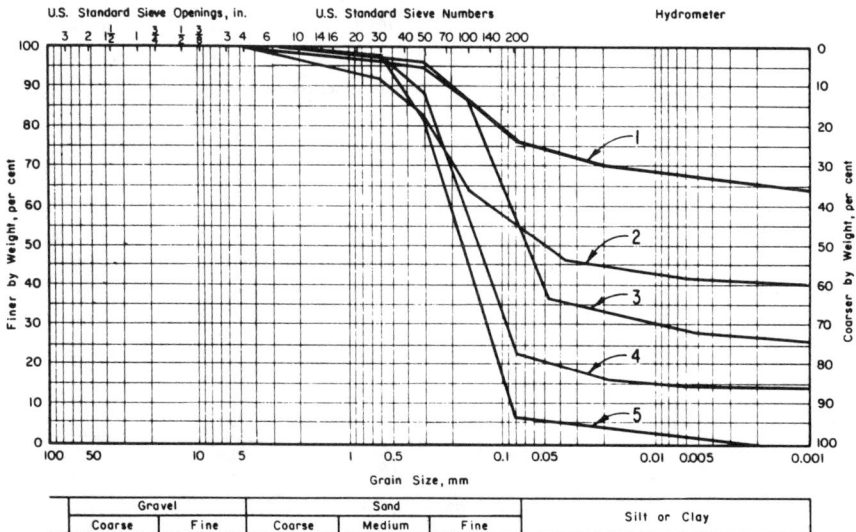

FIG. 2—*Grain-size analyses.*

TABLE 2—*Liquid limit, plastic limit, plasticity index, and shrinkage limit.*

Soil No.	Liquid Limit, %	Plastic Limit, %	Plasticity Index, %	Shrinkage Limit, %
TESTS ON SAMPLES BEFORE PERMEATION				
1	92	19	73	8
2	77	17	60	11
3	52	15	37	12
4	25	17	8	16
5	nonplastic			
TESTS AFTER PERMEATION WITH ACID LIQUOR THROUGH GYPSUM LAYER AND SOIL				
1	103	21	82	8
2	79	18	61	11
3	60	15	45	9
4	22	16	6	15

the 20.3-cm-diameter permeability tests after completion of testing. The samples were obtained as just described under Mineralogical Analyses.

Results of the compaction tests for the four plastic soils and the five silty sand-bentonite mixtures are shown in Table 3.

The calculated permeabilities for the 20.3-cm-diameter tests on the four plastic soils and the silty sand plus 29% bentonite mix using Denver tap water as the permeant fluid are summarized in Table 4, and the test results are shown in Figs. 3–6. A hydraulic gradient of approximately 30 was used.

The calculated permeabilities for the 20.3-cm-diameter tests on the four plastic soils and the silty sand plus 29% bentonite mixture using acid liquor as the permeant fluid are summarized in Table 5, and test results are shown on Figs. 3 through 7. A hydraulic gradient of approximately 30 was used.

TABLE 3—*Compaction test results.*

Soil No.	Compaction Standard ASTM D 698-78	Compaction Standard ASTM D 1557-78	Bentonite Added, %	Maximum Density, pcf[a]	Optimum Moisture Content, %
1	x		0	86.8	27.0
2	x		0	94.8	21.6
3	x		0	105.3	16.6
4	x		0	114.3	14.0
5		x	5	113.5	13.1
5		x	7.5	115.5	13.3
5		x	10	118.9	12.6
5		x	15	117.9	10.5
5		x	29	117.8	10.8

[a]Conversion factor: 1 pcf = 16.01 kg/m^3.

TABLE 4—*The 8-in. (20.3-cm)-diameter permeability test results with Denver tap water as permeant fluid.*

Soil No.	Soil Description	Permeability, cm/s
1	very high plasticity clay	5.2×10^{-9}
2	high plasticity clay	3.5×10^{-9}
3	medium plasticity clay	1.4×10^{-9}
4	low plasticity clayey sand	1.6×10^{-7}
5	sand + 29% bentonite	2.3×10^{-9}

TABLE 5—*The 8-in. (20.3-cm)-diameter permeability tests using "acid liquor" as permeant fluid.*

Soil No.	Soil Description	Permeability, cm/s
1	very high plasticity clay	$<1.0 \times 10^{-10}$
2	high plasticity clay	$<1.0 \times 10^{-10}$
3	medium plasticity clay	$<1.0 \times 10^{-10}$
4	low plasticity clayey sand	1.7×10^{-8}
5	sand + 29% bentonite	2.5×10^{-9}

Some reaction between the iron permeameter walls and the acid liquor was noted and is discussed in paragraphs that follow. Test results reported as $<1 \times 10^{-10}$ cm/s indicate the lower limit of the permeability range believed measurable with this equipment.

The calculated permeabilities for the 5.08-cm-diameter tests on the four plastic soils and sand 5, 7.5, 10, and 15%-bentonite mixtures using acid liquor as the permeant are summarized on Table 6, and test results are shown in Figs. 3–7. A hydraulic gradient of approximately 70 was used.

The results of the mineralogical analyses are presented in Table 7. The results reported are based on petrographic examination and X-ray diffraction and indicated that the medium to very high plasticity clays are low in quartz and higher in clay minerals, while the low plasticity clayey sand was higher in quartz and lower in clay minerals, correlating well with the plasticity index, shrinkage limit, and gradation data which were determined.

The results of the chemical analyses of the soils and soil leaches are given in Table 8. The pH values of the leaches varied from 6.8 to 8.0, about neutral to slightly basic. The total dissolved solids were relatively low, being from 380 to 460 mg/L for the three most plastic clay soils and 990 for the low plasticity clayey sand. Constituents of calcium, magnesium, sodium, potassium bicarbonate, sulphate, and chloride were the most prominent in the leaches. For the dry soil basis of analyses, calcium oxide, iron oxide, and aluminum oxide were prominent with silicon dioxide being the major constituent.

TABLE 6—*The 5.08-cm (2-inch)-diameter permeability tests with acid liquor as permeant fluid.*

Soil No.	Soil Description	Permeability, cm/s
1	very high plasticity clay	1.6×10^{-9}
2	high plasticity clay	1.9×10^{-9}
3	medium plasticity clay	1.6×10^{-9}
4	low plasticity clayey sand	1.6×10^{-8}
5	sand + 5% bentonite	2.5×10^{-7}
5	sand + 7.5% bentonite	2.0×10^{-7}
5	sand + 10% bentonite	1.1×10^{-7}
5	sand + 15% bentonite	1.4×10^{-7}

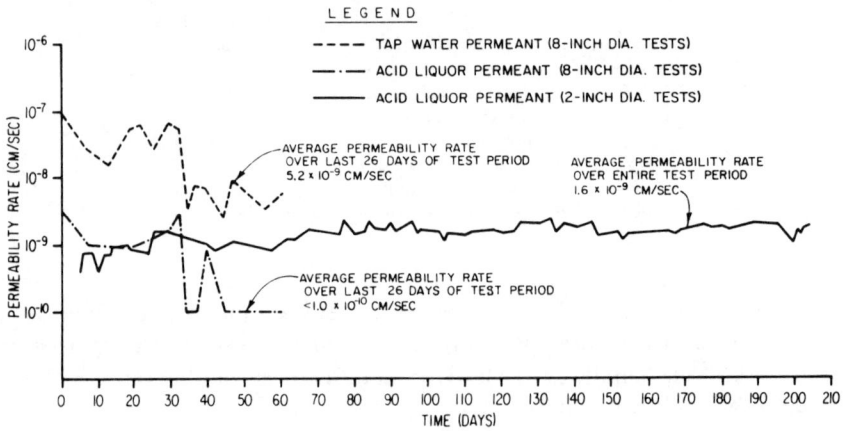

FIG. 3—*Permeability test results for Soil 1—very high plasticity clay (CH).*

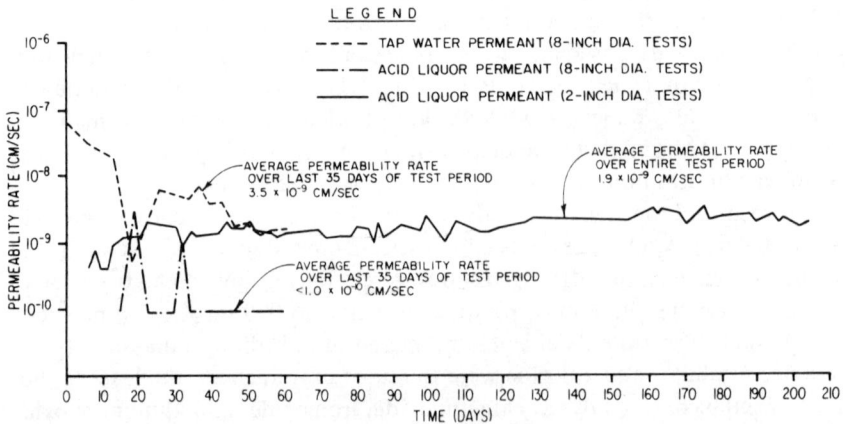

FIG. 4—*Permeability test results for Soil 2—high plasticity clay (CH).*

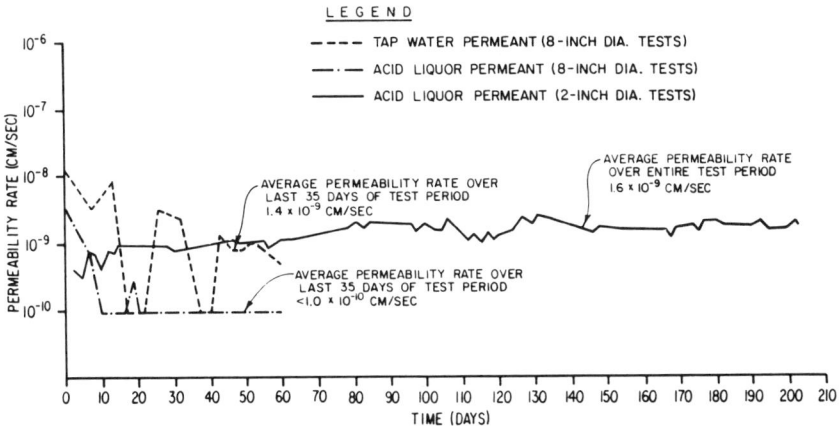

FIG. 5—*Permeability test results for Soil 3—medium plasticity clay (CL).*

The results of the chemical analyses of the acid liquor used as the permeant liquid for the permeability tests is given in Table 9. This liquid was obtained from an operating phosphogypsum field and was believed to represent the type of liquid that will be utilized with the proposed phosphogypsum disposal facility. This liquid has a pH of 2.2 and was high in calcium, calcium oxide, sodium, and chloride and very high in sulphate and fluoride.

Discussion

Plastic Soils

The average permeability rates for the four plastic soil specimens Nos. 1–4 permeated with acid liquor averaged about one tenth that of the specimens permeated with tap water, indicating that the acid liquor enhances the impermeability of the plastic soils planned for this "impermeable" barrier. The permeabilities from the 5.08-cm-diameter permeability tests with acid liquor permeant averaged about one eighth the permeability of the 20.3-cm-diameter permeability tests with tap water permeant, and the 20.3-cm-diameter tests with acid liquor permeant averaged about one tenth the permeability of the 20.3-cm-diameter permeability tests with water permeant. In the 20.3-cm-diameter tests with acid liquor permeant, the values of $< 1 \times 10^{-10}$ cm/s were assumed as 1×10^{-10} cm/s for averaging purposes. The lower average permeabilities for the 20.3-cm diameter tests with acid liquor might be a result of some gas formation in the permeameters, which would have a tendency to reduce flow or to indicate the effect of sample saturation before testing. The 5.08-cm-diameter samples were vacuum saturated before the start of testing, whereas the 20.3-cm-diameter samples were not. The cal-

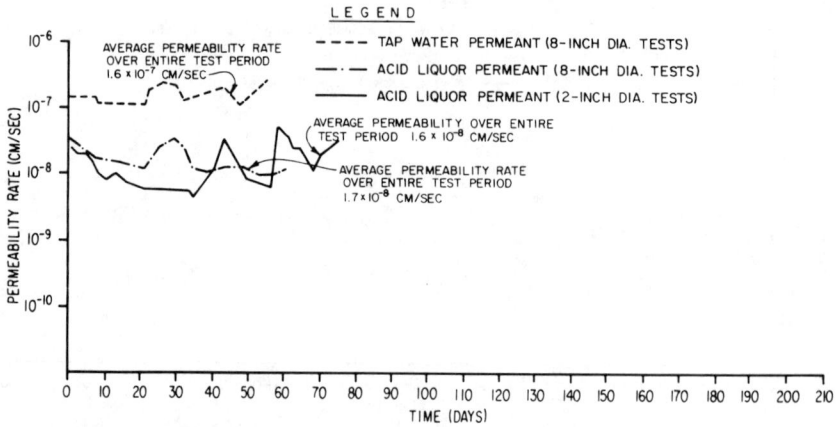

FIG. 6—*Permeability test results for Soil 4—low plasticity clayey sand (SC).*

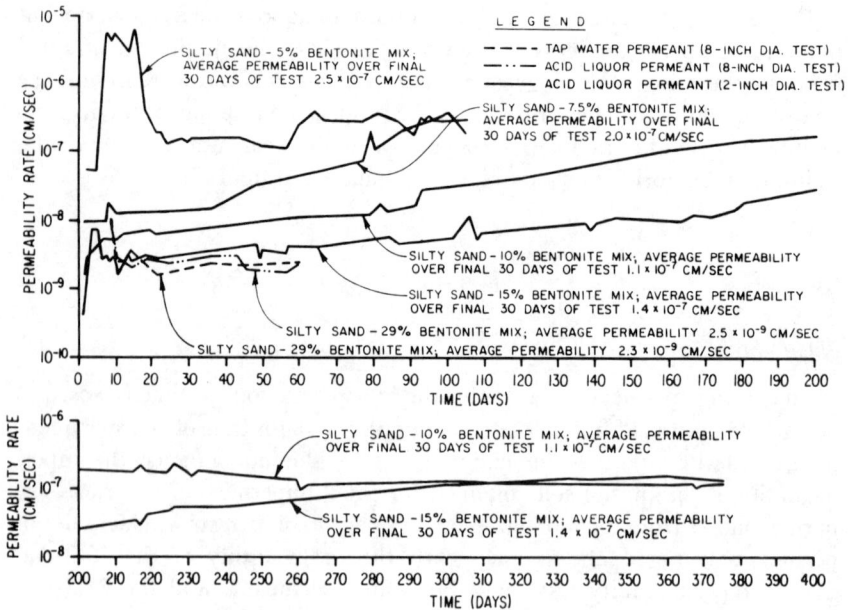

FIG. 7—*Permeability test results for Soil 5—silty sand-Bentonite mixtures.*

culated pore volumes of acid liquor contacting Soils 1, 2, and 3 ranged from about 2.5 to 3.2 pore volumes in the 5.08-cm-diameter tests. No detrimental effects of the acid liquor on these lining materials were observed during the test periods. This is further substantiated by the indication of little, if any, significant change in the other properties considered, including liquid limit, plastic limit, plasticity index, shrinkage limit, and mineralogical analyses.

TABLE 7—*Estimated mineral percentages by volume.*

Component	No. 1	No. 2	No. 3	No. 4
BEFORE PERMEABILITY TEST				
Quartz	23	27	78	89
Attapulgite fibers	2	2	1	trace[a]
Montmorillonite gel	64	57	15	6
Apatite	9	12	3	3
Calcite	2	2	3	2
Hydrated iron oxide	trace	trace	trace	trace
Feldspar	trace	trace	trace	trace
Gypsum	ND[b]	ND	ND	ND
AFTER PERMEABILITY TEST[c]				
Quartz	21	26	81	87
Attapulgite fibers	trace	4	trace	1
Montmorillonite gel	67	58	15	7
Apatite	10	10	2	1
Calcite	2	2	1	0.5
Hydrated iron oxide	trace	trace	trace	trace
Feldspar	trace	trace	trace	trace
Gypsum	ND	ND	1	4

[a]Trace = less than 0.5%.
[b]ND = not detected.
[c]Samples taken from 20.3-cm-diameter permeability tests with acid liquor permeant.

The author feels that the reduction in permeability may be the result of the reaction of phosphoric acid in the acid liquor with calcite to reprecipitate in the pore spaces as gypsum. Another possibility is the solution and reprecipitation of the other mineral species in the pore spaces [2]. An examination of the soil samples after the permeation tests seems to support this hypothesis. The samples permeated with acid liquor generally felt harder and in some cases slightly cemented (based on cutting resistance with a knife) than those permeated with tap water.

The results of the permeability tests on clay lining materials with acid liquor flowing through a gypsum layer showed that the very high, high, and medium plasticity clays, Soils 1, 2, and 3, would produce a compacted lining having permeabilities less than 1×10^{-8} cm/s. It was therefore suggested that these materials and this rate be used for lining design purposes. The low plasticity clayey sand, Soil 4, had a permeability slightly higher than Soils 1–3 in the 20.3-cm-diameter permeability test but was less than 1×10^{-7} cm/s. Also, when this soil was tested with acid liquor permeant in the 5.08-cm-diameter permeameter, the permeability observed was 1.5×10^{-8} cm/s. Since this soil appeared to be borderline, we suggested that it not be used for design because the thickness of the lining would have to be increased to provide an equivalent level of protection.

TABLE 8—*Chemical analyses of clayey soils tested.*

Chemical Component	No. 1	No. 2	No. 3	No. 4
SOIL, DRY BASIS, %				
Calcium (as calcium oxide)	5.3	11.9	2.8	0.60
Magnesium (as magnesium oxide)	1.1	0.88	0.70	0.15
Sodium (as sodium oxide)	0.18	0.33	0.32	0.24
Potassium (as potassium oxide)	0.52	0.33	0.25	0.06
Iron (as iron oxide)	3.4	2.3	1.7	1.4
Aluminum (as aluminum oxide)	7.5	6.4	6.4	3.8
Carbonate	1.3	5.0	2.0	0.41
Chloride	0.03	0.01	0.03	0.01
Sulphate	<0.01	<0.01	1.1	<0.01
Phosphate-phosphorus	<0.01	<0.01	<0.01	<0.01
Silicon dioxide	55.9	54.4	57.1	76.6
1 : 10 LEACH, MG/L				
Calcium	18	33	13	76
Magnesium	20	35	14	30
Sodium	19	9	18	14
Potassium	3.0	1.7	2.0	50
Silicon dioxide	8.3	12	14	6.3
Bicarbonate	180	60	91	30
Carbonate	<0.1	<0.1	<0.1	<0.1
Sulphate	25	48	5	4
Chloride	9.3	75	5.8	<5
Iron	0.10	0.05	<0.05	0.52
Aluminum	4.0	<0.5	3.5	84
Phosphorus	1.7	0.2	0.6	<0.1
Total dissolved solids by evaporation	460	380	460	990
pH	7.9	8.0	8.0	6.8

Silty Sand-Bentonite Mixtures

The results of the permeability tests on mixtures of sand and bentonite with 5 to 15% bentonite, which were conducted in the miniature permeameters with acid liquor flowing through a gypsum layer, showed that the initial permeability varied with changes in the amounts of bentonite and that the permeability increased with time for samples containing between 5 and 15% bentonite. The measured permeabilities of silty sand and bentonite mixtures with 7.5, 10, and 15% bentonite increased to approximately 7, 13, and 41 times, respectively, that of their initial permeabilities over a one-year period. We concluded that a soil-bentonite mixture should not be used in constructing the lining.

The 5.08-cm-Diameter Permeameter

The 5.08-cm-diameter permeameters appear to provide excellent results for the fine-grained soils tested. They are not recommended for use with

TABLE 9—*Chemical analyses of "acid liquor,"*
mg/L.

Component	Analyses
Silicon dioxide	23
Aluminum	105
Aluminum oxide	397
Iron oxide	586
Calcium	1280
Calcium oxide	1644
Magnesium oxide	369
Sulphate	6850
Sodium	1700
Potassium oxide	360
Carbonate	<0.1
Bicarbonate	<0.1
Chloride	992
Hydroxide	<0.1
Fluoride	14 000
Phosphorus	<0.1
Potassium	150
Free ammonia	<0.3%
pH	2.2

coarse-grained soils. The permeameters are highly controllable, and the soil-permeameter side wall contacts can be observed through the Lucite cylinder for signs of piping.

Conclusion

The test results indicated that the clayey soils would be the best for use in constructing the compacted soil lining to minimize seepage losses. The effect of the acid liquor is beneficial in this case because it reduces the permeability. This study demonstrated the need to use site specific soils as well as site specific fluids for permeability testing.

At present there is no ASTM standard for permeability testing for "cohesive" soils. The author and reviewers believe that adoption of a test standard would be beneficial and that the standard should include consideration of permeant fluids other than water.

Acknowledgments

I wish to thank W. G. Holtz for his assistance in developing and analyzing the testing program and for reviewing this article and Mr. S. T. Thorfinnson for his review of this article and for the overall direction he provided for the projects. I also wish to thank H. J. Gibbs, E. G. Hambek, and T. E. Arnold for their assistance in developing, performing, and analyzing the testing program.

References

[1] *Earth Manual*, 2nd ed., U.S. Department of the Interior, Bureau of Reclamation, Washington, DC, 1974, pp. 491–505.
[2] Gee, G. W., Campbell, A. C., Opitz, B. E., and Sherwood, D. R. in *Proceedings*, Third Symposium on Uranium Mill Tailings Management, Geotechnical Engineering Program, Civil Engineering Department, Colorado State University, Fort Collins, CO, 24-25 Nov. 1980, pp. 333–352.

Tuncer B. Edil[1] and Allan E. Erickson[2]

Procedure and Equipment Factors Affecting Permeability Testing of a Bentonite-Sand Liner Material

REFERENCE: Edil, T. B. and Erickson, A. E., **"Procedure and Equipment Factors Affecting Permeability Testing of a Bentonite-Sand Liner Material,"** *Hydraulic Barriers in Soil and Rock, ASTM STP 874*, A. I. Johnson, R. K. Frobel, N. J. Cavalli, and C. B. Pettersson, Eds., American Society for Testing and Materials, Philadelphia, 1985, pp. 155–170.

ABSTRACT: This paper presents the findings of an investigation regarding the effects of procedural and equipmental factors on the laboratory measurement of permeability of a bentonite-sand liner. In the experimental program, permeability tests were performed on compacted specimens of a bentonite-sand mixture to study the effect of (1) permeameter type (rigid- versus flexible-wall permeameters), (2) back pressure application, and (3) hydraulic gradient magnitude on the measured permeabilities. The liner material used in the tests was a mixture of 90% P20-R30 Ottawa sand and 10% bentonite clay. Water was used to prepare the specimens and as the permeant. Specimens were prepared by compaction at moisture contents exceeding optimum using the standard proctor method. A pressure of 380 kPa was used in the tests involving back pressure. The low and high hydraulic gradients used were nominally 29 and 290, respectively. Prior to the termination of a test, a red indicator was passed through the permeameters under the high gradient. The specimens were dissected and inspected after the test for the identification of flow areas and channels.

The test results indicated that specimens continue to hydrate during permeation. Unless wetter specimens are used, this continuing hydration interferes with the inflow-outflow balance, depriving the tester of an important check for leaks. Differential hydration throughout the specimen results in different soil structures and zones of flow. For instance, most of the flow may take place in an annular area surrounding a less hydrated core in the center. Since the total cross-sectional area is used in computing the coefficient of permeability, the values may be underestimated. Back pressure, often used to enhance saturation during testing, appears to have a detrimental effect when applied in the rigid-wall permeameters by increasing the potential for formation of channels and side flow. Test results are affected by hydraulic gradient in different ways depending on the type of permeameter. While gradients as high as 360 did not induce piping in the gap-graded

[1]Professor, Department of Civil and Environmental Engineering, University of Wisconsin-Madison, Madison, WI 53706
[2]Geotechnical engineer, CH2M Hill, Bellevue, WA 98004.

liner material tested, the application of very high gradients to accelerate testing is not desirable for a number of other effects observed.

KEY WORDS: permeability, soil barriers, soil liners, permeability test, laboratory testing, bentonite, hydraulic conductivity, compaction, back pressure, hydraulic gradient, permeameter

Soil liners are commonly used in the construction of low-permeability barriers. During the design of a barrier, the maximum allowable coefficient of permeability must be specified for the soil being used. Prior to construction, the soil that is proposed for use must be tested. The test is intended to show that the soil has the potential to have a coefficient of permeability at or below the specified level in the field. Accurate measurement of the coefficient of permeability has proved to be a difficult task with regard to low-permeability soils (clays and silts). Problems are present within the procedural steps and the equipment used to measure the coefficient of permeability.

This study addresses some of the problems of permeability testing of low-permeability soils. An experimental program was developed to identify causes of problems that are equipment and procedure related. Within the experimental program, three main variables were considered: (1) permeameter type; (2) back pressure application; and (3) hydraulic gradient. Permeability tests were run in permeameters that confined the specimens with rigid or flexible walls. Some specimens were tested with back pressure while others were not. Gradients applied during the tests were also varied. A mixture of sand and bentonite was used as the test soil because bentonite is often used as a permeability-reducing additive in areas with scarce clay sources. The specimens were prepared by impact compaction. Conclusions and a discussion of the test results follow a description of the experimental program.

Experimental Program

Four permeability tests were run in rigid-wall permeameters and three in flexible-wall permeameters to investigate how specimen containment affects the measured permeabilities. Rigid-wall permeameters consisted of standard (944 cm³) steel compaction molds fitted with caps at each end. Flexible-wall permeameters consisted of slightly modified triaxial cells. Figure 1 shows the schematics of each permeameter along with the flow measurement equipment.

Specimens for the flexible-wall permeameter were trimmed to a diameter of 35.6 mm from soils compacted in the compaction molds. They were encased in two latex membranes before installation in the permeameter. Rigid-wall specimens were compacted and tested in the compaction molds so that the diameter of these specimens remained at 101.6 mm. Each permeameter had porous stones at the specimen ends. Two sheets of filter paper were placed between the specimen and the porous stones. Fluid and air pressures were measured with an electronic pressure transducer.

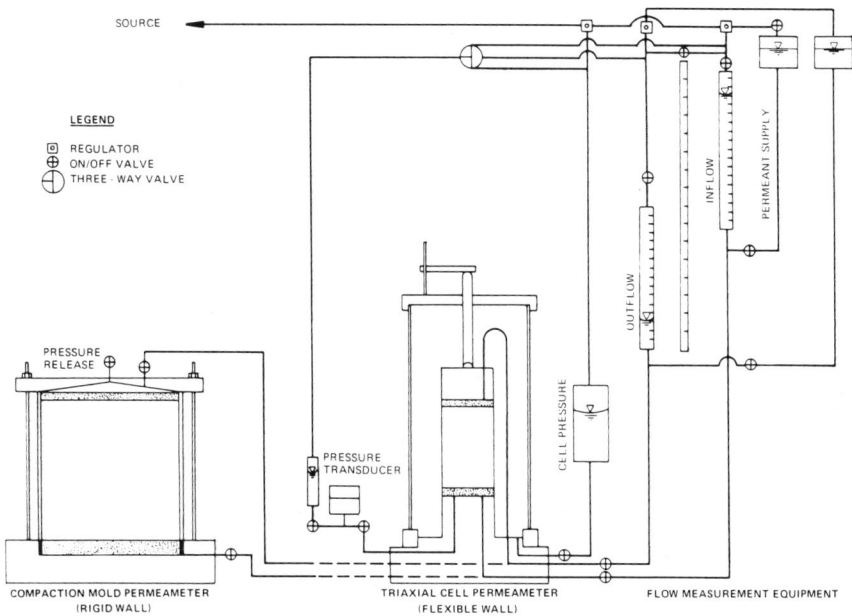

FIG. 1—*Schematic of permeameters.*

Back pressure was applied at the air-water interface in each burette. A back pressure of 380 kPa was used because during preliminary investigations it produced B-values (ratios of the measured pore water pressure to the applied hydrostatic confining pressure increment) exceeding 0.9

Hydraulic gradients were induced by a combination of elevation head differences and air pressure differences between the inflow and outflow burettes shown in Fig. 1. Water levels were monitored and recorded with time. Recorded data were reduced and converted to tables of inflow and outflow water volume versus elapsed time and plotted on graphs to determine average flow rates (Fig. 2). During testing, gradient magnitudes were varied between low and high hydraulic gradients of 29 and 290 as shown in Table 1. Tests were run under falling-head/rising-tailwater conditions, but flow rate and hydraulic gradient changes were small for short time increments. Therefore, for short time increments, permeabilities were similar when computed using variable head or constant head equations. Permeabilities presented herein were computed using constant head calculations. For ease of discussion, specimens are designated by three characters as explained in Table 1.

To provide similar soil characteristics in each test, specimens were prepared in the laboratory under controlled conditions. The soil was a mixture of 90% P20-R30 Ottawa sand and 10% bentonite clay combined on a dry weight basis. Specific gravities of these two materials were measured to be 2.66 and 2.70, respectively. This particular bentonite had a measured liquid limit of 400% with a plasticity index of 350%. A moisture-density relationship for

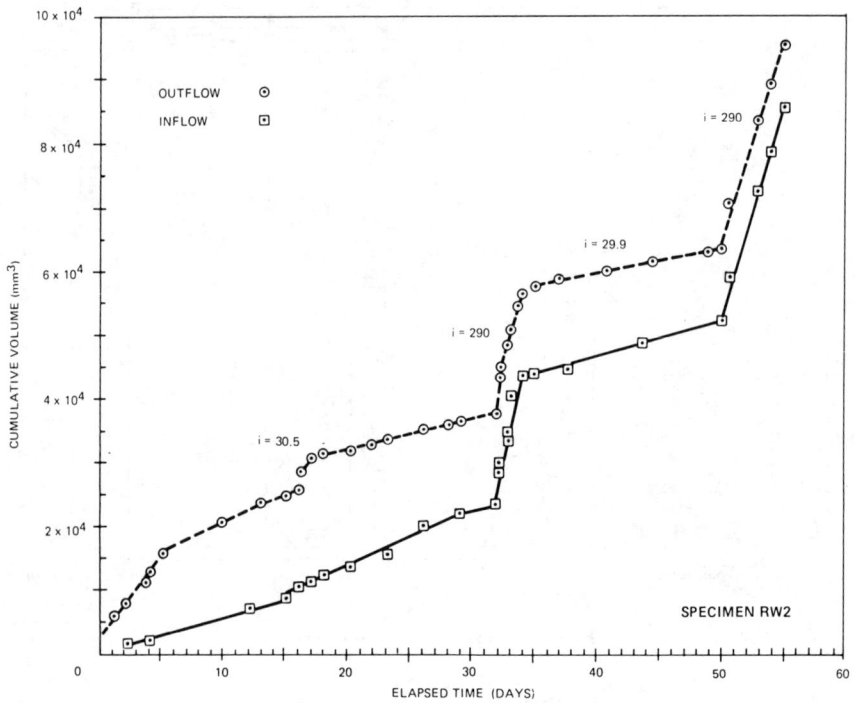

FIG. 2—*Flow versus time.*

TABLE 1—*Experimental program showing gradient schedule, back pressure, and specimen dimensions.*

Test No.[a]	Specimen		Back Pressure, kPa	Gradient Schedule			
	L, mm	D, mm		1st Low	1st High	2nd Low	2nd High
RIGID-WALL PERMEAMETER							
RW1	110	101	0	30.4	300	30.4	
RW2	110	101	0	30.5	290	29.9	290
RB1	110	101	380	33.4	292		
RB2	110	101	380	32.9	292	32.2	
FLEXIBLE-WALL PERMEAMETER							
FW1	30	36	0	48.7	357	45.0	363
FB1	38.6	35.5	380	40.5	288		
FB2	36.6	35.3	380	42.6	299	41.4	

[a]Test/Specimen Designation

  F = Flexible wall permeameter.
  R = Rigid wall permeameter.
  B = With back pressure.
  W = Without back pressure.
  1,2 = Order of testing.

this synthetic soil was developed following the standard Proctor method [ASTM Test for Moisture-Density Relations of Soils and Soil-Aggregate Mixtures, Using 5.5-lb (2.49-Kg) Rammer and 12-in. (304.8-mm) Drop (D 698-78)]. Compaction moisture contents for the specimens were selected to be at least 1% on the wet side of the optimum moisture content of 18% as shown in Fig. 3. Two specimens (RW2 and FW1) were compacted at moisture contents above 19% to see how compaction moisture affects specimen hydration and, consequently, permeability test results. The soil used to make specimens was brought to the desired moisture content at least 24 h prior to compaction. All permeability specimens were compacted following the procedure given in ASTM Method D 698-78.

A basic procedure was developed and followed for each test. First, the specimens were compacted and prepared for the permeameters as just described. Within 2 h after compaction, the specimens were installed in the permeameters. To accelerate the specimen saturation process, the following steps were performed. A vacuum was applied at the top of the specimen, and a low hydraulic head was applied at the bottom of the specimen (following the procedures described in ASTM Test for Permeability of Granular Soils (Constant Head) [D 2434-68 (1974)]. The vacuum was maintained for at least 2 h. After the vacuum was released, water was added to the top of the specimen and air was purged from the system. At this stage, back pressure was applied to designated specimens. Back pressure was increased in steps of 34.5 kPa, allowing 30 min for pressure equilibration at each step. To allow internal pressures

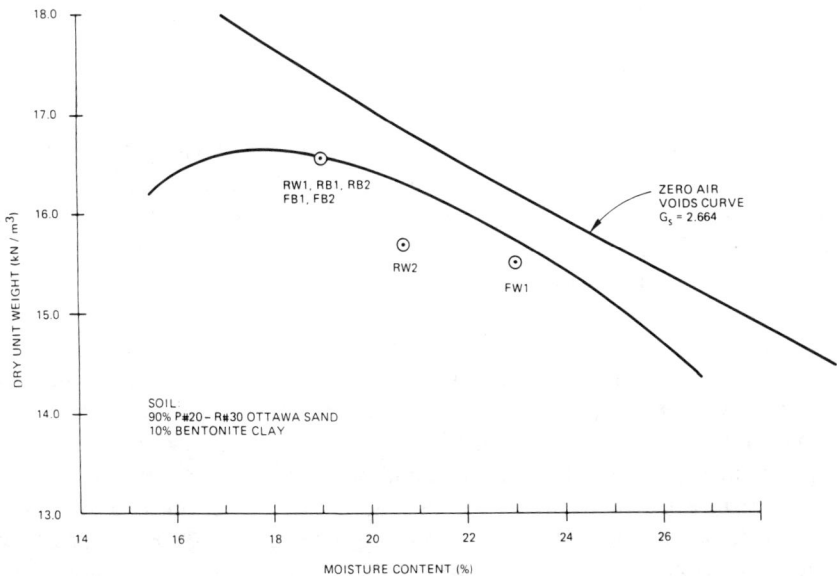

FIG. 3—*Compaction curve of soil barrier (90% P20-R30 Ottawa sand and 10% bentonite clay).*

to stabilize, the specimens were kept for at least 24 h with no hydraulic gradient applied regardless of whether the back pressure was applied. A check of the B-value was made in the case of flexible wall specimens to assess the status of saturation at the end of this period.

Next, hydraulic gradients were applied with the magnitude and progression shown in Table 1. Flow measurements were monitored and recorded daily. At each magnitude, the applied gradients were maintained until inflow and outflow rates became constant. Gradients remained nearly constant due to maintaining a constant air pressure difference between the inflow and outflow burettes. The relatively small rate of flow resulted in small changes in head when considered over a short increment of time.

During the last portions of Tests RW1, RW2, RB2, FW1, and FB3, a red indicator (phenol sulfon phthalein) was added to the permeant to highlight areas of the specimens that were transmitting more flow. The permeameters were then disassembled and the specimens dissected, measuring the moisture contents at various locations within. A discussion of test and inspection results follows.

Results and Discussion

During these experiments certain problems arose that were common to all tests performed regardless of equipment or procedure. These problems can be placed in one of two categories, equipment or specimen. In some cases, solutions to these problems are readily available; in other cases, more research is required.

Equipment-Related Problems

In measuring the coefficient of permeability of a low-permeability soil, the most common equipment-related problem is system leaks. With such low-permeability soils, even a small leak could substantially increase the measured inflow rate or decrease the measured outflow rate. Therefore, a method of checking for leaks is needed. One way of checking that there are no leaks in the system is to perform a water budget analysis. This involves a comparison of inflow and outflow volumes to show that the inflow rate equals the outflow rate. If the inflow rate equals the outflow rate, then continuity is maintained and it may be assumed that there is no leakage. However, if inflow does not equal outflow, it is not necessarily true that there is a leak. For instance, equipment-related flow imbalances could be caused by evaporation.

If flow measurement burettes allow evaporation that is not accounted for during data reduction, inflow rates will appear to be higher than outflow rates. Evaporation from the dropping water level of an inflow burette increases the apparent inflow rate. Evaporation from the rising water level of an outflow burette decreases the apparent outflow rate. Evaporation was accounted for in these experiments by measuring evaporation with a separate burette. Water was measured to be evaporating at an average rate of 5.5 mm³ per day. The measured inflow and outflow rates for each test were adjusted by subtracting evaporation from inflow rates and adding evaporation to outflow rates. In these tests, evaporation was less than 5% of the lowest measured inflow rate.

Another reported equipment-related cause of nonequal inflow and outflow rates is air bubbles in the outflow burette [1]. Producing hydraulic gradients with air pressure differences may drive air into the water at the high pressure inflow end. As water passes through the specimen, pressure reduces and air comes out of the permeant to produce air bubbles. During this study, air bubbles were present in the outflow water of Tests RW1, RW2, and RB2. High gradient flow through these specimens was induced by a 320-kPa air pressure difference between inflow and outflow burettes. As air bubbles collected and expanded in the outflow lines, the outflow rates appeared to increase. Air-saturated permeant was replaced with freshly de-aired water, but bubbles continued to collect. When air bubbles were collecting, measured outflow rates exceeded inflow rates in Tests RW1, RW2, and RB2.

Figure 2 shows a plot of the cumulative volume of permeant that entered and exited Specimen RW2 versus elapsed time. This is a typical figure from which inflow and outflow rates were measured for calculating the coefficients of permeability. Bubbles were present in the outflow water of this test throughout the duration; therefore, the outflow volumes appeared to exceed inflow volumes. The presence of bubbles in these tests indicates that if hydraulic gradients are produced by air pressure differences, the driving pressure should be as low as possible. One possible way of reducing the amount of bubbles collecting in the outflow burette is to apply hydraulic gradients with elevation differences rather than air pressure. Equipment-related problems did not appear to be the only cause of flow rate imbalances in these tests. Specimen-related problems also affected the measured flow rates.

Specimen-Related Problems

The moisture content of the specimens tested seemed to affect how the permeant passed through them. During the testing of two specimens with higher initial moisture contents (RW2 and FW1), measurable outflow appeared soon after the application of the first low-hydraulic gradient. During testing of the specimens compacted at moisture contents just above optimum (RW1, FB1, and FB2), measurable outflow did not appear during application of the first low gradient. The dry unit weights of the wetter specimens were about 6% lighter than the dry unit weights of the other specimens tested. It is reported that decreases in dry unit weights do not cause significant increases in permeability for clay soils compacted wet of optimum [2]. Therefore, the increased outflow is not attributed to the lower initial dry unit weights of Specimens RW2 and FW1.

In these tests, the production of measurable outflow appeared to be linked to the initial state of hydration of the clay particles in the specimens. This is illustrated by comparing the outflow-to-inflow ratios shown in Table 2. The wetter specimens, RW2 and FW1, produced outflow rates that were 60% and 90% of their respective inflow rates. This indicates that those specimens adsorbed lower percentages of inflowing water. One reason why the clay of the wetter specimens adsorbed less permeant may be related to matrix suction. The highly active bentonite may have a threshold moisture content below which all water entering the specimen is adsorbed by the clay. This threshold appears to be related more to initial moisture content than to the degree of saturation. The initial and final degrees of saturation for all specimens were within 5% of each other. Both flexible-wall specimens that were tested with back pressure had B-values exceeding 0.9. From these results it appears that using specimens prepared at least 2% wetter than optimum helps in performing water balances that check for leaks in the system. This intentional deviation from simulating design/construction moisture content in laboratory specimens shortens the time for inducing long-term field moisture conditions in the specimens. Other results of this study indicate that there may be another reason for beginning the permeability tests with wetter, more hydrated specimens.

Specimens compacted near optimum appear to continue to hydrate as permeability tests progress. As the clay adsorbs more inflowing permeant, zones of differential hydration may develop within the specimen. Wetter zones may transmit a higher percentage of the total flow through a specimen, therefore violating the assumption (based on Darcy's law) that flow is uniform throughout the specimen. If more permeant is flowing through these more hydrated zones and the cross-sectional area (A_h) of these zones is smaller than the measured specimen area ($A_m = \pi r^2$), the actual permeability would be unconservatively higher than the measured value. In Darcy's law, $Q = kiA$, where k equals the coefficient of permeability and Q equals the volumetric flow rate flowing through cross-sectional area A as driven by hydraulic gradient i. Ac-

TABLE 2—*Ratios of outflow rate/inflow rate during the first application of low gradients.*

Specimen No.[a]	RW1	RW2	RB1	RB2	FW1	FB1	FB2
Compaction, w%	19.0	23.0	19.0	19.0	20.7	19.0	19.0
Outflow rate, $\dfrac{\text{Outflow rate}}{\text{Inflow rate}}$	NM[b]	0.6	NS[c]	NS	0.9	0.0	0.0

[a]See Table 1.
[b]NM = not measurable.
[c]NS = not significant, permeant flowing between soil and permeameter wall.

tual flow rate and hydraulic head are physically measured in the permeability test, and area is assumed to be equal to the specimen area (A_m). From the relationship just given, if the actual hydrated cross-sectional area of flow (A_h) equals 10% of the measured (assumed) cross-sectional area (A_m), the actual permeability may be ten times higher than the test results indicate [1].

The condition just described may have been present during at least one of the tests of this study. Inspection of Specimen FB2 showed that a small-diameter zone of wetter soil extended up from the bottom of the specimen. Figure 4 is a photo of the cross section of Specimen FB2 showing a wetter zone (lighter colored) near the center of the specimen. Moisture content of this wet zone was measured at 23% and the surrounding soil at about 20%. While this occurrence was assessed in only one specimen, it demonstrates a possible problem that may affect the testing of hydrating barrier materials. The locations where moisture contents were measured are shown in Fig. 5. The presence of this wetter zone indicates that a significant portion of the permeant entering this specimen was flowing through a cross-sectional area that was smaller than the measured area. Therefore, the actual permeability of this soil was probably higher than the test results indicate.

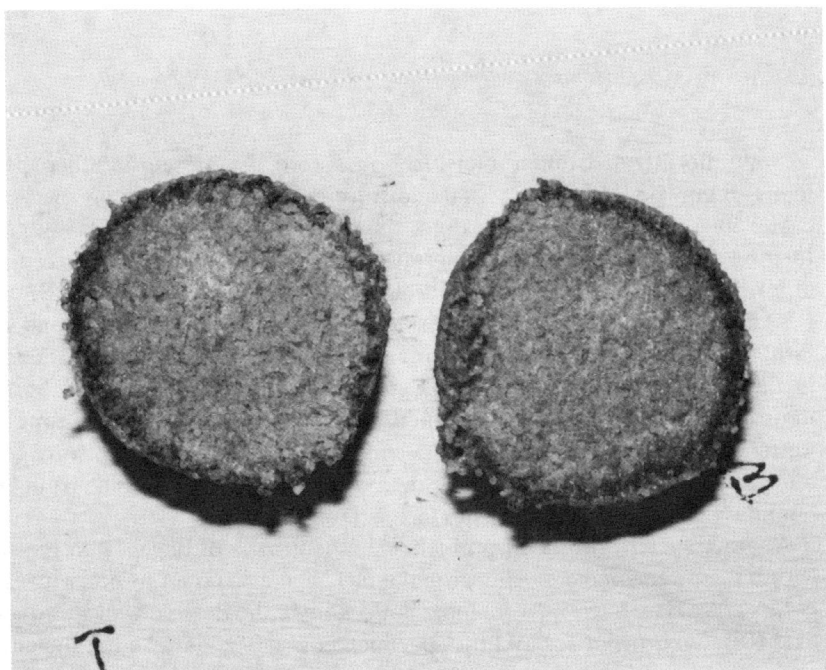

FIG. 4—*Photograph of cross section of Specimen FB2 at the midpoint (after testing). Note darkening around outer edges and lightening near center.*

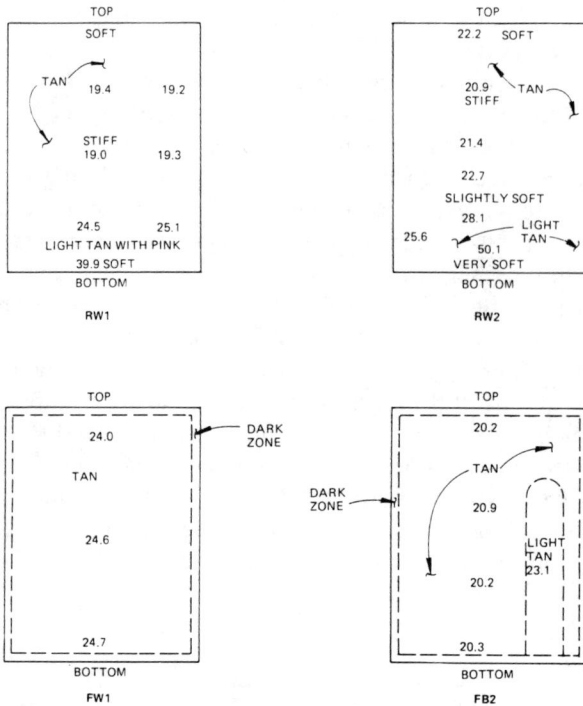

FIG. 5—*Moisture content distributions in specimens (after testing).*

Nonuniformly distributed moisture zones were also present in other specimens. Figures 5 and 6 show that moisture contents varied along specimen center lines and edges. After testing, the rigid-wall specimens were wetter at their bases and along the specimen sides. The moisture contents at the centers of RW1 and RW2 equaled their initial values. Moisture contents in RB1 and RB2 did not change because the permeant flowed between the mold and soil rather than through the soil. Final moisture contents of the specimens tested in the flexible-wall permeameters were uniformly higher than their initial moisture contents. However, specimens FW1 and FB2 also had thin zones of wetter soil along the edges and at the ends. These variations in moisture content may cause the permeant to flow under conditions different from the boundary conditions assumed in Darcy's law.

Another specimen-related problem was the growth of bacteria on the soil within the permeameters. This problem has been also noted by other investigators [3,4]. The bacteria were most noticeable in the flexible-wall specimens of this study, growing around the specimen edges. The dark soil at the outer edges of the specimen shown in Fig. 4 is representative of the two other flexible-wall specimens. This bacterial growth did not seem to have a significant effect on the results of this study, perhaps because of the relatively small areas

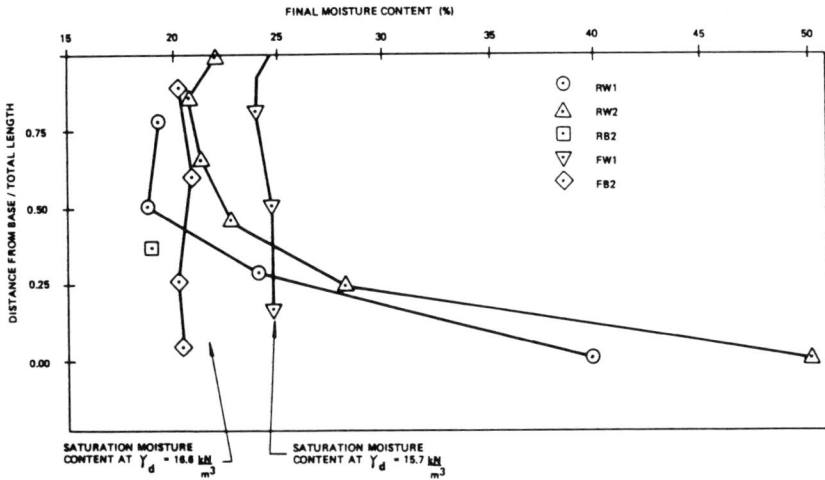

FIG. 6—*Final moisture contents along each specimen's centerline versus dimensionless distance from specimen base.*

affected by it. However, if gas is produced by the bacteria, inflow and outflow measurements may be adversely affected, as discussed previously.

Effect of Back Pressure

Application of back pressure has been promoted as an effective procedure for improving the degree of saturation of a specimen during permeability tests [5–7]. Measuring B-values above 0.9 is reported to indicate that most of the air initially trapped within the system (burettes, lines, porous stones, specimens) has been forced into solution. This study attempted to explore the effects of back pressure by performing tests with and without back pressure.

It was apparent that the application of back pressure did drive the air in the burettes, lines, and porous stones into solution. However, there was no physical proof that the degree of saturation in the specimens was actually increased by the back pressure. For instance, the final calculated degrees of saturation for Specimen FB2 and FW1 were about 99% and 98%, respectively (based on the density of specimen as determined by the specimen dimensions after test). During application of back pressure in flexible-wall permeameters, burette water levels were not seen to decrease once the back pressure exceeded approximately 210 kPa. Effective stresses within the specimen were held at 34.5 kPa during back pressure application. Since specimens were not consolidated to pressures above 34.5 kPa and back pressures did not force more water into the specimen, back pressuring does not appear to have actually increased the degree of saturation of flexible-wall specimens. A comparison of permeabilities measured during these tests shows that flexible-wall permeameters produced results with similar orders of magnitude for FW1, FB1, and FB2 (Table 3). Insufficient data were generated by this study to support a stronger statement for or against the application of back pressure in flexible-wall permeameters; however, there was strong evidence indicating that back pressure should not be applied in rigid-wall permeameters similar to those used herein.

Back pressuring in the rigid-wall permeameters adversely affected test results. The increases in permeant pressure at the top and bottom of Specimens RB1 and RB2 during back pressuring appeared to facilitate the formation of channels between the specimen and the compaction mold. Channels allow permeant to flow around the soil rather than through it. Measured permeabilities in RB1 and RB2 were as much as 1000 times higher than the permeabilities measured in any other tests of this study. A channel was visible in the soil during inspection of Specimen RB2 after passage of the red indicator through the permeameter. This channel is visible as a light line on the photo of Specimen RB2 in Fig. 7. Once the channel had formed, the permeant did not flow through the soil; it passed around it. More research is needed, but these tests indicate that the application of back pressure is appropriate for flexible-wall permeameters and not for rigid-wall permeameters.

Hydraulic Gradient

Measuring permeability of barrier soils can be a long process because of the low flow rates and the long time required to reach steady-state conditions. In an attempt to reduce the length of time required to finish a permeability test, hydraulic gradients in excess of 100 have been used and investigated [2]. Darcy's law seems to apply to clay soils under gradients exceeding 20 [2], but there are concerns that these high gradients may adversely affect test results [5].

TABLE 3—Coefficients of permeability (cm/s) $\times 10^7$ (based on inflow rates).

| | Coefficients of Permeability | | | |
Test No.	Leg 1, 1st Low i	Leg 2, 1st High i	Leg 3, 2nd Low i	Leg 4, 2nd High i
RW1	NM[a]	1.3	0.5	
RW2	4.3	4.2	3.0	3.1
RB1	7.2	380.0		
RB2	2750.0	5100.0	5490.0	
FW1	2.4	1.6	2.9	1.5
FB1	4.1	1.7		
FB2	3.3	1.4	3.4	

[a]NM = not measurable.

FIG. 7—*Photograph of Specimen RB2 (after testing). Note channel and large voids on sides.*

The results of this study indicated that the use of higher hydraulic gradients under certain conditions may cause physical changes in the specimens that change the measured permeability. The permeabilities measured in flexible-wall permeameters (Table 3) decreased as much as 58% when hydraulic gradients were increased. To increase the gradient from 29 to 290 and still maintain a 34.5-kPa effective consolidation pressure, the cell pressure had to be increased from 11 to 103.5 kPa. The pressure at the outflow burette remained constant, therefore allowing the effective consolidation pressure at the outflow end of the specimen to increase from 34.5 to 138 kPa. This increase in pressure at the top of the specimen may have resulted in a differential consolidation of the specimen.

A first interpretation of this result was that the entire upper portion of the specimen consolidated, which reduced the porosity and permeability. However, this may not necessarily be the case. Specimens were densely compacted bentonite-sand mixtures with void-bound fabrics because of the relatively low clay content [8]. Since the sand grains were compacted into a very dense con-figuration, it is doubtful that the specimens would experience significant dif-ferential consolidation from stress differences between the top and bottom of the specimens. However, inspection of flexible-wall permeameter specimens showed that they were surrounded by thin zones of more hydrated soil. Per-meant probably was flowing through this highly hydrated area near the speci-men edges. Soil here may have had a clay matrix fabric due to extensive swell-ing of bentonite [8]. This hydrated area may have consolidated sufficiently to reduce the measured permeability of the specimen under high confining pres-sures. Table 3 shows that the reduction in measured permeability reversed slightly when gradients were returned to their initial levels. Partial reversibil-ity of the permeability decrease when the gradient was lowered indicates that the physical changes may have occurred near the specimen's edges [1]. The problem just described may have been a result of the effects of nonuniform zones of specimen hydration.

Other results produced by the variation of gradients showed that the high gradients applied did not induce piping but did cause air bubbles in the out-flowing permeant. Some migration of the bentonite from these gap-graded specimens was expected, but no noticeable piping was found. This is exclud-ing the flow channels that formed in RB1 and RB2 because of the application of back pressure, as was previously discussed. Air bubbles were especially prevalent during high-gradient portions of the tests when driving air pres-sures exceeded 100 kPa. Possible solutions to these problems may involve shortening the specimen, increasing the initial consolidation pressure, or both. By shortening the specimen, higher gradients could be produced with lower driving pressures. Increasing the consolidation pressure would mean the variation in specimen pressure caused by the application of high gradients from bottom to top would be less significant.

Permeameters

Flexible-wall permeameters have been described as the best type of equip-ment for performing permeability tests [5]. They model in situ pressure con-ditions and provide a better seal along the specimen edges. It appears from the results of this study of bentonite-sand liner soils that rigid- or flexible-wall permeameters, when carefully used, may produce similar results unless the specimen is shrinking in some way.

In these tests it seemed that the permeant was flowing through soil speci-mens in both the rigid- (RW1 and RW2) and flexible-wall (FW1, FB1, FB2) devices and through the specimen-permeameter interface in Specimens RB1 and RB2. Figure 5 shows that the moisture contents of specimens RW1 and RW2 increased by over 15% in their lower halves. Inspection of Specimen RW1 after the test showed that the red indicator was dispersed throughout. Coloring due to the indicator was concentrated more heavily near the specimen edges, perhaps due to a higher localized flow of water adjacent to the interface, but a pink hue was also detectable in the center of the specimen. The apparent flow of permeant near the specimen edges in the flexible-wall equipment appeared to be more a function of zones of differential hydration than of the equipment.

Conclusions

The seven permeability tests performed and described herein were designed to show some of the specific problems that are present in soil barrier permeability testing. Each of the seven specimens was subjected to different test procedures, and the accuracy of the results cannot be uniquely defined. However, certain valuable observations can be made based on these carefully performed tests. Rigid- and flexible-wall permeameters were found to produce measured permeabilities with the same order of magnitude for the bentonite-sand specimens tested in this program. High gradients seem to cause problems within the specimens that change the measured permeabilities. Although back pressure is known to improve the results of permeability tests, it should not be used with rigid-wall permeameters. Compaction of specimens at the moisture content expected in the field makes good geotechnical sense but can be troublesome because of inadequate hydration in testing bentonite-sand barrier materials.

In conclusion, it seems that adequate equipment has been developed to perform permeability tests on barrier soils, but the exact procedures have not been defined that will produce accurate results each time. It is hoped that this paper will provide a step toward the final goal of developing consensus standards for measuring the coefficient of permeability of low-permeability soil barriers.

Acknowledgment

The authors acknowledge the assistance provided by Norman H. Severson in setting up the laboratory tests in this study. The support of the University-Industry Research Program of the University of Wisconsin during this study is also acknowledged.

References

[1] Erickson, A. E., "Equipment and Procedures for Determining the Coefficient of Permeability for Low Permeability Soils," Master of Science Report, Department of Civil and Environmental Engineering, University of Wisconsin-Madison, Madison, WI, 1983.
[2] Mitchell, J. K., Hooper, D. R., and Campanella, R. G. *Journal of the Soil Mechanics and Foundations Division*, American Society of Civil Engineers, Vol. 21, No. SM4, Proceedings Paper 4392, July 1965, pp. 41-65.
[3] Allison, L. E., *Soil Science*, Vol. 63, 1947, pp. 439-450.

[4] Olson, R. E. and Daniel, D. E. in *Permeability and Groundwater Contaminant Transport, ASTM STP 746*, T. F. Zimmie and C. O. Riggs, Eds., American Society for Testing and Materials, Philadelphia, 1981, pp. 18–46.

[5] Zimmie, T. F. in *Hazardous Solid Waste Testing: First Conference, ASTM STP 760*, R. A. Conway and B. C. Malloy, Eds., American Society for Testing and Materials, Philadelphia, 1981, pp. 293–304.

[6] Matyas, E. L. in *Permeability and Capillarity of Soils, ASTM STP 417*, American Society for Testing and Materials, Philadelphia, 1967, p. 160

[7] Black, D. K. and Lee, K. L., *Journal of the Soil Mechanics and Foundations Division*, American Society of Civil Engineers, Vol. 99, No. SM1, Proceedings Paper 9484, 1973, pp. 75–93.

[8] Sowers, G. F. in *Soil Mechanics and Foundations: Geotechnical Engineering*, MacMillan, New York, 1979, p. 38.

Yalcin B. Acar,[1] Abu-bakr Hamidon,[2] Stephen D. Field,[1] and Lisa Scott[2]

The Effect of Organic Fluids on Hydraulic Conductivity of Compacted Kaolinite

REFERENCE: Acar, Y. B., Hamidon, A., Field, S. D., and Scott, L., **"The Effect of Organic Fluids on Hydraulic Conductivity of Compacted Kaolinite,"** *Hydraulic Barriers in Soil and Rock, ASTM STP 874*, A. I. Johnson, R. K. Frobel, N. J. Cavalli, and C. B. Pettersson, Eds., American Society for Testing and Materials, Philadelphia, 1985, pp. 171–187.

ABSTRACT: The effects of four organic fluids on hydraulic conductivity of compacted kaolinite are presented. Permeation fluids were 0.1 and 100% solutions of nitrobenzene, acetone, phenol, and benzene, which represent a wide range of dielectric constants. Full saturation hydraulic conductivities were obtained in flexible wall permeameters under continuous back pressure, at hydraulic gradients of less than 100 and effective stresses of 69 kPa (10 psi). Reference hydraulic conductivity values were determined with 0.01 N calcium sulfate (CaSO$_4$) solution.

The effect of the testing scheme was evaluated by measuring the hydraulic conductivity of acetone in both a flexible wall permeameter at variable effective stresses and in a rigid wall permeameter. Dramatic hydraulic conductivity increases were observed in rigid wall permeameters. The results indicate that increases in the hydraulic conductivity measured in rigid wall permeameters can only be explained by side leakages due to shrinkage of the specimen.

All tests with chemicals at low concentrations resulted in slight decreases of hydraulic conductivity. Hydraulic conductivity with pure solutions slightly increased with acetone and phenol and significantly decreased with benzene and nitrobenzene. Diffusion through the cell membrane was found to be a considerable source of error in assessing the full strength of organic fluids. The direction of variations in the liquid limit and the free swell of kaolinite with organic fluids was observed to be inversely related to changes in absolute values of hydraulic conductivity at low effective stresses. These results together with fabric studies indicated that changes in hydraulic conductivity with organic fluids can be explained by the variations in the surface forces of interaction on clay particles affecting the flow characteristics.

[1]Assistant professor, Dept. of Civil Engineering, Louisiana State University, Baton Rouge, LA 70803.

[2]Graduate research assistant, Dept. of Civil Engineering, Louisiana State University, Baton Rouge, LA 70803.

KEY WORDS: hydraulic conductivity, organic fluids, compaction, kaolinite, free swell, liquid limit, testing procedure, groundwater

Containment of organic compounds in shallow land waste disposal sites has prompted the need to study the effects of these compounds on the engineering properties of soils. The hydraulic conductivity of compacted soil liners used in these facilities is one of the most significant parameters necessary to predict the time-dependent release of the contaminants [1]. Several studies have assessed the effect of low level electrolyte concentrations on permeability of compacted clay [2,3]. The more recent studies have indicated that organic fluids might increase the hydraulic conductivity of compacted soil liners and have a detrimental effect on the structural integrity of the waste disposal facility [4,5].

Presently, soils with high activities (activity is defined as the plasticity index divided by percent weight of particles finer than 2 μm fraction) are used to achieve the low permeability of the liner material. Although compacted soils with higher activities are associated with lower permeabilities [6], the structure of these soils are more sensitive to variations in the attractive and repulsive electrical forces between clay particles.

This study presents the effect of four organic fluids with a wide range of dielectric constants and their 0.1% aqueous solutions on the hydraulic conductivity of compacted Georgia kaolinite. Calcium and sodium montmorillonite were also used in tests for index properties and free swell. The mineralogical composition, index properties, and compaction parameters of these clays are presented in Table 1. Table 2 presents the characteristics of organic fluids used in this study together with their hazard classification and maximum reported concentration in leachates collected from waste disposal facilities [7].

Apparatus and Procedure

Considering the present state of the art in permeability testing with leachates [8,9] and various shortcomings of other methods of testing [10], five constant head flexible wall permeameters were set up. Figure 1 presents a schematic view of the test setup. This setup allows testing of 3.50- to 7.62-cm diameter specimens under continuous back pressure. The constant head is maintained by mariotte bottles and the gradient is monitored continuously throughout the testing by individual differential pressure transducers. All couplings and tubings were made of Teflon to avoid adsorption and corrosion. Provisions are taken to record both the inflow and outflow.

Hydraulic conductivity testing of fine-grained soils with leachates often requires the testing to continue until the effluent concentration reaches the influent level (100% breakthrough). Since the transport of the leachate is gov-

TABLE 1—*Composition and characteristics of clays used in this study.*[a]

Characteristic	K	Ca-M	Na-M
Mineralogical composition, wt%			
Kaolinite	98	...	...
Illite	2	8	T
Ca-montmorillonite	...	92	...
Na-montmorillonite	...	...	100
Index properties, % (ASTM D 4318)[b]			
Liquid limit	64	88	425
Plastic limit	34	54	58
Plasticity index	30	34	367
Specific gravity (ASTM D 854)[c]	2.65	2.70	2.70
Finer than 2-μm size, %	90	12	80
Activity	0.32	2.8	4.5
Proctor compaction parameters			
Maximum dry density, tons/m^3	1.37	1.15	1.15
Optimum water content, %	31.0	25	25

[a] K = Georgia kaolinite.
Ca-M = ca-montmorillonite.
Na-M = na-montmorillonite.
T = trace quantity.
[b]ASTM Method for Liquid Limit, Plastic Limit, and Plasticity Index of Soils (D 4318-83).
[c]ASTM Method for Specific Gravity of Soils [D 854-58 (1979)].

TABLE 2—*Characteristics of organic fluids.*[a]

Compound	Formula	ε	μ	γ	DM	pH	Class.	Max. Concentration, mg/L	SBL, g/L at 25°C
Water	H$_2$O	80.4	1.0	0.98	1.83	7.0	...	...	...
Ethylene glycol	C$_2$H$_6$O$_2$	38.66	21.0	1.11	2.2	6.4	...	...	...
Nitrobenzene	C$_6$H$_5$NO$_2$	35.74	2.03	1.20	4.22	3.9	H,T,S,P	0.74	1.90
Acetone	C$_3$H$_6$O	20.7	0.54	0.79	1.66	6.8	T	42.4	∞
Phenol	C$_6$H$_6$O	13.13	12.7	1.06	1.45	3.5	H,T,P,S	17.0	86.34
Aniline	C$_6$H$_5$NH$_2$	6.9	4.40	1.02	1.55	9.9	T,S	1.9	34.0
Xylene	C$_8$H$_{10}$	2.50	0.81	0.87	0.40	5.6	T,S	60	0.20
Benzene	C$_6$H$_6$	2.28	0.65	0.88	0	5.7	H,T,P,S	7.4	1.77
P-dioxane	C$_4$H$_8$O$_2$	2.21	1.44	1.03	0.45	5.8	H,T	...	∞
Heptane	C$_4$H$_{16}$	1.0	0.41	0.68	0	5.4	...	...	0.003

[a] ε = dielectric constant (20°C).
γ = unit weight, g/cm^3 (20°C).
μ = viscosity, cP (20°C).
DM = dipole moment (debyes).
Class. = hazard classification [7].
H = hazardous.
S = Section 311 compound.
T = toxic.
P = priority pollutant.
Max Conc. = maximum concentration reported in leachates (mg/L).
SBL = water solubility.

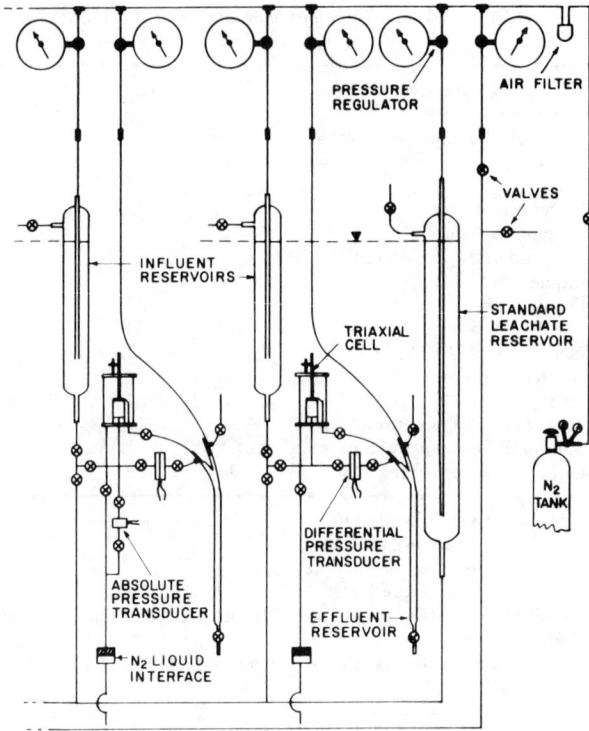

FIG. 1—*A schematic view of the test setup.*

erned by advective-dispersive-reactive transport mechanisms [*11*], it is necessary to continue testing more than one pore volume in order to reach full breakthrough and assess the effect of the full strength leachate. Unless the specimen pore volume is decreased, testing might then take several months even at considerably high gradients. Consequently, specimen dimensions were restricted to 3.55 cm in diameter and 3.8 to 5.1 cm in height, resulting in pore volumes of 10 to 15 cm^3.

Kaolinite was cured for a week at a water content of 32% and was compacted in a Harvard miniature compaction mold [*12*] at an energy level corresponding to standard Proctor compaction.

The reactive and corrosive nature of organic contaminants adversely affect the commercially available latex membranes used in flexible wall permeameters. Figure 2 presents the effect of benzene on deformation of different types of membranes. In such extreme cases of deformation, interconnected vertical and horizontal wrinkles were observed on the membrane, making the testing susceptible to side leakages. It was concluded that latex membranes are not compatible to organic fluids with dielectric constants of less than 10 [*13*]. This latex membrane deterioration was effectively remedied when specimens were

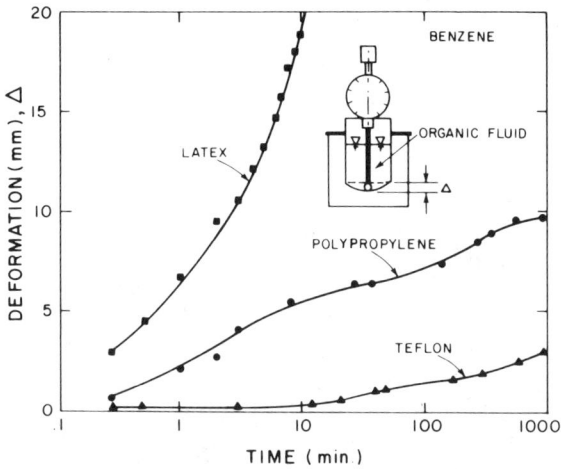

FIG. 2—*The effect of benzene on different types of membrane* [13].

first wrapped by two rounds of an 0.03-mm sheet of Teflon. The membranes were then placed and coated with a contaminant-resistant (silicon base) grease to decrease the possibility of chemical diffusion. Even with such provisions a significant amount of contaminant was found in the cell water after prolonged testing times.

Back pressures of 414 to 449 kPa (60 to 65 psi) were used to fully saturate the specimens, which had initial saturations of 88 to 90% [14]. The average effective stress was increased to 69 kPa (10 psi) by incrementally raising the cell pressure and the back pressure after measurement of B values [15]. To promote saturation, a slight gradient was applied during back pressuring. Approximately one pore volume of 0.01 N $CaSO_4$ solution was permeated to obtain the reference hydraulic conductivity and to homoionize the specimens. The influent was then switched to organic fluids without any change in the back pressure. The effluent was collected and continuously monitored for pH, K^+, Ca^{+2}, and organic solvent concentration. Organic fluid concentrations were determined using a gas chromatograph equipped with a packed column (Tenax 80/100[3]), flame ionization detection, and an auto integrator. Ca^{++} and K^+ concentrations were determined by atomic absorption techniques using a spectrophotometer. Testing continued until hydraulic conductivity and the effluent concentration were stabilized. Initial and final unit weights and fluid contents of specimens were determined.

It is known that particle cloggings and uncloggings together with consolidation and swelling of the specimen at high gradients result in an apparent non-Darcian flow in fine-grained soils [16,17]. Therefore, it was first decided to determine the effect of gradients on the velocity of flow. Figure 3 presents

[3]Teklab, Inc., Baton Rouge, LA 70814.

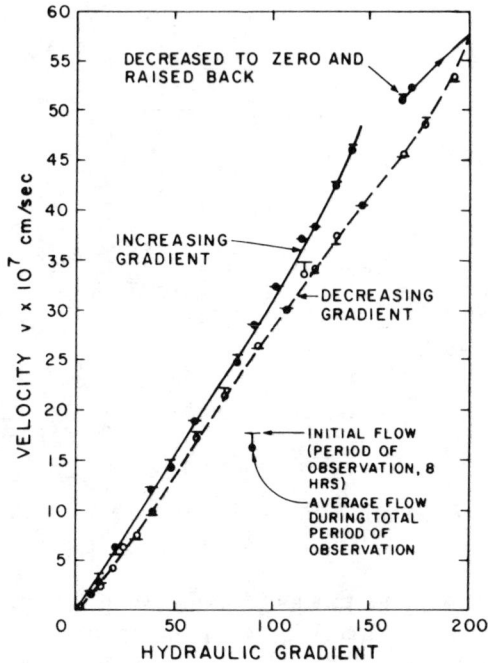

FIG. 3—*The effect of gradients on velocity of flow in compacted Georgia kaolinite.*

the results of this test with Georgia kaolinite compacted at the wet of optimum water content. This test indicated a minimum amount of hysteresis for gradients less than 100, and the hydraulic conductivity varied between 3.0 to 3.2×10^{-8} cm/s. Consequently, gradients of less than 100 were used while simultaneously testing a control specimen for comparison with the tests using organic permeants. It is noted that when a gradient of 100 is applied on a specimen of 4 cm long, the reference effective stress at the middle of the specimen is decreased at the bottom and increased at the top by almost 20 kPa (2.8 psi), resulting in effective stress variations along the specimen. In order to obtain identical changes, the specimen heights were kept the same in each batch.

Analysis of Results

Figure 4 presents the results obtained with 1000 ppm (0.1%) acetone solution. The decreasing trend in hydraulic conductivity is observable. Local variations were attributed to experimental and reading errors and were smoothed by taking moving averages of readings in 0.20 pore volumes [18]. The 0.1% solutions of organic fluids were prepared in 0.01 N CaSO$_4$ solution; Fig. 4 indicates that although Ca^{+2} almost reached the influent concentration and

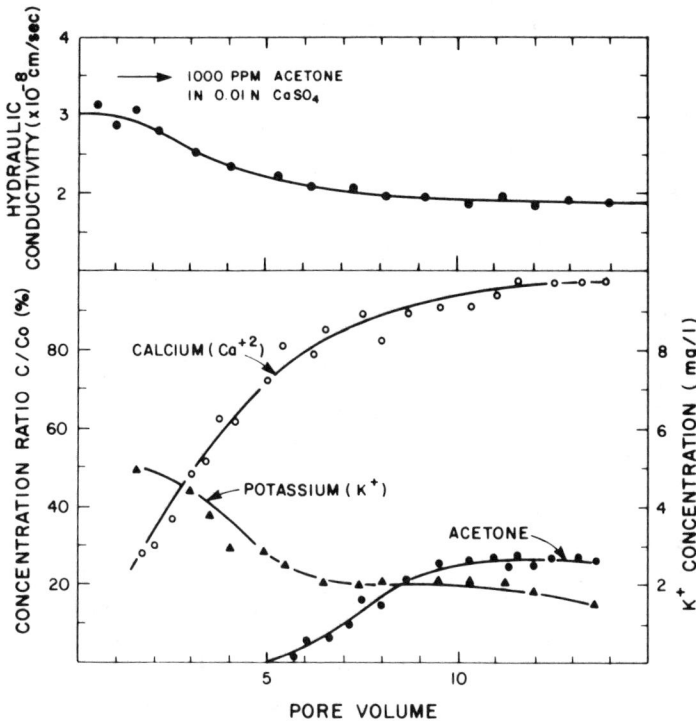

FIG. 4—*Hydraulic conductivity of compacted kaolinite to low level acetone solution.*

hydraulic conductivity was stabilized, the acetone concentration was only 25% of the influent. A significant amount of chemical was found in the cell water. Unless the concentration in the cell water was monitored, this observation may have misled to the conclusion that organic fluid was adsorbed on the clay. Adsorption tests showed no measurable adsorption of the organics onto the kaolinite. Table 3 summarizes the chemical characteristics of the effluent and the cell water. A mass balance analysis showed that only 20% of the total amount of acetone permeated was in the effluent; the rest was either adsorbed on the latex membrane or diffused into the cell water. The fact that Ca^{+2} concentration reached 96.2% and pH values dropped indicated that the organic fluid was removed from the system by adsorption on the membrane and subsequent diffusion into the cell water.

Figure 5 presents the changes in hydraulic conductivity with all low level solutions tested. The control specimen was simultaneously subjected to similar gradients. The slight decrease in hydraulic conductivity of control specimen was attributed to the gradient variations and changes in void ratio due to consolidation with seepage forces [18]. As presented in Fig. 5, a decrease in hydraulic conductivity was observed in all tests with the low level solutions. It

TABLE 3—*Effluent characteristics in low level tests.*[a]

Influent (1000 ppm Solutions)	No. of Pore Volumes	K[+], mg/L		Ca[+2], %	Organic Fluid, %	pH		Cell[b] Water, mg/L
		C_o	C_f	C_f/C_o	C_f/C_o	C_o	C_f	
Nitrobenzene	8	0.35	2.70	67.5	20.7	6.9	4.7	52.5
Acetone	14	0.35	1.6	96.2	24.6	6.7	4.7	64.7
Phenol								
(1)	9	0.2	3.4	70.0	22.3	7.1	4.4	...
(2)	9	0.4	2.6	96.9	76.6	7.0	3.8	...
Benzene	9	0.10	2.8	68.8	30.0	6.6	4.6	...

[a]C_o = concentration when the influent is switched to organic fluid.
C_f = final concentration.
[b]Volume of cell water is approximately 1.1 L.

should be noted that these tests exhibited the effect of final effluent concentrations given in Table 3.

The results of the tests with pure organic fluids are presented in Fig. 6. Upon switching to organic fluids, all tests indicated an immediate decrease in hydraulic conductivity. This decrease was followed by increases for phenol and acetone. The final values of hydraulic conductivity were relatively constant. Tests with nitrobenzene and benzene resulted in two orders of magnitude decrease in hydraulic conductivity, while acetone and phenol doubled the hydraulic conductivity. Since hydraulic conductivity with benzene and nitrobenzene reduced to 1×10^{-10} cm/s, the gradients were increased to increase the outflow. This resulted in a slight increase in hydraulic conductivity. The tests with benzene and nitrobenzene were discontinued immediately after the breakthrough of the organic fluid. Table 4 compares the final mean relative and absolute hydraulic conductivity values for pure organic solvents and the 0.01 N CaSO$_4$ solution.

One concern accompanying the use of high effective stresses in flexible wall permeability testing is that the effective stresses in excess of in situ conditions might veil changes in hydraulic conductivity by restricting any possible structural changes due to organic fluid permeation. This possibility, together with the need to assess the susceptibility of the test to possible side leakages in rigid wall permeameters, necessitated the investigation of the effect of effective stresses and the testing scheme on the results. Two specimens were first identically saturated and consolidated to an average effective stress of 69 kPa (10 psi). One of these specimens was rebounded to 13.8 kPa (2 psi). Since the coefficient of recompression is smaller than the coefficient of compression in consolidation, this procedure resulted in a void ratio change of only 0.01. Both specimens were then permeated with 0.01 N CaSO$_4$ solution and then pure acetone. Figure 7 compares the results of these flexible wall permeameter tests with variable effective stresses and the rigid wall permeameter test.

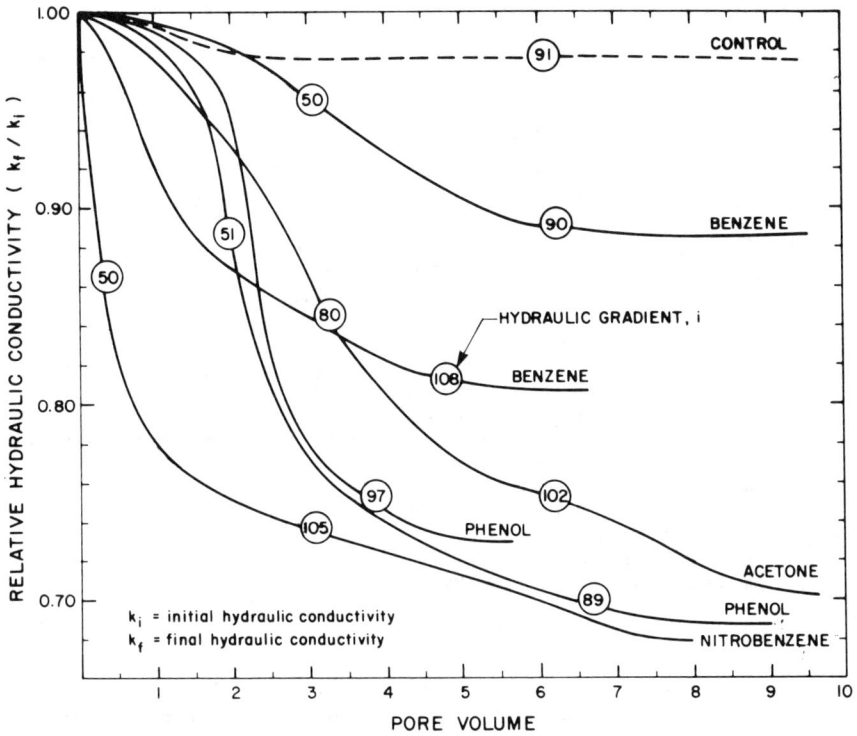

FIG. 5—*The effect of low level solutions on hydraulic conductivity of compacted kaolinite.*

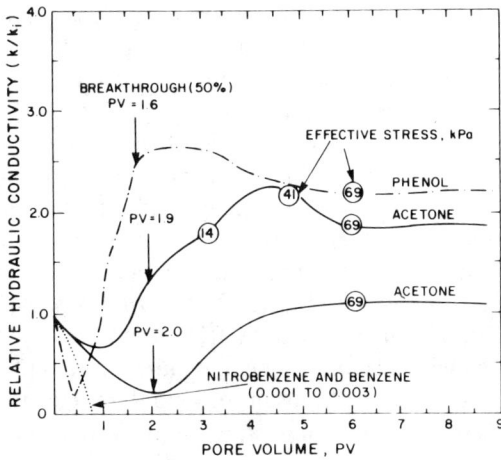

FIG. 6—*The effect of pure organic fluids on hydraulic conductivity of compacted kaolinite.*

TABLE 4—*A comparison of relative and absolute hydraulic conductivity of compacted kaolinite.*[a]

| | Hydraulic Conductivity | | | | |
| | Relative $\times 10^{-8}$ cm/s | | Absolute $\times 10^{-15}$ cm^2 | | |
Permeant	k_i	k_f	K_i	K_f	K_f/K_i
Nitrobenzene					
(1)	5.60	0.025	560	1.51	2.7×10^{-3}
(2)	5.20	0.015	520	0.91	1.8×10^{-3}
Acetone					
(1)	5.60	6.00	560	1470	2.63
(2)[b]	5.00	11.00	500	2690	5.38
Phenol	6.00	14.00	600	119.0	0.20
Benzene	5.10	0.010	510	1.38	2.7×10^{-3}

[a] k_i = reference hydraulic conductivity with 0.1 N CaSO$_4$.
k_f = final hydraulic conductivity.
K = $k\mu/\gamma$ = absolute hydraulic conductivity.
γ = unit weight of permeant.
μ = viscosity of permeant.
[b] Test at average effective stress of 13.8 kPa (2 psi).

The specimen for the rigid wall permeameter was prepared at the same compaction effort, molding water content, and saturation procedure and was tested at a gradient of 100. The hydraulic conductivity in the rigid wall permeameter test stabilized at a value of 2×10^{-6} cm/s, while the two tests at variable effective stresses in flexible wall permeameters rendered final acetone hydraulic conductivity values ranging between 6 to 9×10^{-8} cm/s. The greater hydraulic conductivity obtained with the 13.8 kPa (2 psi) test, as compared with the 69 kPa (10 psi) test, indicated that effective stresses affect the results. However, the rigid wall permeameter test resulted in two orders of magnitude increase in hydraulic conductivity. Furthermore, although the breakthrough in the two flexible wall tests replicate each other, the influent concentration was reached earlier and abruptly in the rigid wall permeameter. Since a specimen compacted in a rigid wall permeameter was also subjected to effective stresses of unknown magnitude, these anomalous results can only be explained by side leakages in this testing scheme. Side leakage in a rigid wall permeameter implies that the specimen shrinks during acetone permeation. It is then expected that the swell-shrink behavior of compacted soil is different with organic fluids and 0.01 N CaSO$_4$ solution.

The swelling behavior of compacted kaolinite specimens were recorded with 0.01 N CaSO$_4$ and organic fluids in consolidation cells. Figure 8 presents the results of this test. This figure indicates that the volume increase with 0.01 N CaSO$_4$ solution and subsequent shrinkage with acetone permeation was the mechanism initiating the side leakages in rigid wall permeameters. It should

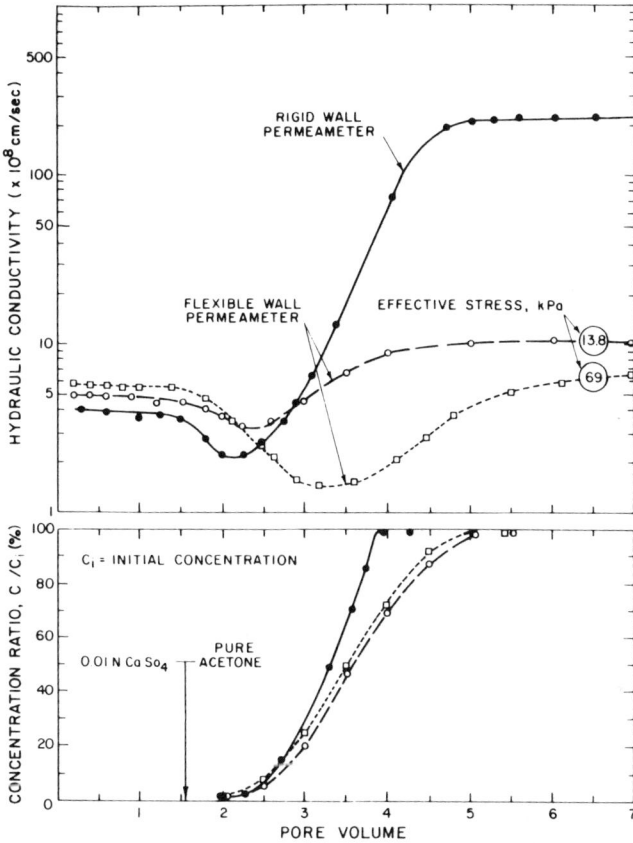

FIG. 7—*The effect of effective stresses and testing scheme on the hydraulic conductivity with acetone.*

be noted that any side leakage in a test would result in "apparent" hydraulic conductivity values of greater than 1×10^{-7} cm/s [16]. Such tests should be avoided when permeants are leachates of different composition than the initial pore fluid.

The previous discussion indicates that slight volume changes will occur during permeation. However, it was not possible to detect such small changes during testing or from the volume of specimens before and after permeation. Furthermore, such slight changes in volume would not explain the changes in absolute values of hydraulic conductivity. Consequently, variations in absolute values of hydraulic conductivity indicate differences in the structure of the compacted soil. Structural changes are initiated by changes in the surface forces of interaction due to variations in the attractive and repulsive forces between clay minerals.

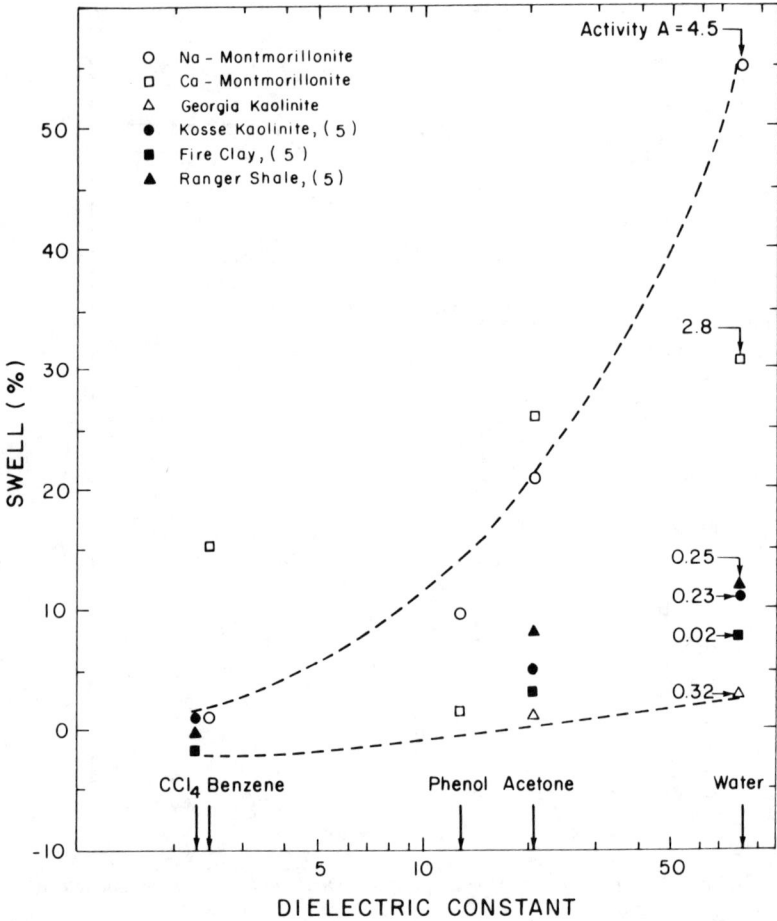

FIG. 8—*The effect of organic fluids on the swelling behavior of compacted soils.*

Repulsive forces are primarily attributed to the interaction between diffuse double layers. The system variables that control these forces are defined in the development of the Gouy-Chapman theory of diffuse double layer around the clay minerals [*19,20*]. An approximate quantitative indication of the thickness of diffuse double layer is given in Ref *21*

$$H = f\left[\left(\frac{DT}{h_0^2 v^2}\right)^n\right]$$ (1)

where

H = relative thickness of double layer,
D = dielectric constant of the medium,

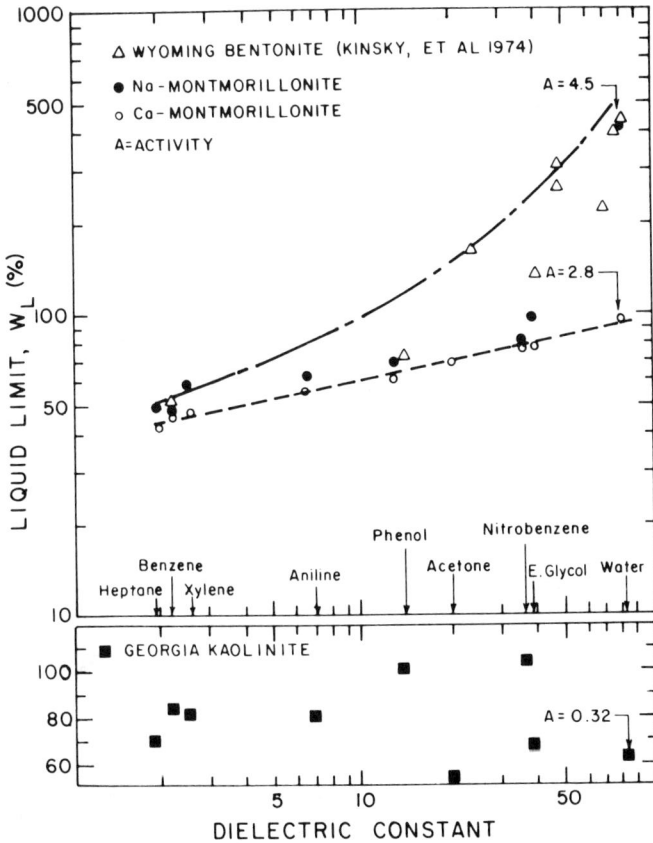

FIG. 9—*The effect of organic fluids on the liquid limit of clay minerals.*

T = temperature,
h_0 = electrolyte concentration,
v = cation valence, and
n = a constant ($\cong$ 1/2).

Changes in the thickness of the diffuse double layer are directly related to the forces of repulsion between the clay minerals.

The principal contribution to attractive forces arise from van der Waal's forces. These forces could either be ion-dipole or dipole-dipole interactions. Other factors that contribute to attractive forces are Coulombic attractions between negative surfaces and positive edges, cation linkages, and hydrogen bondings [6]. Moore and Mitchell [22] developed expressions for both the electrostatic and electrodynamic forces of interaction between clay particles. These expressions indicated that the net forces of interaction are functions of the static dielectric constant of the pore fluid. Shear strength tests on resedi-

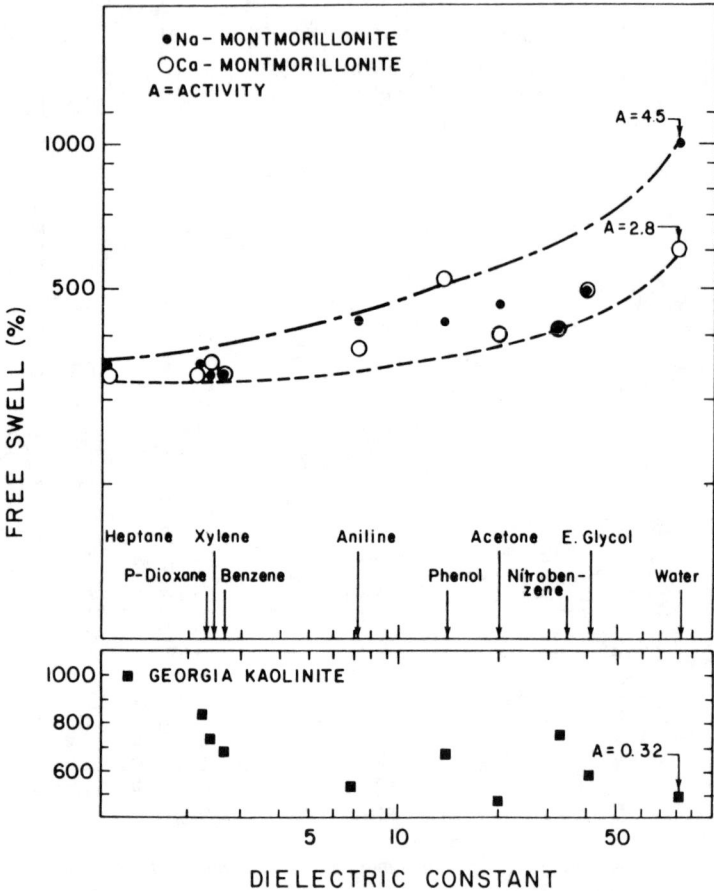

FIG. 10—*The effect of organic fluids on the free swell of clay.*

mented and miscible organic fluid permeated Georgia kaolinite verified the validity of the analytical expressions. Consequently, it is well recognized that changes in the dielectric constant of the medium would reestablish the net forces of interaction resulting in variations in the structure of the soil. Since structure of the soil implies the arrangement of particles and pore spaces (fabric) together with the effects of composition and interparticle forces, these changes will result in differences in the geotechnical properties [23-26].

Free swell tests and tests for index properties of a clay with a specific fluid would show the extent of diffuse layer around the particles giving a qualitative indication of expected changes in the repulsive forces. Figures 9 and 10 give the results of these tests with a variety of organic fluids. The liquid limit and free swell of highly active smectic minerals increased with an increase in the dielectric constant of the pore fluid. Kaolinite did not follow this trend; the liquid limit and free swell was increased significantly with nitrobenzene and benzene, slightly increased with phenol, and decreased with acetone. Such observations are possible in lowly active kaolinite minerals since the positive charges at the broken edges of this mineral are unaccounted in the development of Eq 1, although they have a major effect on the forces of interaction. However, the direction of changes in the liquid limit and free swell of kaolinite are consistent with the changes in the absolute values of hydraulic conductivity presented in Table 4.

In its classical sense of explanation, increases in liquid limit and free swell of a fine-grained soil with a specific fluid demonstrate that the repulsive forces between particles will increase, resulting in dispersion and in a subsequent decrease in hydraulic conductivity. Conversely, decrease in repulsive forces will promote flocculation, increasing the hydraulic conductivity. If this point of view is taken, it is also expected that the intercluster pore size distribution would change at a relative constant total porosity. However, mercury intrusion tests before and after permeation with organic fluids indicated that the distribution of pores greater than 80 Å are not significantly changed and do not explain the variations in hydraulic conductivity [27]. Since pores of sizes less than 80 Å are not expected to have a major contribution to total flow, one possible explanation for differences in hydraulic conductivity would be that changes in the physicochemical properties of the pore fluid close to the clay surfaces lead to variations in flow volume and behavior. Tests with calcium-montmorillonite are in progress to further investigate this observation.

Summary and Conclusions

The following results and conclusions were obtained in this study:

1. When organic fluid permeation through fine-grained soils is of concern, the testing scheme is a major factor affecting the validity of results. Side leakages due to shrinking of the specimen with some organic fluids might give excessive increases in hydraulic conductivity values obtained in rigid wall permeameters.

2. When low concentration organic fluids are used, diffusion of chemicals through the latex membrane may prevent an assessment of the effect of actual strength leachate in testing with flexible wall permeameters.

3. Some full strength organic fluids adversely affect the membranes used in flexible wall permeameters. This latex membrane deterioration is effectively remedied by wrapping specimens with Teflon for the organic solutions tested.

4. All tests at low concentrations of nitrobenzene, acetone, phenol, and benzene indicated slight decreases in hydraulic conductivity. Absolute values of hydraulic conductivity with pure solutions increased with acetone and slightly decreased with phenol; three orders of magnitude decrease were observed with benzene and nitrobenzene. Free swell and liquid limit tests with

the specific leachate provide qualitative explanations of these changes in the hydraulic conductivity.

5. Hydraulic conductivity changes with pure organic fluids can be explained by variations in flow characteristics initiated by differences in the surface forces of interaction due to changes in the chemical properties of the pore fluid.

Acknowledgments

Ivan Olivieri is acknowledged for his help in the index and in the swell and rigid wall permeability tests. The investigation presented in this study was funded by the U.S. Environmental Protection Agency (EPA) under assistance agreement EPA cooperative agreement CR 809714010 to the Hazardous Waste Research Center, Louisiana State University. However, the study does not necessarily reflect the views of the EPA and no official endorsement is inferred. Dr. Roger K. Seals and the Department of Civil Engineering at Louisiana State University are acknowledged for supplemental financial support of this investigation.

References

[1] Zimmie, T. F. and Riggs, C. O., Eds., *Permeability and Groundwater Contaminant Transport, STP 746*, American Society for Testing and Materials, Philadelphia, 1981.

[2] Mitchell, J. K., Hooper, D. R., and Campanella, R. C., *Journal of the Soil Mechanics and Foundations Division*, American Society for Civil Engineers, Vol. 91, No. SM4, 1965, pp. 41-65.

[3] Hardcastle, J. H. and Mitchell, J. K., *Clays and Clay Minerals*, Vol. 22, 1974, pp. 143-154.

[4] Anderson, D. and Brown, K. in *Proceedings*, Seventh Annual Research Symposium on Land Disposal of Hazardous Wastes, EPA-600/9-81-002b, U.S. Environmental Protection Agency, Cincinnati, OH, 1981, pp. 119-130.

[5] Green, W. J., Lee, G. F., and Jones, R. A., *Journal of the Water Pollution Control Federation*, Vol. 53, 1981, pp. 1347-1354.

[6] Lambe, T. W. in *Symposium on Permeability of Soils, ASTM STP 163*, American Society for Testing and Materials, Philadelphia, pp. 56-67.

[7] Shuckrow, A. L., Pajak, A. P., and Touhill, C. J., "Management of Hazardous Waste Leachate," No. SW871, U.S. Environmental Protection Agency, Office of Water and Waste Management, Washington, DC, 1980.

[8] Zimmie, T. F., Doynow, J. S., and Wardell, J. T. in *Proceedings*, 10th International Conference in Soil Mechanics and Foundation Engineering, Stockholm, Balkema Publishers, Rotterdam, Vol. 2, 1981, pp. 403-406.

[9] Zimmie, T. F. in *Hazardous Solid Waste Testing: First Conference, STP 760*, American Society for Testing and Materials, Philadelphia, 1982, pp. 293-320.

[10] Olson, R. E. and Daniel, D. E. in *Permeability and Groundwater Contaminant Transport, STP 746*, American Society for Testing and Materials, Philadelphia, 1981, pp. 18-64.

[11] Gilham, R. W. and Cherry, J. A., *Geological Society of America*, Special Paper 189, 1982, pp. 31-62.

[12] Wilson, S. D. in *Special Procedures for Testing Soil and Rock for Engineering Purposes, ASTM STP 479*, American Society for Testing and Materials, Philadelphia, 1970, pp. 34-36.

[13] Rad, N. S. and Acar, Y. B., *Geotechnical Testing Journal*, Vol. 7, No. 2, 1984, pp. 104-106.

[14] Lowe, J. and John, T. C. in *Proceedings*, Conference on Shear Strength of Cohesive Soils, American Society of Civil Engineers, Boulder, CO, 1960, pp. 819-836.

[15] Skempton, A. W., *Geotechnique*, Vol. 4, No. 4, 1954, pp. 143-147.

[16] Mitchell, J. K. and Younger, J. S. in *Symposium on Permeability and Capillary of Soils, STP 417*, American Society for Testing and Materials, Philadelphia, 1967, pp. 106-141.

[17] Seals, R. K., "The Effect of Pore Water Flow Characteristics on the Consolidation of Fine-Grained Soils," Ph.D. thesis, Civil Engineering Department, North Carolina State University, Raleigh, 1967.

[18] Hamidon, A., "Organic Leachate Effects to Hydraulic Conductivity of Compacted Kaolinite," M.S. thesis, Department of Civil Engineering, Louisiana State University, Baton Rouge, 1984.

[19] Gouy, G., *Anniue Physique* (Paris), Series 4, Vol. 9, 1910, pp. 457-468.

[20] Chapman, D. L., *Philosophical Magazine*, Vol. 25, No. 6, 1913, pp. 475-481.

[21] Mitchell, J. K., *Fundamentals of Soil Behavior*, Wiley, New York, 1976.

[22] Moore, C. A. and Mitchell, J. K., *Geotechnique*, Vol. 24, No. 4, 1974, pp. 627-640.

[23] Green, W. J., Lee, G. F., Jones, R. A., and Pallt, T., *Environmental Science and Technology*, Vol. 17, 1983, pp. 278-282.

[24] Sridharan, A. and Rao, V. G., *Geotechnique*, Vol. 23, No. 3, 1973, pp. 359-382.

[25] Kinsky, J., Frydman, S., and Zaslavsky, D. in *Proceedings*, 4th Asian Conference on Soil Mechanics and Foundation Engineering, 1974, pp. 367-372.

[26] Ladd, C. C. and Martin, R. T., "The Effects of Pore Fluid on the Undrained Strength of Kaolinite," Civil Engineering Research Report R67-15, Massachusetts Institute of Technology, Cambridge, 1967.

[27] Acar, Y., Olivieri, I., and Field, S., this publication, pp. 203-212.

A. Gwen Eklund[1]

A Laboratory Comparison of the Effects of Water and Waste Leachate on the Performance of Soil Liners

REFERENCE: Eklund, A. G., "**A Laboratory Comparison of the Effects of Water and Waste Leachate on the Performance of Soil Liners,**" *Hydraulic Barriers in Soil and Rock, ASTM STP 874*, A. I. Johnson, R. K. Frobel, N. J. Cavalli, and C. B. Pettersson, Eds., American Society for Testing and Materials, Philadelphia, 1985, pp. 188–202.

ABSTRACT: A study was conducted to determine the performance of two compacted native soils and the soils plus a commercial additive when exposed to specific paper mill waste leachates. Fluids produced from actual paper mill wastes were used in accelerated (high-pressure), anaerobic permeability tests with two disposal site soils and the soils with a beneficient added. Comparison tests measuring hydraulic conductivity were performed to determine behavioral differences of the candidate soil liners with the waste leachate and water. Test results were used to evaluate waste impacts on soil and on the soil plus a commercial additive. Emphasis was placed on identifying fatal flaws in expected field liner performance.

Test results indicated that the two soils tested were effective in containing the paper mill waste leachates. One soil physically contained the waste leachates by providing an impermeable barrier ($K = \ll 10^{-7}$ cm/s). The other soil, while not providing as strong a physical barrier, ($K = 10^{-7}$ cm/s), did contain the leachate through chemical attenuation of the trace contaminants in the leachate. While the commercial additive improved the impervious barrier characteristics of this soil, it was determined to be optional. Both mechanisms of containment will be effective in providing protection of groundwater resources below a nonhazardous waste landfill.

KEY WORDS: permeability, soil liners, liner compatibility

The use of compacted soils is frequently used to retard leachate and waste liquid components from leakage and subsequent pollution of groundwater. In this study two soils and the soils plus a commercial additive were tested for permeability performance with water and leachate from paper mill wastes.

[1]Senior scientist, geochemistry and waste characterization, Physical Chemistry Div., Radian Corp., Austin, TX 78766.

The major objectives of this study were to (1) provide an approach for evaluating in the laboratory a soil liner's performance expected in the field, and (2) to determine if a soil additive was necessary for good liner performance. The approach was comprised of tests allowing rapid identification of liner failure mechanisms.

Background

The purpose of lining a waste disposal facility is to retard waste contaminants from leaving the site and entering natural groundwater systems near the site. In their performance, liners function by two mechanisms: (1) impedance of flow of the pollutant carrier (usually water) into the subsoil, or (2) attenuation of suspended or dissolved pollutants so that the exiting leachate contains contaminants within acceptable ranges for groundwater [1], or both (1) and (2).

The first liner function is attained by constructing a material with low permeability. The sorptive or attenuative capability relies on the chemical composition and mass of the liner. Soil liners generally function by both mechanisms. Soils generally have a large capacity to sorb materials of different types, but some soils do not provide an impermeable boundary. These properties can sometimes be enhanced by the use of soil additives. A greater thickness of a soil layer beneath the disposal impoundment also can result in a lower flux of a contaminant through the liner. A good soil liner will contain a minimum of 25 to 28% clay-size particles by weight [1]. The clay-size fraction (<0.002 mm) has a relatively large surface area, and the mechanical behavior of a clayey soil is dependent on the physiochemical interactions between soil particle surfaces.

Soils used as liners also must have acceptable engineering properties. The soil consistency differs with moisture content, and the degree of hydration of clayey soils has a direct effect upon the liner's compressive strength. The plastic limit and the liquid limit are also valuable in assessing the engineering characteristics of soils [2].

The mechanical soil properties can be altered by mechanical compaction, which affects the permeability. (The mechanical behavior is also controlled by the stresses that act on soil, for example, overburden pressure.) For a given method of compaction and compactive effort, each soil has a unique laboratory moisture/density relation, that is, Proctor compaction curve, which defines optimum moisture and maximum dry density for design considerations. Soil permeability is a numerical measure of the ability of the soil to transmit a fluid. The permeability is dependent on both the soil properties and leachate properties. A good liner will usually possess a hydraulic conductivity coefficient (K) of 1×10^{-7} cm/s or less when it is tested with a simulated (or actual) waste fluid.

Prior to contacting the groundwater, the contaminants in the wastes must migrate through soil liners where some degree of attenuation (reduction in concentration) of many species will take place; however, not all contaminants are attenuated by soil. Even after the contaminants enter the groundwater, soil attenuation mechanisms often continue to reduce the contaminant levels. For nonhazardous wastes a desirable disposal practice is to utilize the natural attenuation processes of the soil media beneath the disposal site. These attenuation processes include precipitation by changes in the aqueous environment, adsorption onto soil surfaces, filtration of contaminants by tightly packed soils, and dispersion and diffusion within the unsaturated zone and in the groundwater. Some species, for example chloride, often migrate through soil at the same velocity as the transport liquid; no attenuation occurs in this case.

Approach

Each waste and candidate soil liner was characterized for the physical and chemical properties described in following paragraphs. However, the main thrust of the approach to liner evaluation was determining the interactions of liners and wastes. The major focus of the liner evaluation involved short-term and long-term compatibility studies. The liner evaluation criteria for waste characterization, soil properties, and compatibility studies are as follows.

Waste Leachate Properties

The leachate and the contaminants it contains can be aggressive to liner materials. Water, itself, can be an aggressive constituent, causing a liner material to swell. The initial whole sample compositions of actual waste composites were used in the liner study to identify possible reactive chemical constituents that might preclude the use of clay soils as liners, thus requiring synthetic liners. Waste leachates generated using proposed ASTM Method A[2] procedure were used in the short-term and long-term liner compatibility studies. The compositions of column leachates were used to screen for long-term leachate composition changes accompanying anaerobic saturated conditions and to provide soil attenuation data as necessary.

Soil Properties

Each candidate liner was characterized for the soil properties listed in Table 1. Included in the table are synopses of the test objectives.

[2]ASTM Method A, proposed by Subcommittee D34.0201.

TABLE 1—*Soil properties determined for candidate natural liners.*

	Test Method	Purpose
Soil Property		
Particle size distribution	sieve and hydrometer [ASTM D 421-58 (1978)][a]	estimates percent sand, silt, and clay
Compaction test	compact liner at different moisture contents and measure compacted density (ASTM D 558-82)[b]	determines the moisture content which results in maximum compacted density at a given compactive effort
Atterberg limits	measure of plastic and liquid limits [ASTM D 423-66 (1972)[c] and ASTM D 424-59 (1971)[d]]	defines effects of varying water content on soil consistency
Chemical Properties		
Organic matter, %	titration [4]	organic matter in liners may attenuate organics from leachates
Soil pH	electrometric [5]	alkalinity of a soil will affect its ability to attenuate chemical contaminants
Cation exchange capacity	sodium saturation followed by ammonium extraction; analyze for sodium [4]	estimates the extent to which the liner may exchange (remove) cations from the leachate
Major exchangeable bases	ammonium extraction; analyze for sodium, potassium, magnesium, and calcium [4]	estimates the liner's capacity to exchange ions on base sites and identifies major ions absorbed on clay before interaction
X-ray powder diffraction	X-ray diffractometer	identifies clay minerals in liners

[a] Method for Dry Preparation of Soil Samples for Particle-Size Analysis and Determination of Soil Constants.
[b] Method for Moisture-Density Relations of Soil-Cement Mixtures.
[c] Method for Liquid Limit of Soils.
[d] Method for Plastic Limit and Plasticity Index of Soils.

Short-Term Waste Leachate/Soil Compatibility

The initial compatibility tests consisted of a set of laboratory experiments designed to determine the initial behavioral differences in the soil when it is interacted with a waste leachate. Short-term compatibility of liners with waste leachate was evaluated by comparing the liners' properties after wetting with water to their properties after wetting with waste leachate. Following wetting with each fluid, the liners were tested for permeability, internal structural changes, and plasticity.

Permeability—A test method for examining effects of waste leachates on liner permeabilities, developed for the U.S. Environmental Protection Agency (EPA) by Matrecon, Inc., was modified in this study for use in liner evaluation [1]. Depths of 152 mm (6 in.) of liner materials were compacted in transparent permeameters [76 mm (3-in. diameter)] at maximum density at optimum moisture content [ASTM Methods for Moisture-Density Relations of Soils and Soil-Aggregate Mixtures, Using 5.5-lb (2.49-kg) Rammer and 12-in. (304.8-mm) Drop (D 698-78)]. The appropriate fluid (water or waste leachate) was placed in the permeameter over the compacted soil and forced through the soil at slightly elevated head pressure by using nitrogen at a pressure of 20 to 41 kPa gage (3 to 5 psig).

Internal Structural Changes—Internal structural changes were investigated using X-ray diffraction. First, to identify the clay minerals present in each candidate liner, a powder pattern of a representative sample of each bulk soil was obtained. These patterns also were used to identify regions showing dioctahedral structures susceptible to possible interaction with organic species in the waste leachate. A second set of X-ray diffraction experiments was performed to determine the initial effect of waste leachate on the candidate liners. If a substantial interlayer spacing decreased for the second set of data, shrinkage cracks may be anticipated for that particular liner-waste combination. (This method was performed again at the end of long-term accelerated batch leaching to determine changes in soil crystallinity due to extended shaking with a 10:1 ratio of leachate with soil.)

External Property Changes—Indication of external property change was evaluated using Atterberg limits. The plasticity (Atterberg limits) was evaluated for each candidate liner with the appropriate waste leachate, as well as with water; large differences in the plasticity with water and waste fluid would indicate possible expected failure in the field behavior of the liner due to changes like shrinking or swelling.

Long-Term Compatibility

Long-term waste leachate/liner compatibility was assessed by continuing to monitor soil permeability and failure mechanisms observed in soils undergoing permeability tests, by performing accelerated batch leaching tests of the soils with leachate, and by monitoring chemical breakdowns of the soil and internal structural changes in the clay minerals in the soils.

Permeability and Failure Mechanism Observation—Constant head permeability measurements in ASTM Method for Permeability of Granular Soils (Constant Head) [(D 2434-68 1974)] were taken once per week to note changes due to fluid-liner interactions. At the end of six months of testing, the soil cores were visually examined for indications of the degradation.

Accelerated Batch Leaching—The accelerated long-term compatibility involved leaching of the soil liners with water or waste leachate by batch extraction. The soils were agitated for two months at approximately a 10:1 liquid-to-solid ratio in closed polyethylene bottles. Aliquots were removed monthly for analysis of aluminum and silicon concentration to assess chemical degradation of the clays in the soils. At the end of the two months of leaching, the soils were separated from the liquor and examined for physical or structural changes using X-ray powder diffraction.

Leachate/Soil Interaction—The approach taken in this project for estimating the degree of soil attenuation was to simulate the leaching and soil attenuation processes through column studies (separate from the permeameter study) using actual samples of solid waste and soil. This phase of investigation was performed on natural soil only. The concentration of contaminants found in the column leachates and subsequent soil attenuate served as an indicator of the soils' ability to reduce contaminant levels entering the groundwater.

The columns for leachate/soil interaction were constructed of Pyrex glass with Teflon fittings to prevent organic contamination or sorption. Deionized and deaerated (with nitrogen) water was percolated through the columns in an upflow mode. Water flow was maintained constant by adjusting the hydraulic head pressure. Wastes were packed at 0.6 porosity and the soil at 0.4 porosity.

Leachate aliquots were collected at the end of each day, filtered, and refrigerated. For every litre of leachate collected, the following index parameters were analyzed: pH, conductivity, redox potential, and total organic carbon (TOC). The first aliquots from the column were characterized for inorganic species. Subsequent aliquots were analyzed only for those inorganic and organic species found in measurable concentrations in the first aliquot.

Discussion of Liner Test Results

The discussion of results is separated into sections on waste character, soil properties, initial compatibility, and long-term compatibility. These tests were designed to evaluate two site soils in the disposal area. Waste 1 was evaluated with Soil 1 and Soil 1 plus beneficient, and Waste 2 was evaluated with Soil 2 and Soil 2 plus beneficient.

Waste and Leachate Character

Waste 1 contained mainly calcium, magnesium, and silicon compounds with the presence of some trace elements. Waste 2 contained mainly calcium compounds.

Waste Leachates—Leachates were obtained from the wastes by ASTM Method A,[3] and these chemical compositions are given in Table 2. (This method uses a 20:1 liquid-solid ratio with no pH adjustment and deionized

[3]Proposed by ASTM Subcommittee D34.02.01 on Batch Extraction Methods.

TABLE 2—*Chemical analysis of ASTM extract of liner study wastes, mg/L, unless specified.*

Species	Concentrations	
	Waste 1	Waste 2
Major/Minor Elements		
Calcium	62	3.0
Iron	<0.008	1.2
Potassium	24	13
Magnesium	0.059	0.72
Sodium	94	350
Trace Elements		
Silver	0.005	<0.002
Aluminum	0.11	2.3
Arsenic	<0.003	<0.003
Boron	0.21	<0.01
Barium	0.42	0.035
Cadmium	<0.002	<0.002
Chromium	0.005	0.048
Copper	0.012	0.045
Mercury	0.0006	<0.0002
Manganese	<0.001	0.46
Molybdenum	<0.002	<0.002
Nickel	<0.003	0.007
Lead	<0.002	<0.08
Antimony	<0.03	<0.03
Selenium	<0.08	<0.003
Zinc	<0.003	<0.003
Water quality parameters, hydrogen sulfide (H_2S)	13	14
Total dissolved solids (TDS)	480	1100
pH	11.2 SU[a]	10.1
Total organic carbon (TOC)	7	134
Total organic halogen (TOX)	0.10	1.3
Fluoride (F)	0.1	0.2
Nitrate (NO_3)	0.01	<0.01
Color	5 SCU[b]	1500
Tannin	3	31

[a]SU = standard units.
[b]SCU = standard color units.

water as the liquid.) Leachate from Waste 1 had a total dissolved solids (TDS) content of 480 mg/L and the pH was 11.2. Calcium, sodium, potassium, and hydrogen sulfide were the major dissolved species. Waste 2 had a TDS content of 1100 mg/L and contained major concentrations of sodium and potassium with a pH of 10.1.

Soil Properties

A summary of the properties of the soil liners is given in Table 3. These properties were identified to aid in interpretation of soil-waste leachate inter-

TABLE 3—*Results of soil property tests on candidate liners.*

Soil Property	Soil 1	Soil 1 + Beneficient	Soil 2	Soil 2 + Beneficient
Particle size distribution				
% Sand	13	8	12	12
% Silt	70	74	48	48
% Clay	17	18	40	40
Compaction moisture/density relation				
Optimum moisture, %	13	16	18	18.9
Maximum dry density, lb/ft³ [a]	116	106	110	106
Atterberg limits				
Liquid limit, %	30	50	38	58
Plastic limit, %	21	19	17	24
Plasticity index, %	9	31	21	34
Organic matter content, %	0.4	0.4	0.6	0.6
Soil pH	7.1	7.5	6.2	7.2
Cation exchange capacity, meq/100g	12	20	25	31
Major exchangeable bases, meq/100g	12	17	20	25
Clay minerals present	ca-substituted nontronite (iron-rich montmorillonoid) kaolinite microcline (alkaline feldspar) muscovite (dioctahedral mica) quartz dolomite		ca-substituted nontronite (iron-rich montmorillonoid) kaolinite microcline (alkaline feldspar) muscovite (dioctahedral mica) quartz dolomite	

[a] Conversion factor: 1 pcf = 16.01 kg/m³.

actions. Tests were performed on soil samples recovered from borings at expected excavated soil liner levels.

Volclay Bentonite Saline Seal 100 was used as the soil beneficient. Saline Seal 100 is a chemically treated sodium-based bentonite formulated to contain wastes with high levels of dissolved salts, acids, or alkalis. (For example, Saline Seal previously has been used successfully as an additive to Wisconsin beach sand to contain sulfite liquor from a paper mill pulping operation.) The beneficient was added at a level of 7% by weight to the soils.

Particle Size Distribution—Results from particle size testing showed that Soil 2 (from a different site) contains much more clay than Soil 1. Although a higher clay content is desired, the presence of silt with clay in the natural soil should reduce the shrink/swell problems experienced with clays while still providing fairly large surface area for attenuation reactions. Adding a soil beneficient to Soil 1 increased the clay and silt fraction only slightly. Addition of the beneficient to Soil 2 had no noticeable effect on particle size distribution.

Compaction—Data correlate with the general concepts that the more granular the soil (as with Soil 1), the higher the maximum density and the lower the optimum moisture; and the finer the soil (as with Soil 2), the less defined the maximum density and the lower the slope of the curve of the density as a function of the moisture content [1,2]. For Soil 2, increasing the moisture requirements during compaction causes the density to become more critical for the montmorillonite clay; for example, a small decrease in density (1%) may result in an increase in permeability by one order of magnitude. Addition of soil beneficient increased the optimum moisture requirement and slightly reduced the maximum dry density for both soils.

Atterberg Limits—Preferred Atterberg limits for liner performance require a liquid limit between 35 and 60% water and a plasticity index above the A-line as shown in Fig. 1 [1]. While Soil 2 falls in the preferred zone, Soil 1 has liquid limits too low to lie within the preferred zone. Both soils with beneficients fall into the preferred zone.

Organic Matter Content—Soil properties including absorption and retention of water, reserves of exchangeable bases, and stability of soil structure are dependent to some degree on the quantity of organic matter present. The organic matter contents for the two soil borings down to 15 m (50 ft) were expectedly low (<1%) with Soil 2 showing slight superiority. The addition of beneficient did not alter organic matter content.

Soil pH—The soil pH of Soil 1 was practically neutral, while Soil 2 was slightly below neutral (6.20). The pH of a soil generally decreases with increasing concentration of neutral salts, for example, sodium chloride (NaCl)

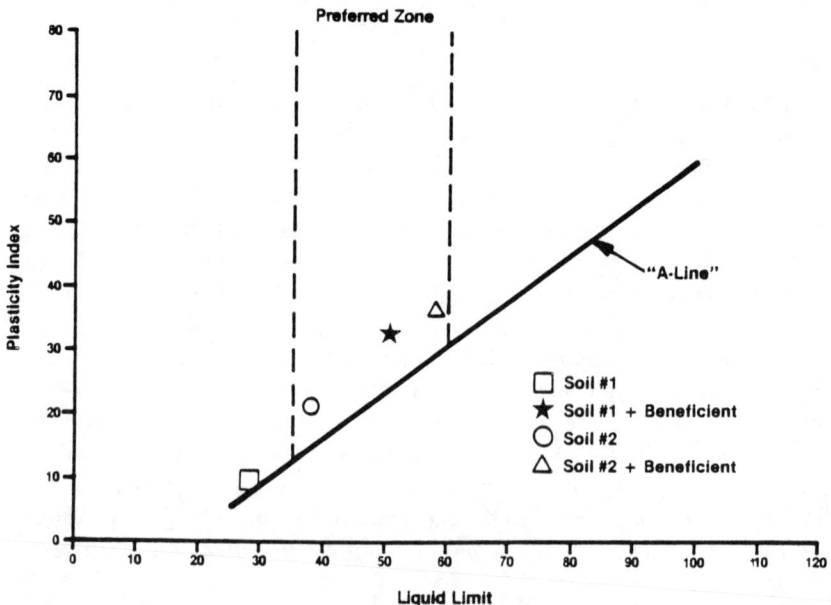

FIG. 1—*Plasticity of soil liner candidates.*

and calcium sulfate ($CaSO_4$). The pH increased slightly with beneficient addition.

Cation Exchange Capacity—Cations held on the surface of soil minerals and within the crystal framework can undergo exchange reactions reversibly in salt solutions and acids. The cation exchange capacity (CEC) is defined as the sum of the exchangeable cations of a soil. A high CEC generally denotes a high clay content and potentially a high capacity to attenuate contaminants. Soils vary in CEC from <1.0 to >100 meq/100g. The CEC of Soil 2 is at a medium level for soils, and Soil 1 is low. Cation exchange capacity of both soils was improved by beneficient addition.

Major Exchangeable Bases—Exchangeable bases are defined as the alkali and alkaline earth metals (principally calcium, magnesium, potassium, and sodium) that are attached to the clay and organic constituents of the soils. These ions can be exchanged with each other and with other positively charged ions in the soil solution. The major exchangeable bases (MEBs) in Soil 1 and beneficiated Soil 1 accounted for almost the entire cation exchange capacity with slightly higher values for the beneficiated soil. The MEBs do not account for the entire CEC for Soil 2 with or without beneficient.

Clay Mineral Identification—Clay minerals were identified in the soils using X-ray diffraction in order that any microstructural changes due to subsequent leachate interaction could be monitored. Originally, both natural soils contained kaolinite, microcline (alkali feldspar), muscovite (dioctahedral mica), and quartz. The clay fraction present behaved as a dioctahedral expanding smectite, probably nontronite (the dioctahedral structure is indicated by the 060 peak at approximately 62°, or 1.50 Å). X-ray fluorescence spectroscopic scan of elements in the samples (magnesium, aluminum, silicon, potassium, calcium, titanium, and iron) and the sample color implied the presence of nontronite, an iron-rich montmorillonoid. The X-ray diffraction of these samples indicates calcium substitution. No major changes in structure were observed upon addition of soil beneficient; however, adding a soil beneficient to the natural soils intensified montmorillonite clay peaks, especially for Soil 1.

Initial Compatibility Results

The initial compatibility testing involved comparison of each soil's behavior after wetting with water to its properties after wetting with the appropriate waste leachate.

Short-Term Permeability Behavior—Permeability measurements after one week of testing with water as the testing fluid were affected by beneficient addition for Soil 1 only. Soil 1 had a permeability coefficient (K) of 7.4×10^{-8} cm/s, while Soil 1 plus beneficient had "K" equal to $<1 \times 10^{-10}$ cm/s. Soil 2 with and without beneficient had an initial "K" of $<1 \times 10^{-10}$ cm/s.

Internal Structural Changes—The initial reaction of waste leachate with the natural soil liners did not produce marked differences from exposure to water. X-ray diffraction results showed some minor changes which included shifts indicating possible swelling of Soil 2 and magnesium to calcium exchange on dolomite. These results indicate that no immediate internal structure changes are a result of leachate interaction and that the soils are stable in water and initial waste leachate interaction. These soils showed good orientation and high intensity due to stability of suspensions in the X-ray powder pattern. With beneficients minor swelling occurred, but initial interaction with waste leachate did not result in major internal structural changes.

External Property Effect—To evaluate the effect of waste leachate interaction on the external engineering properties of the candidate liners, Atterberg limit tests were performed and the results compared to liner performance with water. In no case did the waste leachates produce major difference in the plasticity properties of the liner candidates.

Long-Term Compatibility

To evaluate the long-term compatibility of the candidate liners with waste leachate, permeability was monitored weekly for six months. At the end of six months, visual observations of the liners in the permeability tests were recorded.

Accelerated batch leaching tests were performed with waste leachate and the soils and beneficiated soils. Under constant agitation in excess leachate (10:1 liquid–solid ratio), the liners were tested for long-term compatibility by (1) measurement of aluminum and silicon in leachate which would imply structural clay breakdown, and (2) microstructural changes in the clay minerals that would show up during X-ray diffraction analyses.

Long-Term Permeability Behavior—The final permeability for Soil 1 with Waste 1 leachate was 8.5×10^{-8} cm/s; this is the same range as the initial measurement showing good stability in behavior. The Soil 1 test cylinder did not show visual evidence of shrinking or swelling; it did show slight back diffusion (suspended clay in leachate above liner). (While back diffusion on a major scale can indicate that the liner's clay affinity is much greater for suspension in leachate in the waste body as opposed to staying in a compacted layer beneath the waste, slight back diffusion should not result in failure of the liner.) The final permeability coefficient for Soil 1 plus beneficient was 2.1×10^{-8} cm/s. However, major visual changes had occurred. Major swelling had occurred as well as lateral crack formation in the liner test cylinder. If swelling and cracking occurs in a liner under field conditions, a liner failure or breach is likely to result. The clay appeared to be back diffusing and piping, and circular white formation on the liner exterior was occurring (possibly bacterial action). Under field conditions, piping and back diffusion indicate that the clay may migrate out of the liner, weakening the barrier.

The final permeability for Soil 2 with Waste 2 leachate was $4.4 \times$

10^{-9} cm/s. Soil 2 is highly impermeable and provides a good physical barrier. With Waste 2 leachate, Soil 2 showed slight shrinkage over six months. Soil 2 plus beneficient had a final permeability of $<1 \times 10^{-10}$ cm/s, and the test cylinder had developed some cracks due to swelling. Slight back diffusion of clay into the leachate above the liner was also occurring.

Accelerated Batch Tests—Aluminum and silicon concentrations measured in the accelerated batch leach testing were analyzed monthly. These concentrations were extremely low and imply no chemical breakdown in any of the liner candidates. These results agree with the X-ray diffraction results, which also indicate that no major changes in the clay structures are occurring.

Auxillary Studies (Column Attenuation)—Because Soil 1 alone did not provide an impervious barrier, the soil's ability to attenuate contaminants in the waste became an important factor in evaluating the possible effectiveness of the candidate liners.

Two leaching columns were operated for Waste 1; one column contained waste only and the other contained a layer of waste and a layer of Soil 1. In general, the concentrations of the parameters in leachate from the waste again decreased with increased leaching. Total organic carbon, total dissolved solids (as indicated by conductivity), sodium, and barium were attenuated by the soil layer, particularly in the initial pore volume displacement (see Fig. 2). The pH, which was 10 or higher in the waste leachate, was at or near 7 in the waste/soil leachate. Total sulfur, on the other hand, was not appreciably reduced when the leachate was exposed to Soil 1.

Conclusions

The Soil 1 liner candidate is acceptable. While Soil 1 is lower in clay content than preferred and has a lower liquid limit and plasticity index than preferred, it has acceptable permeability properties and shows good stability over six months of testing; it also attenuated contaminants from the waste leachate. While Soil 1 plus beneficient had superior soil properties and good initial compatibility behavior, long-term compatibility tests indicated possible problems with swelling. If the liner swells under field conditions, cracks may form and leachate may breach the liner.

Soil 2 exhibited good performance with Waste 2 leachate. The excellent physical barrier provided by this soil's high clay content affords expected reliable field performance. Soil 2 plus beneficient also performed well with the exception of possible swelling properties.

Tests in this study were performed to identify fatal flaws for risk assessment. While laboratory tests are not directly applicable to determining extent or expected length of acceptable performance, these tests do provide state-of-the-art indication of soils that should have acceptable field behavior. For future liner studies, the laboratory testing matrix for liner compatibility can be limited to permeability studies, which indicate failure mechanisms. Soil iden-

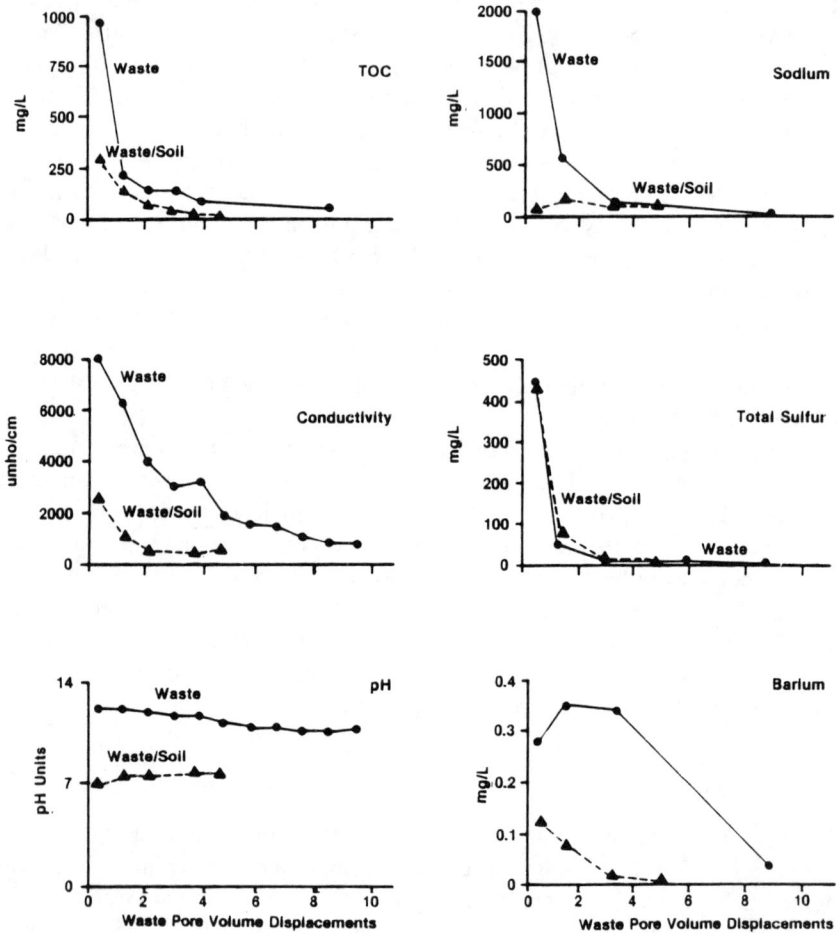

FIG. 2—*Leachate/soil interactions for Soil 1 with Waste 1 (concentration versus waste pore volume displacement).*

tification for definition of liner properties can be limited to particle size distribution, compaction, Atterberg limits, soil pH, and mineral composition. This matrix provides enough information to assess interaction with wastes having similar compositions. Chemical attenuation studies are necessary to document migration control only in the event that the candidate soil cannot provide an adequate physical barrier.

References

[1] Matrecon, Inc., *Lining of Waste Impoundment and Disposal Facilities*, U.S. Environmental Protection Agency, Cincinnati, OH, 1980.

[2] Sowers, G. B. and Sowers, G. F., *Introductory Soil Mechanics and Foundations*, 3rd ed., Macmillan, New York, 1970.
[3] Hillel, D., *Soil and Water Physical Principles and Processes*, Academic Press, New York, 1971.
[4] Allison, L. E. et al in *Diagnosis and Improvement of Saline and Alkali Soils*, L. A. Richards, Ed., U.S. Department of Agriculture, Washington, DC, Aug. 1969.
[5] Black, C. A., Ed., *Methods of Soils Analysis: Part 2—Chemical and Microbiological Properties*, American Society of Agronomy, Madison, WI, 1965.

DISCUSSION

J. H. Kleppe (written discussion)—What is the rationale of the liner material "preferred zone" in the Atterberg limits plasticity chart?

J. A. Mundell (written discussion)—What constitutes an acceptable "Atterberg limit" range? How were these ranges determined?

A. G. Eklund (author's closure)—Using the unified soil classification and plotting plasticity index versus liquid limit for laboratory classification of fine-grained soils, the toughness and dry strength increase with increasing plasticity index when soils are compared at equal liquid limit. Soils which belong to the groups CL or CH should be considered to be the most suitable soil liners with the most favorable soils with the liquid limit between 35 and 60, placed above the A-line in the "preferred zone" in the Atterberg limits plasticity chart, according to a study by Matrecon for the Environmental Protection Agency (EPA) [1, p. 41].

J. A. Mundell (written discussion)—What procedure would the author recommend regarding the *design* of natural attenuation landfills? Which geochemical properties are most pertinent?

A. G. Eklund (author's closure)—The design of soil containment in natural attenuation landfills should be based on leachable waste constituents and species likely to be mobile in the environment that will require attenuation. Then, a soil type likely to be capable of attenuation can be chosen. For organic wastes, a soil containing clay materials likely to undergo organic reactions would be ideal. For inorganic wastes, a soil containing a clay with a high cation exchange capacity would be advantageous.

The most pertinent geochemical property of soils in landfill design is probably permeability or hydraulic conductivity. If the soil forms a good physical barrier, chemical attenuation is not necessary.

Y. Acar (additional closure)—Ion exchange capacity is most important. Calcium-based bentonites, such as Mississippi bentonites, can be mixed into the soil and will reduce the permeability. The advantage of such a liner is that it is an open system; water mounding does not occur or only at a small level, but heavy metals and some organics will attenuate on the clay, thus filtering the water.

[1]Geotechnical engineer, Hart-Crowser and Associates, Seattle, WA.
[2]ATEC Associates, Indianapolis, IN 46220.

Yalcin B. Acar,[1] *Ivan Olivieri,*[1] *and Stephen D. Field*[1]

The Effect of Organic Fluids on the Pore Size Distribution of Compacted Kaolinite

REFERENCE: Acar, Y. B., Olivieri, I., and Field, S. D., **"The Effects of Organic Fluids on the Pore Size Distribution of Compacted Kaolinite,"** *Hydraulic Barriers in Soil and Rock, ASTM STP 874*, A. I. Johnson, R. K. Frobel, N. J. Cavalli, and C. B. Pettersson, Eds., American Society for Testing and Materials, Philadelphia, 1985, pp. 203-212.

ABSTRACT: This paper determines that permeating organic fluids through compacted kaolinite does not significantly change the distribution of different pore sizes. The mercury intrusion method was used to quantify the pore size and frequency distribution of specimens before and after permeation with nitrobenzene, acetone, phenol, and benzene. Porosity quantification parameters obtained from available hydraulic conductivity models were then correlated with experimental values of absolute permeability. Acetone and phenol resulted in changes of absolute hydraulic conductivity within one order of magnitude, that of 0.01 N calcium sulfate ($CaSO_4$) solution, while benzene and nitrobenzene led to a decrease of two orders of magnitude. Mercury intrusion tests on all specimens indicated that the size and distribution of pores within the 0.008 to 10 μm range were not significantly changed. The conclusion is that changes in hydraulic conductivity of compacted kaolinite are not due to a redistribution of pore sizes due to changes in forces of interaction. The results suggest that the decrease in hydraulic conductivity with benzene and nitrobenzene are due to the low solubility of these chemicals in water.

KEY WORDS: hydraulic conductivity, porosity, pore size distribution, organic fluid, kaolinite, compaction, mercury intrusion, dielectric constant, groundwater

The practice of containing hazardous wastes in shallow land waste disposal facilities has recently stimulated investigations of the effects of the generated leachates on the hydraulic conductivity and the structural stability of the compacted soil liners used to prevent and/or retard release of the contaminants to the environment [1-3]. The present state of knowledge on engineering behavior of compacted soils indicates that hydraulic conductivity is most affected by soil structure [4]. The arrangement of particles and pore spaces or

---

[1]Assistant professor, graduate research assistant, and assistant professor, respectively, Department of Civil Engineering, Louisiana State University, Baton Rouge, LA 70803.

the fabric of compacted soil together with the effects of composition and interparticle forces constitute the structure. Although the structure of a compacted soil could well be controlled during construction by controlling the composition, compaction scheme, and molding water content [5,6], the interparticle forces would be greatly affected by postconstructional changes in the chemistry of the pore fluid. It is well established that such changes in the interparticle forces would change the arrangement of particles and pore spaces, altering the fabric and hence the engineering behavior of a soil [7–8].

The mercury intrusion test is a well established method of quantifying the size and distribution of pores greater than 0.008 μm. Recent studies indicate that the changes in soil fabric caused by altering the compaction variables are quantitatively well reflected by the pore size distribution [9]. Furthermore, pore size parameters from theoretical hydraulic conductivity prediction models that relate the pore size distribution to hydraulic conductivity are found to correlate with experimental values. Previous studies indicated dramatic increases in the hydraulic conductivity of compacted soils to organic fluids [1,2] and suggested that such changes might be induced by major structural variations such as microcracks formed by syneresis of the clay particles. This paper presents the results of a study to assess the nature and extent of fabric changes by an analysis of pore size and frequency distribution.

Procedure

Mercury intrusion tests were conducted on compacted kaolinite specimens before and after permeation with water and organic fluids. The freeze drying method [10] was used as the pore fluid removal technique.

Benzene, phenol, acetone, and nitrobenzene were used as the organic fluids that represent a wide range of dielectric constants. Hydraulic conductivity to 0.01 N $CaSO_4$ solution and organic fluids was determined in a constant head setup with triaxial cells. After measurement of B values, specimens were fully saturated with 0.01 N $CaSO_4$ solution, raising the effective stresses to 69 kPa (10 psi) with incremental increases in the back pressure and the cell pressure [3]. Full saturation hydraulic conductivity values were obtained. Subsequently, organic fluids were permeated under continuous back pressure. Characteristics of the kaolinite and the organic fluids and the details of the hydraulic conductivity testing and procedure are given in an accompanying paper in this publication [3].

The most important operation in analysis of fabric of soils with high water contents is the preparation of the specimen. The drying of specimens must not result in excessive swelling or shrinkage. Freeze drying was used to remove the pore fluid before mercury intrusion tests. The procedure presented in other studies [10–12] was adopted in this study. Three to four specimens of approximately 10 mm long were cut from the compacted kaolinite samples before and after permeation with the organic fluids and the 0.01 N $CaSO_4$

solution. The specimens were then subjected to quick freezing in liquid nitrogen followed by drying by vacuum sublimation. This quick freezing of the pore fluid causes it to be transferred into an amorphous solid without recrystallization. Consequently, due to the low expansion of this phase, minimal strains are induced on the structure. After freeze drying, the specimens were split with a blade to expose fresh planes for the mercury intrusion tests. Volumetric methods were used to determine the effectiveness of freeze drying.

Several studies exist in the literature explaining the apparatus and procedure for mercury intrusion tests [7,13–16]. In this study, a Micromeritics pore sizer with a pressure capacity of 6.9 kPa to 207 MPa (1 to 30 000 psi) was used. The porosimeter measures volumetric increments of 0.001 mL. Consequently, it was possible to obtain the size and distribution of pores within the range of 0.008 μm (80 Å) to greater than 100 μm.

Cumulative porosity and pore size distributions were determined. Two models which were previously developed to predict hydraulic conductivity from pore size distribution [9] are used together with experimental values in order to assess whether changes in hydraulic conductivity with organic fluids are due to variations in pore size and distribution. Table 1 presents the basic parameters used in these models.

Results

Figure 1 presents the compaction curve and the location of specimens for the kaolinite. All specimens except one were compacted at standard Proctor

TABLE 1—*Summary of permeability models* [9].[a]

MARSHALL MODEL	
Equation	$K = C_s d^2$
Pore size distribution parameter	$d^2 = \sum_i^n \sum_j^n d^2 f(d_i) f(d_j)$
	where $d = d_i$ if $d_i < d_j$
	or $d = d_j$ if $d_i \geq d_j$
	and $\sum f(d_i) = \sum f(d_j) = n$
HYDRAULIC RADIUS MODEL	
Equation	$K = C_s R_H^2 n$
Pore size distribution parameter	$R_H = \dfrac{1}{4 \sum_i^l \dfrac{f(d_i)}{d_i}}$
	where $\sum_i f(d_i) = 1$

[a] K = absolute permeability, L^2.
   n = porosity.
   C_s = shape factor.
   d_i = pore diameter, L.
   $f(d_i)$ = volumetric frequency of d_i.

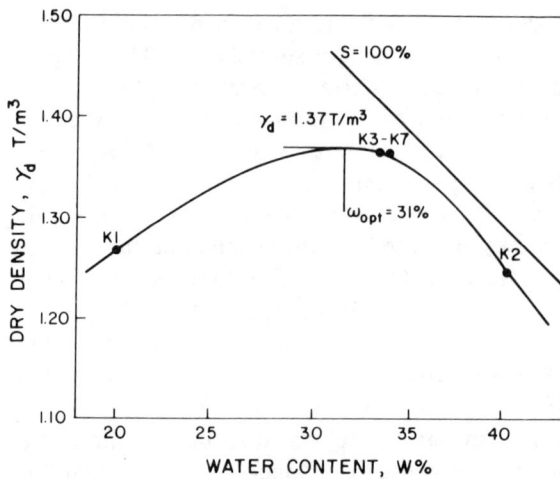

FIG. 1—*Compaction curve for Georgia kaolinite compacted at standard proctor efforts.*

effort and the wet side of optimum water content. Consequently, the initial fabric of these specimens were the same. The specimen at the dry side of optimum water content was prepared in order to compare the changes in fabric with the molding water content.

Table 2 presents the effect of freeze drying on the specimens with different pore fluids. The results indicate that the volume change experienced in specimens with a pore fluid of 0.01 N CaSO$_4$ solution was comparable with other studies [17]. It is interesting to note that volume changes less than that of 0.01 N CaSO$_4$ solution were observed in specimens with pores filled with organic fluids. This implies that freeze drying was more effective in these specimens.

Figure 2 presents a comparative plot of pore size distribution of the speci-

TABLE 2— *Volumetric changes in freeze dried compacted kaolinite.*

Pore Fluid	Dielectric Constant	Swell, %	
		Mean[a]	Variance[a]
0.01 N CaSO$_4$	84	5.22	0.28
1000 ppm solutions			
Nitrobenzene	80	3.43	0.57
Phenol	80	3.57	0.03
Pure solutions			
Nitrobenzene	35.7	4.83	0.07
Acetone	20.7	0.77	0.06
Phenol	13.1	2.25	0.20
Benzene	2.3	1.23	0.09

[a]Mean and variances are for three specimens.

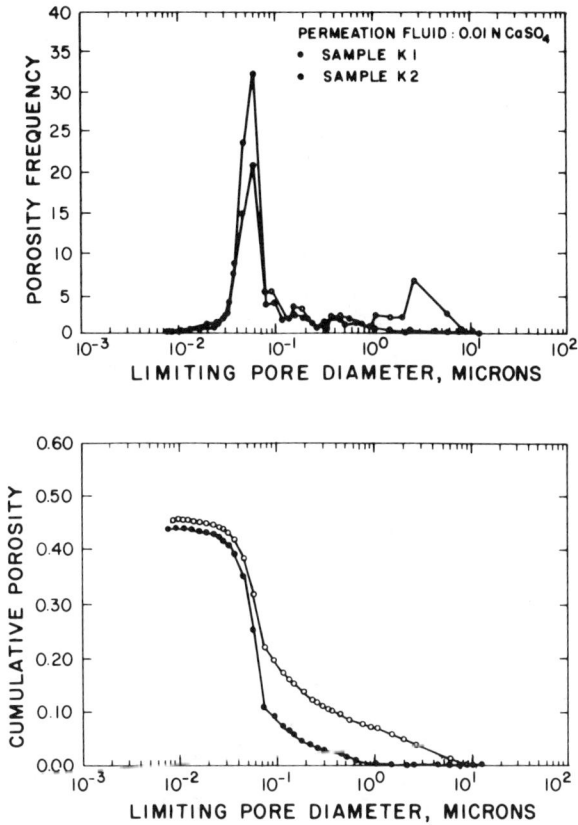

FIG. 2—*Differential and cumulative pore size distribution for kaolinite specimens compacted at dry and wet of optimum (Specimens K1 and K2).*

mens compacted at the wet and dry sides of optimum. As it is determined in studies with compacted silty clays [9], pore size distributions of compacted kaolinite are also bimodal; a large pore mode at 10 μm to 1 μm and a small pore mode at 0.1 to 0.008 μm. Changes in molding water content from the dry of optimum at the same compaction effort eliminates the large pore mode. Cumulative distributions indicate that approximately 90% of the pore spaces are accounted for with the available equipment. It should also be pointed out that the mean pore size in the small pore mode is approximately 0.06 μm (600 Å) and the variance is less than that observed with compacted silty clay [9].

Hydraulic conductivity tests indicated that absolute values of hydraulic conductivity slightly increased with acetone, slightly decreased with phenol, and a decrease of three orders of magnitude was observed with benzene and nitrobenzene [3]. Figures 3 and 4 present the pore size distribution of speci-

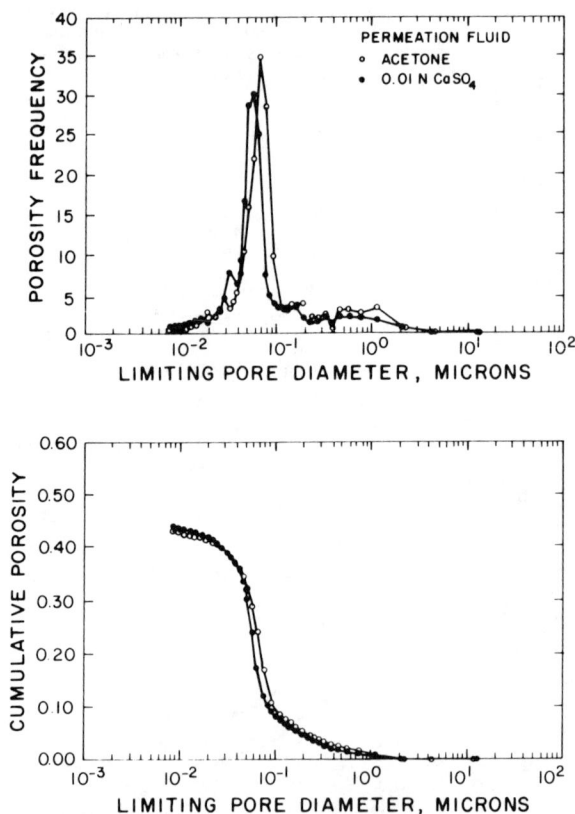

FIG. 3—*Differential and cumulative pore size distribution curves for the acetone permeated specimen (Specimen K3).*

mens before and after permeation with acetone and nitrobenzene. These two specimens were chosen for presentation since they represent the two extremes in changes in hydraulic conductivity. As depicted from these figures, major changes in the pore size distribution are not observed. The same observation was valid for other specimens. Table 3 summarizes the pore size parameters obtained from these distributions and the experimental values of absolute hydraulic conductivity. Figure 5 presents a comparative plot of these values together with the results of a previous correlative study between hydraulic conductivity and pore size parameters.

Table 3 and Fig. 5 indicate that changes in pore size distribution from the dry of optimum to the wet of optimum water content are well reflected by the change in absolute hydraulic conductivity and these points plot close to the regression line given by Bengochea and Lovell [9]. The same conclusion is not deducible for specimens permeated with organic fluids. Hydraulic conductiv-

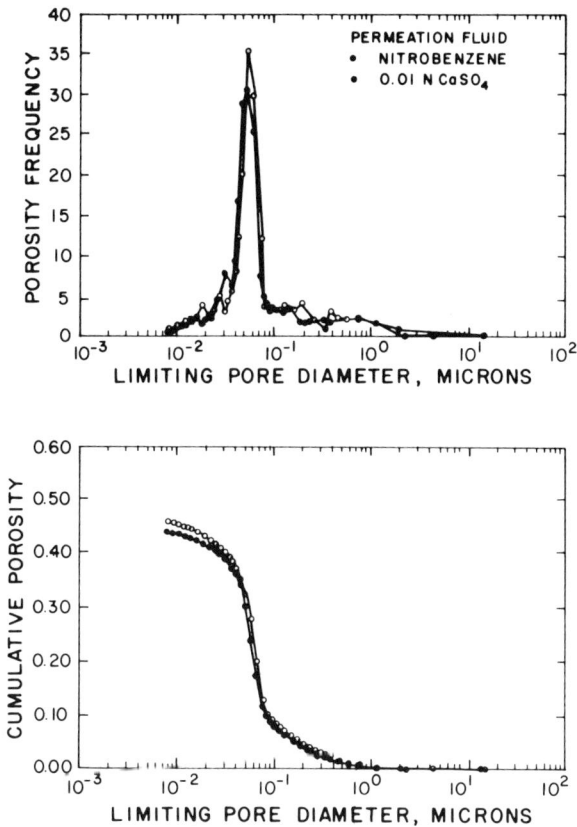

FIG. 4—*Differential and cumulative pore size distribution curves for the nitrobenzene permeated specimen (Specimen K5).*

TABLE 3—*Hydraulic conductivity and pore size parameters of compacted kaolinite.*[a]

Specimen Type	Hydraulic Radius Model, $R_h^2 \times n \times 10^{-15}$ μm²		Marshall Model, $d^2 \times 10^{-2}$ μm²		Absolute Hydraulic Conductivity, $\times 10^{-13}$ cm²
	μ	σ[b]	μ	σ	
Dry of optimum	14.75	0.22	4.75	0.18	55.30
Wet of optimum					
0.01 N CaSO₄	8.01	0.35	2.10	0.30	5.40
Nitrobenzene	7.25	0.22	1.78	0.05	0.015
Acetone	8.66	0.70	2.25	0.15	14.70
Phenol	8.69	0.60	2.27	0.25	1.19
Benzene	9.66	0.78	1.87	0.07	0.014

[a] σ = mean.
μ = standard deviation.
[b] Means and deviations are for three specimens.

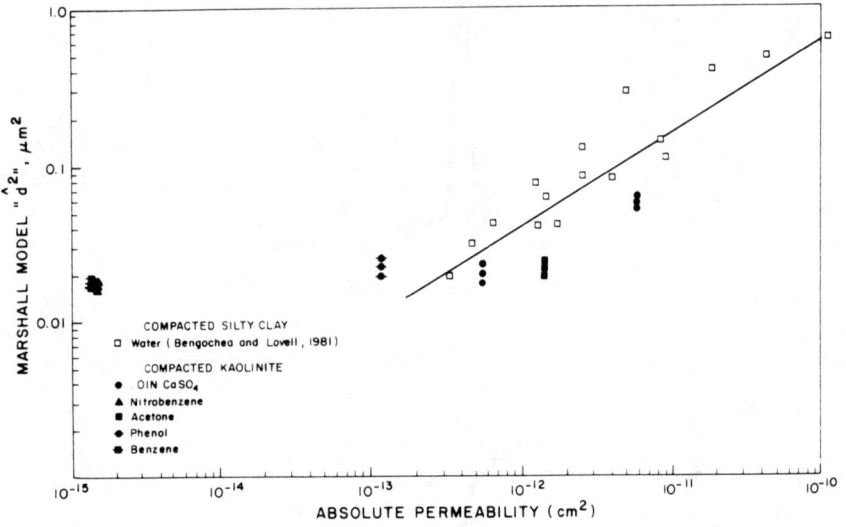

(a)

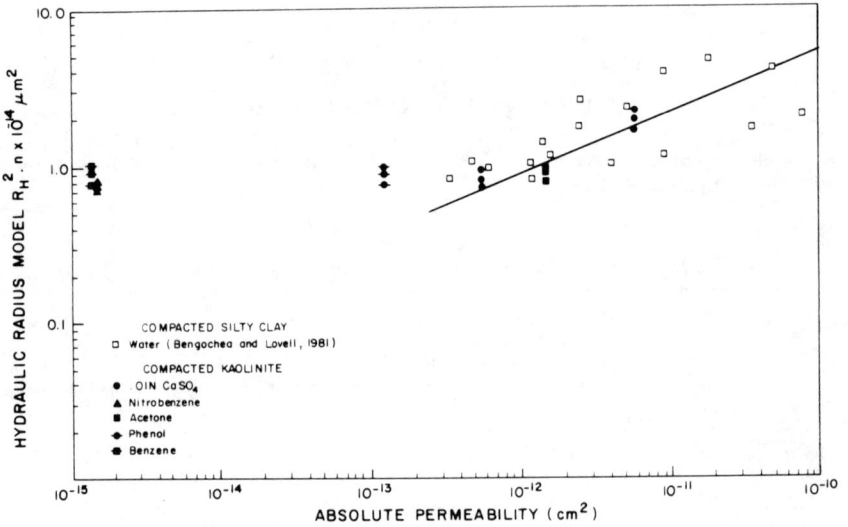

(b)

FIG. 5—*Hydraulic conductivity versus the pore size distribution parameters.*

ity with acetone and phenol varies within one order of magnitude of 0.01 N $CaSO_4$. Benzene and nitrobenzene resulted in three orders of magnitude decrease in absolute values of hydraulic conductivity [3], while the pore size parameters did not significantly change. These results suggest that changes in hydraulic conductivity of compacted kaolinite with organic fluids tested cannot be explained by the redistribution of pore sizes. However, it is noted that benzene and nitrobenzene have very low solubilities in water. It is then expected that these fluids will not be able to fully displace the water in the smaller pores due to the higher flow initiation pressures required to overcome the surface tension at the interface of the two fluids. Consequently, the resulting decrease in flow volume will lead to lower hydraulic conductivities.

Summary and Conclusions

The results of this study on the pore size distribution of organic fluid permeated compacted kaolinite indicate:

1. As previously detected with compacted silty clays, compacted kaolinite also displays bimodal to unimodal distributions with increases in molding water content. The pore size parameters reflecting these changes correlate well with experimental values of hydraulic conductivity with 0.01 N $CaSO_4$ solution.

2. The size and distribution of pores within 0.008 to 10 μm are not significantly changed with organic fluid permeation. This indicates that variations in hydraulic conductivity with organic fluids are not due to major structural changes.

3. The three orders of magnitude change in hydraulic conductivity with benzene and nitrobenzene at the same size and distribution of pores indicates that these changes cannot be explained by the redistribution of pore sizes due to changes in forces of interaction. The decrease in flow volume due to the low solubility of these two fluids in water would explain these decreases in hydraulic conductivity.

Acknowledgments

The investigation presented in this study was funded by the U.S. Environmental Protection Agency (EPA) under assistance agreement EPA Cooperative Agreement CR 809714010 to the Hazardous Waste Research Center, Louisiana State University. However, it does not necessarily reflect the views of the agency and no official endorsement is inferred.

References

[1] Anderson, D. and Brown, L., "Organic Leachate Effects on the Permeability of Clay Liners," in *Land Disposal: Hazardous Wastes*, 7th Annual Research Symposium, EPA-600/9-81-0026, D. W. Schultz, Ed., U.S. Environmental Protection Agency, Cincinnati, OH 45268, 1981.

[2] Green, W. J., Lee, G. F., and Jones, R. A., *Journal of Water Pollution Control Federation*, Vol. 52, 1981, pp. 1347-1354.
[3] Acar, Y., Hamidon, A., Field, S., and Scott, L., this publication, pp. 171-187.
[4] Mitchell, J. K., *Fundamentals of Soil Behavior*, Wiley, New york, 1976.
[5] Seed, H. B., Mitchell, J. K., and Chan, C. K., *Proceedings*, Conference on Shear Strength of Soils, Boulder, CO, American Society of Civil Engineers, New York, 1960, pp. 887-961.
[6] Lambe, T. W. in *Proceedings*, Conference on Shear Strength of Soils, Boulder, CO, American Society of Civil Engineers, New York, 1960, pp. 555-580.
[7] Mitchell, J. K., *Highway Research Board*, Vol. 35, 1956, pp. 693-713.
[8] Moore, C. A. and Mitchell, J. K., *Geotechnique*, Vol. 24, No. 4, 1974, pp. 627-640.
[9] Bengochea, J. G. and Lovell, C. W. in *Permeability and Groundwater Transport, ASTM STP 746*, American Society for Testing and Materials, Philadelphia, 1981, pp. 137-150.
[10] Tovey, N. K. and Yan, W. K. in *Proceedings*, International Symposium on Soil Structures, Gothenburg, Sweden, 1973, pp. 59-68.
[11] Gillot, J. E., *Journal of Sedimentary Petrology*, Vol. 39, No. 1, 1969, pp. 90-105.
[12] Zimmie, T. F. and Almaleh, L. J. in *Soil Specimen Preparation for Laboratory Testing, ASTM STP 599*, American Society for Testing and Materials, Philadelphia, 1975, pp. 202-215.
[13] Bhasin, R. N., "Pore Size Distribution of Compacted Soils after Critical Region Drying," Ph.D. thesis, Purdue University, West Lafayette, IN, 1975.
[14] Ahmed, S., "Pore Size and its Effects on the Behavior of a Compacted Clay," M.S.C.E. thesis, Purdue University, West Lafayette, IN, 1971.
[15] Reed, M. A., "Frost Heaving Rate of Silty Soils as a Function of Pore Size Distribution," M.S.C.E. thesis, Purdue University, West Lafayette, IN, 1977.
[16] Diamond, S., *Clays and Clay Minerals*, Vol. 18, 1970, pp. 7-23.
[17] Bengochea, I. G., "The Relation Between Permeability and Pore Size Distribution of Compacted Clayey Soils," Joint Highway Research Project JHRP-78-4, Purdue University-Indiana State Highway Commission, 1978.

Chester W. Jones[1]

Effects of Brine on the Soil Lining of an Evaporation Pond

REFERENCE: Jones, C. W., "**Effects of Brine on the Soil Lining of an Evaporation Pond**," *Hydraulic Barriers in Soil and Rock, ASTM STP 874*, A. I. Johnson, R. K. Frobel, N. J. Cavalli, and C. B. Pettersson, Eds., American Society for Testing and Materials, Philadelphia, 1985, pp. 213-228.

ABSTRACT: Information is needed concerning the long-term effects of brine on the permeability of clay linings for salt-gradient solar and salt evaporation ponds. A case history is presented for the soil lining at Anderson Lake, New Mexico, which was operated as a brine evaporation pond from 1963 to 1976. The soil lining was sampled and tested in March 1982. At that time, the brine level was below the deposited salt surface and near the top of the lined area in the lake. From a comparison of the 1982 tests on the lining and preconstruction tests in 1962, the soil density appears to have increased slightly since lining construction, possibly due to effects of the brine on the soil. From approximate measurements of the drop in brine surface in 1982, the seepage through the lining appeared to be very low and was approximately one order of magnitude below the seepage determined during the first year of pond operation.

KEY WORDS: earth linings, seepage, brine disposal, evaporation reservoirs, soil tests, salt-gradient solar ponds

The Bureau of Reclamation is currently conducting feasibility studies of salt-gradient solar ponds as a source of energy [1]. One of the requirements of a pond that contains brine is a lining to conserve the brine and to prevent contamination of groundwater. Studies show that for the pond to be cost-effective, the lining needs to be reasonably low in cost. If suitable soil for lining is readily available, the cost of a compacted soil lining is often the lowest cost type of lining. However, no information has been found on the long-term effects of brine composed mostly of a high concentration of sodium chloride on the permeability of soil, which for linings usually contains a significant amount of clay.

This paper describes an investigation of a soil lining in an abandoned brine pond known as Anderson Lake, located near Carlsbad, New Mexico. It in-

[1]Soils engineering technical specialist, Bureau of Reclamation, Denver, CO 80225.

cludes data on (1) preconstruction tests made on soil used for the lining, (2) specifications for lining construction, (3) seepage from the monitoring of the brine pond operation by the U.S. Geological Survey (USGS), and (4) 1982 field and laboratory tests on the soil to evaluate present soil conditions and pond seepage.

Background

Starting in the early 1950s, the USGS studied the geology of Malaga Bend on the Pecos River, near the southeast corner of New Mexico. Of particular concern was (a) contamination of the river at this location by brine seeping into the river from a large brine aquifer and (b) possible ways to alleviate the condition. The most feasible method appeared to be lowering the level of the aquifer by pumping from wells and diverting the brine into an evaporation pond. To this end, the USGS conducted field investigations supplemented by laboratory tests in several natural depressions in the area where an evaporation pond could be located [2,3]. The location finally selected was known as the Northeast Depression, later renamed Anderson Lake, which is about 1 km from the Pecos River.

Preconstruction Investigation

In 1962, prior to construction of the soil lining for Anderson Lake, near-surface disturbed samples of soil from three areas (Fig. 1) in the depression proposed for lining were tested at the U.S. Bureau of Reclamation Engineering and Research Center in Denver. The soil from Area 1 at the center of the depression was a lean clay (CL) of medium plasticity, while the soil from Area 2 was a silty sand (SM) to silty clay (CL); the soil from Area 3 was a sandy silt (ML). Laboratory gradation and compaction tests were performed, and the results of these tests are presented in later sections of this paper along with the 1982 test on samples of the soil lining.

Lining Construction

The lining was constructed by scarifying or plowing the soil in place to a depth of approximately 460 mm and by compaction of the scarified surface with a vibratory sheepsfoot roller. The specifications [4] provided that the density of the lining to 460-mm depth be a minimum of 98% of the standard Proctor maximum density [5, pp. 466–478] up to elevation 894.90 m and to 95% to the top of the lining.

No soil moisture limits above and below optimum were specified for compaction, but brine was to be used where necessary to raise the moisture to near the optimum. The specifications provided for a 150-mm layer of uncom-

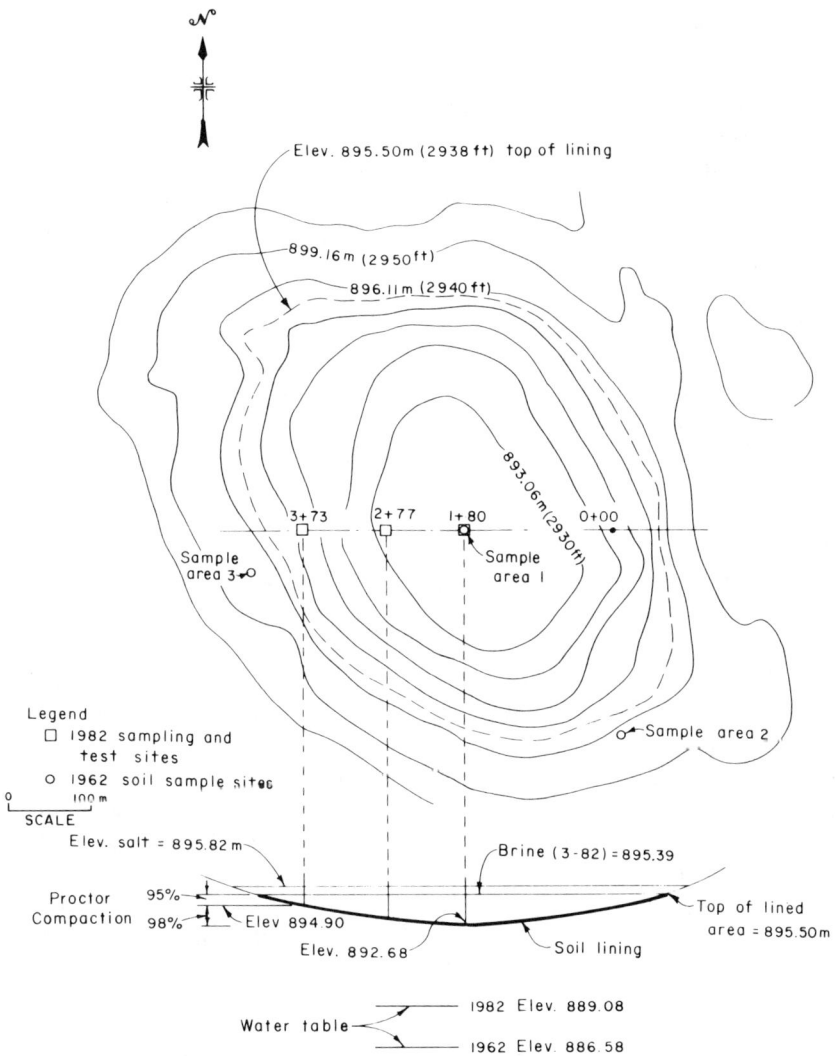

FIG. 1—*Plan and profile of Anderson Lake with locations of 1962 and 1982 soil sampling and testing sites.*

pacted soil to be selected by the construction engineer for placement on the compacted lining; apparently this was to reduce possible shrinkage and cracking of the lining by drying before the pond was filled. This top layer was to be tilled by harrow, disk, or other means to a depth of 100 mm and left in a loose and finely broken condition. No record of construction operations, inspection, or density and moisture control of the soil during construction has been found.

Pond Operation

The brine diversion system was operated by the Red Bluff Water Power Control District under the Pecos River Commission from July 1963 to 1976, when diversion was discontinued. The USGS monitored the pond operation and recorded (1) brine inflow that had 334 000 mg/L total dissolved solids, (2) the liquid level and chloride content in numerous wells surrounding the pond and at various points in the Pecos River, and (3) weather conditions with precipitation, evaporation, and other data to calculate inflow into the lake [6,7,8]. From these data, estimates of seepage from the pond were made. During the first phase of pond operation, some alleviation of the salt problem in the river was accomplished, but this proved to be temporary as the river salt content increased after three years of pond operation. After the first year of operation, the brine level in the pond was maintained above the top of the soil-lined area and seepage increased accordingly.

Soil Sampling (1982)

In March 1982, undisturbed samples of soil lining from three test sites (Fig. 1) in the area specified for 98% compaction were obtained for laboratory tests, and field permeability tests adjacent to the sampling sites were made in the lining. Offsets for sample locations were designated north (N) and south (S) from the east-west survey line. Disturbed samples were also obtained around the lake perimeter. At the time, the lake surface was a level expanse of salt (Fig. 2) at elevation 895.82 m, and the maximum depth of salt at the lake center was 3.14 m. There was brine in the voids of the porous salt up to a level about 430 mm below the surface of the salt, which was about 110 mm below the top of the lined area.

Prior to soil sampling at each location within the pond, a 600-mm-diameter hole was made through the salt to the soil lining with a bucket auger. The brine in the hole served as a drilling fluid for rotary drilling adjacent 108-mm-

FIG. 2—*View of Anderson Lake.*

diameter holes through the salt for soil sampling. Seventy-six millimetre-diameter, thin-wall, open-drive tube samplers were used for sampling [5, pp. 349–386].

With the exception of one sample that slipped out of the tube before it could be recovered, the samples were tight inside the tube with the bottoms of the samples at or protruding from the ends of the tubes. It is possible that for the samples showing partial recovery some soil consolidation occurred; however, it is believed that after a portion of the sample entered the tube, in spite of bit clearance ratios of 0.5 and 1.0%, the friction on the inside of the tube was sufficient to form a plug of soil near the end of the tube that pushed aside less compacted and wetter soil below. From examination of the samples prior to testing, there was no evidence of disturbance due to soil compaction in the tube. Also, any compaction would tend to reduce permeability and, as will be seen later, there was good agreement between laboratory and field permeability test results.

At the center of the lake, a hole was air-drilled to the water table, which was located at 3.60 m below the soil surface; this was 2.50 m above the 1962 level.

Soil Tests

Laboratory Tests

The tests were performed using procedures in the *Earth Manual* [5] unless otherwise noted.

Gradation—The results of 1962 and 1982 gradation analyses are shown on Fig. 3. The gradation analyses of the soil at the center of the lake, as determined in 1962 in Area 1 and in 1982 at Station $1+80$, are similar. Better comparisons can be made between 1962 and 1982 soil conditions here than at other places in the lake where locations for the separate investigations do not correspond.

Compaction and Density—Of the Proctor compaction tests performed in 1962 on soil proposed for lining (Fig. 4), one of the tests on soil from Area 1 was made with Denver tap water and the other with synthetic brine made up with chemicals to represent the brine to be discharged into the lake. The higher maximum density for the test with brine can be at least partly accounted for by the salt added in the brine.

Laboratory tests were made on the 1982, 76-mm-diameter, drive-tube samples after they were brought to the laboratory. The 1982 laboratory density test results are plotted on Fig. 4 with the 1962 compaction curves. The density tests were performed by cutting the soil-filled tube into sections and for each section determining the mass of the soil and the volume of the tube, with allowances being made for bit clearance ratio. The water content tests on the soil samples containing brine in the voids left the precipitated salt from the brine in the soil. Therefore, the resulting dry density calculated on the basis of

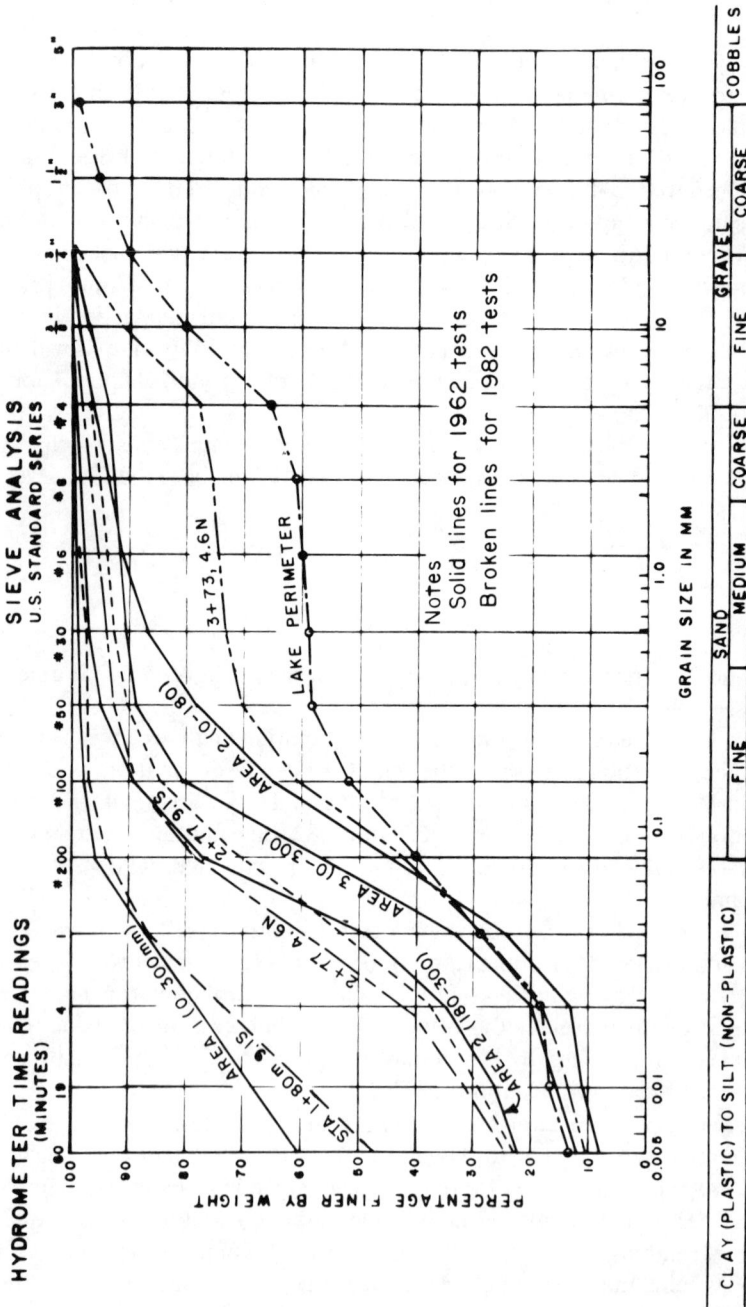

FIG. 3—Gradations of soil proposed for lining of Anderson Lake in 1962 and on 1982 samples from the lining.

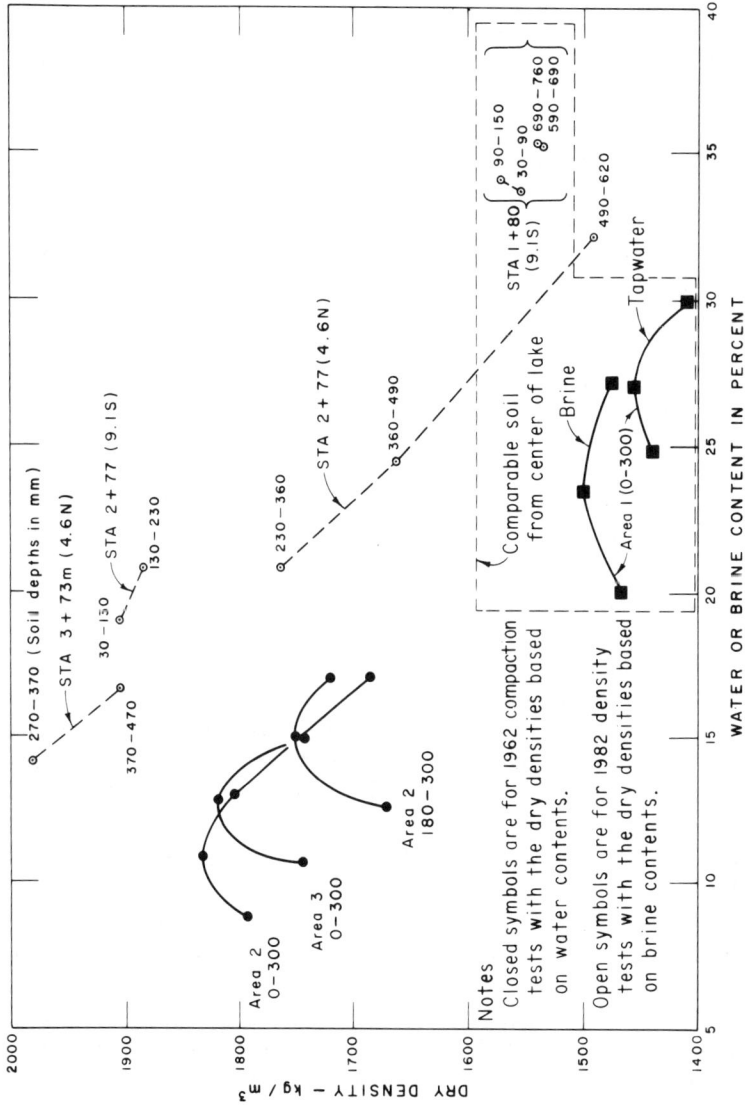

FIG. 4—*Proctor compaction curves on soil before lining in 1962 and 1982 density test results on samples from lining.*

water content represented a density of soil plus salt precipitated from brine rather than for the soil alone. Because of this, the dry density based on water content was reduced by deducting the amount of salt in the brine. This could be done, in an approximate way, by calculations based on the proportion of salts to water in the brine, assuming that the concentration of salts in the brine of the soil voids was the same as that for the brine in the voids of the salt deposit. The total dissolved solids for the brine in the voids of the salt deposit was 315 g/L, and the density of the brine was determined to be 1203 g/L at 22.5°C. With these values and the water content test results, the dry densities based on the water content tests could be corrected to represent soil dry densities without the added salt and with corresponding brine contents.

Since the 1962 samples from Area 1 and the 1982 density tests at Station 1+80 were both near the center of the lake and the soil gradations are similar (Fig. 3), the test data from these locations provide the best comparison that can be made between the properties of the soil before and after lining. The dry density of the soil in the lining averages 3 percentage points, about 50 kg/m^3 higher than the maximum density for the compaction test with brine. The brine content of the lining is about 10% higher than the optimum brine content for the 1962 compaction tests.

Figure 5 shows plots of the variation of soil density and brine content with depth, together with permeability test values, minus No. 200 fractions, liquid limits, plastic indexes, and soil classification. The data show no significant differences between the top 150 mm of soil, specified as loose soil cover, and the underlying compacted soil lining. Within the depth specified for 98% compaction, there was a general decrease in density and an increase in brine content with depth. However, the changes in density with depth were least for the lean clay near the center of the lake. Vibratory surface compaction of clay soil would be expected to be less effective with depth than for less cohesive soils. The top portion of the sampling hole at Station 2+77 (4.6 N) was cased, and an unsuccessful attempt made to retrieve a sample below the level where compaction would be expected. The sample fell back in the hole before the tube was withdrawn to the surface of the salt deposit, and a thin coating of wet soil clung to the inside wall of the tube.

Petrographic and Chemical—Since the petrographic and chemical tests were performed after the samples were air dried, the samples contained salts deposited from brine in the soil voids.

The soil from the lining contained 15 to 20% illite, 10 to 15% montmorillonite, 2 to 5% kaolinite, 25 to 40% quartz, 5 to 10% feldspar, 10 to 15% calcite, a trace to 3% gypsum, 5% hematite, and 5 to 10% halite.

From chemical analyses, the lean clay lining at the pond center (Station 1+80) contained about 22 mg/g of sodium, 5 mg/g of calcium, and 3 mg/g of magnesium. At the other two test locations, the chemical test results were similar except the sodium was about 5 mg/g less than at the pond center.

Electrical conductivity tests were made on a filtered leach of soil and dis-

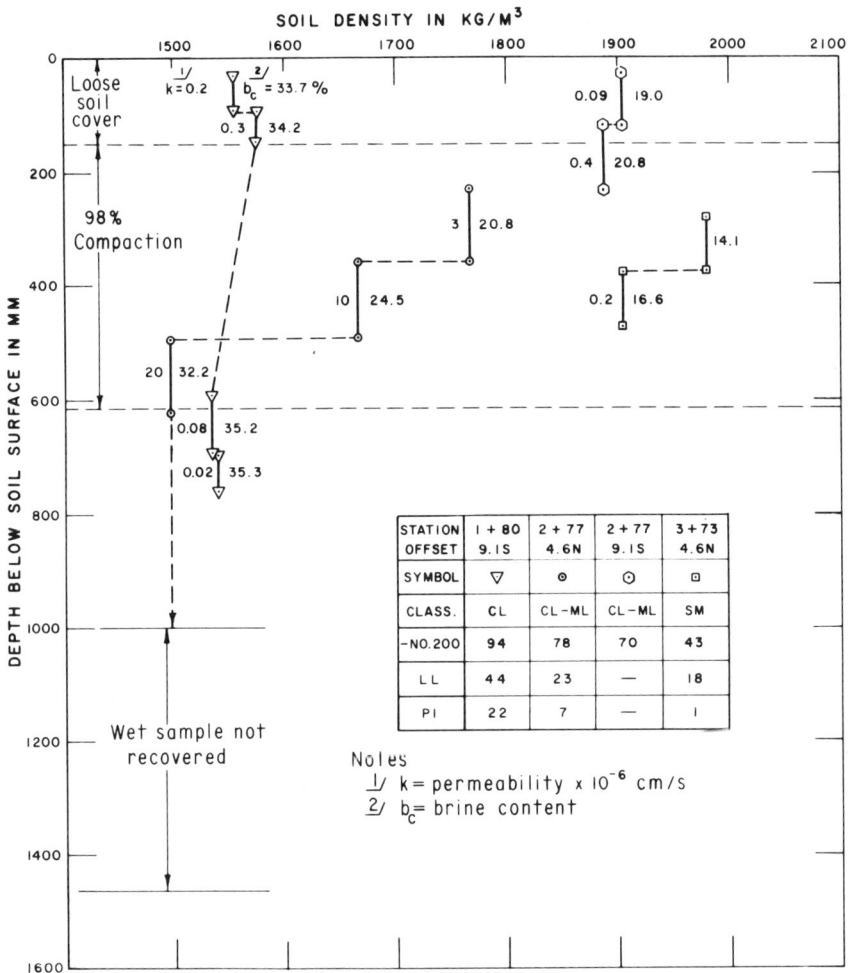

FIG. 5—*Variation of soil density, brine content* (b_c), *and laboratory permeability* (k) *with depth from the soil surface.*

tilled water for a relative comparison of dissolved salts in the lining. The conductivity for the clay at the center of the pond averaged 11 000 μS/cm compared to 8500 for the soil at the other two locations.

The sum of the major ions in the brine from a 1982 sample was 307 000 mg/L with the constituents in milligrams per litre as follows: calcium, 1540; magnesium, 868; sodium, 113 000; potassium, 977; bicarbonate, 6; sulfate, 4110; chloride, 186 000. The brine had a pH of 7.8.

Electron Microscopy—Electron micrographs obtained by a scanning electron microscope revealed sodium chloride in the soil voids of some of the lining samples, but it is uncertain whether or not the salt was (1) in the soil originally, (2) precipitated when brine was flowing through the soil, (3) or precipitated when the sample was dried for the electron microscopy.

Permeability—In 1982, laboratory permeability tests with Anderson Lake brine were performed by the falling-head method on segments of the open-drive tube samples from the lining. The sample segments were trimmed and sealed in plastic permeameter cylinders with modeling clay. The laboratory permeability test results (Fig. 5) with brine ranged from 0.02×10^{-6} to 20×10^{-6} cm/s with an average of 3×10^{-6} cm/s. For comparison of permeability and seepage, the original laboratory results, which were calculated on a 20°C basis, have been corrected for viscosity to that of the field temperature of 16°C recorded for the brine in March 1982. This has been done by the formula

$$K_F = \frac{V_L K_L}{V_F}$$

where

K_L = laboratory permeability at laboratory temperature,
V_L = kinematic viscosity of a concentrated solution of sodium chloride at the laboratory temperature (0.0056 m²/s), and
V_F = kinematic viscosity at field temperature (0.0062 m²/s [9]).

The permeability of the lean clay at the center of the lake remained low (below 0.3×10^{-6} cm/s) regardless of depth. In fact, it was lowest (0.08×10^{-6} cm/s) below the depth specified for 98% compaction. For the soil at the other test sites, there was a tendency for the permeability to increase with soil depth. This was most noticeable for the soil at Station 2+77 (4.6 N) where there was 100% sample recovery. The permeability rates here were the highest, ranging from 3×10^{-6} to 20×10^{-6} cm/s within the specified 98% compaction depth. It is significant that the permeability of the soil in the specified 150-mm-thick loose soil cover layer was about the same as that of samples in the underlying compacted layer.

Field Permeability Tests

The permeability of the lining was determined by the piezometer test for hydraulic conductivity as used in drainage investigations [10, pp. 67–74]. This is applicable to soil layers, particularly water barriers, below the water table. A 25-mm pipe with a 22-mm-diameter ship's auger was advanced in the soil to a point immediately above the depth interval to be tested, which was at the center of the 460-mm lining depth specified for compaction. With the auger, a cavity about 100 mm long was formed below the bottom of the pipe. After the brine in the bottom of the pipe was bailed out, the brine flowed into the pipe and, at appropriate time intervals, the rising liquid level in the pipe was measured with an electrical indicator. From these data and the static water level above the cavity, the coefficient of permeability was determined. The permeability rates were as follows:

Station, m	Offset, m	Brine Depth, m	Permeability, cm/s
1 + 80	4.6 S	2.72	0.06×10^{-6}
2 + 77	4.6 S	2.58	4×10^{-6}
3 + 73	4.9 N	1.23	0

Seepage

During Operation of the Lake

During the period from July 1963, when the operation of the brine evaporation pond was started, through 1968 and at times until 1977, USGS personnel monitored seepage from Anderson Lake. Inflow from pumpage, rainfall, runoff, and evaporation was recorded. The difference between inflow and evaporation was recorded as "other losses," which were considered to be mostly seepage from the depression. The average seepage during the first year of pond operation was about 3 mm/day. Thereafter, the brine level was operated above the top of the lining, and the seepage rate increased to a maximum of nearly 8 mm/day in 1968.

Brine Level Measurements in 1982

The measurement of brine level in a well installed in brine-saturated salt registers the level of brine in the voids of the salt. Therefore, to arrive at a calculated rate of seepage through the soil at the lake bottom where the area for seepage can be considered to be the total soil surface area, a measurement of brine level drop needs to be corrected for the porosity of the salt within the zone of brine level fluctuation. The average porosity of the salt, as measured by the USGS [8, p. 10], was 15%. During the 1982 field tests, a chunk of salt within about 150 mm of the surface was broken out and sawed into a rectangular specimen with a measured volume of 620 cm^3. From the specimen volume, the relative mass density of a pure sodium chloride crystal, and the dry weight of the specimen, the calculated porosity of the salt specimen was about 34%. A higher porosity for the salt near the surface could be expected because of some dissolution of the highly soluble sodium chloride during infiltration of rainwater. This would account for a difference between the Bureau and USGS porosity measurements.

Figure 6 shows (1) the elevation of the top of the salt surface, (2) the specified top of the soil-lined area, (3) brine surface elevations measured from establishment of the wells on 26 March to 16 July 1982, and (4) precipitation recorded from 1 January through 30 June 1983, at the closest official weather stations in the general area. Measurements from the salt surface to the brine level made on 22 January and 19 March 1982, showed that the brine level dropped about 80 mm. Assuming, as a conservative value, that the porosity of the salt is 30% within the zone of fluctuation for the brine, a calculated value of seepage—if any contribution from rainfall during the period is disregarded—would be about 0.4 mm/day. The rainfall shown for February and March was relatively low, and it is not known how much rainfall would have infiltrated the salt to the brine surface. A major portion of the small amount of the rainfall undoubtedly would have been held by capillarity in the voids of the salt before it could reach the brine level and be lost by evaporation.

Measurements in the cased wells on 26 March and 13 April showed that the brine level dropped about 22 mm, and the seepage rate, calculated with a 30% porosity for the salt, was about 0.5 mm/day during the 18-day period. This was during a period of insignificant rainfall. In these calculations of seepage rates, no allowance has been made for evaporation from the brine surface through the pore space between the brine level and the salt surface. This could be a significant part of the water loss and would further reduce the seepage values calculated. Figure 6 shows that during the latter part of April and during May and June there were larger amounts of rainfall in the area, and during the first part of May 1982 the brine level rose above the specified top of the soil-lined area at elevation 895.50. Although not shown on Fig. 6, on 10 August 1982, the brine level had risen to elevation 895.768, and on 21 April 1983, it had dropped to elevation 895.551, which is very near the specified top of the lining.

At a seepage rate of 0.4 mm/day and a depth (610 mm) from the soil surface to the bottom of the layer specified for 98% compaction, a calculated coefficient of permeability, neglecting a possible negative head below the lining, would be about 0.1×10^{-6} cm/s. According to Blight [11], it is possible that for a large, soil-lined pond negative pressures could develop in the soil below the lining. He believed that aeration of the soil below the lining would reduce the negative head and resulting seepage. On the other hand, Kisch [12], based on a theoretical analysis, concluded that "If the saturated permeability of the soil under the blanket is large compared with the permeability of the clay in the blanket, it is found that the characteristics of the underlying soil have practically no effect in determining the seepage from the reservoir." In any future investigations of seepage from soil-lined ponds, measurements of pressures below the lining would be of value.

Investigations are needed to develop soil linings for salt ponds with seepage rates still lower than that at Anderson Lake. Among practicable measures that could be tried are (1) surface compaction with modern equipment com-

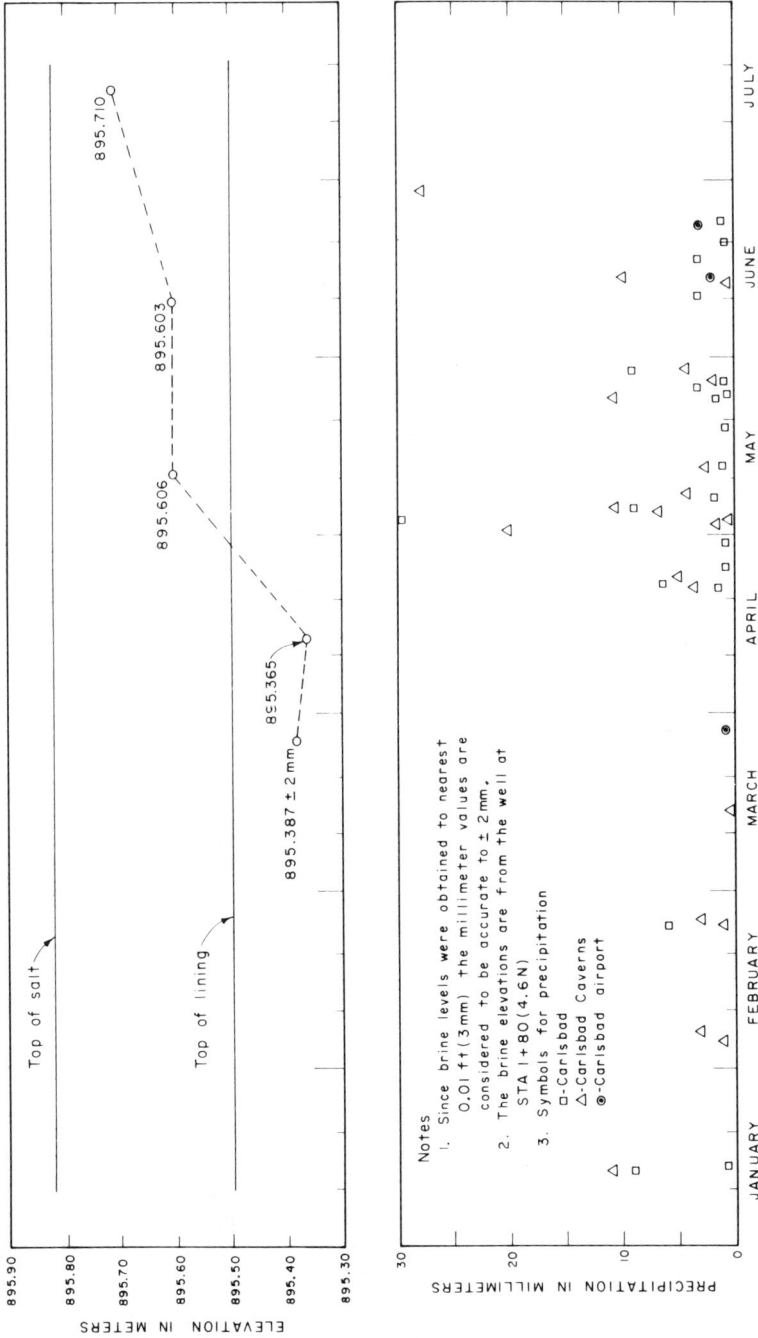

FIG. 6—*Brine level variations at Anderson Lake and precipitation at the nearest weather stations for the first half of 1982.*

pared with the conventional method of constructing the lining in a series of compacted layers and (2) treatment of the soil with additives such as chemical dispersants compared with treatment with brine. For whatever measures are tried, a basic understanding is needed of the physical and chemical properties involved.

Literature Review

In order to fully understand the effects of brine on the permeability of a compacted soil lining, an in-depth study of site-specific brine and soil properties and chemical and physical interactions would be necessary. Although there are references to the effects of some salts on the properties of clay soils, none have been found directly related to long-term effects on compacted soil lining in a pond containing concentrated natural brine.

Lee [13] has described an example where a freshwater, clay-lined lagoon at Treasure Island in San Francisco Bay, California, was sealed by the admittance of seawater to the lagoon. The 250-mm-thick lining in the 30 000 m² lagoon was made up of two layers of clay compacted by a flat roller with a mass of 12.7 metric tons to a density 60 to 90% of the laboratory maximum. The seawater (33 394 ppm of dissolved solids) reduced seepage from the lagoon from 25 to 2.5 mm/day.

It is possible under certain conditions for soil in contact with salt solutions to be compacted by osmosis. Turk [14] studied interactions between brine and unconsolidated sediment at the bottom of solar evaporation ponds in playas. He found that shrinkage and cracking of the sediments occurred that caused leakage from the ponds. He attributed the sediment shrinkage either to osmosis between the brine and the fresher water beneath the sediment, or syneresis, which is a mechanism wherein liquid is extracted from a gel by contraction.

In considering the possible long-term changes in soil properties of a soil lining, such as that at Anderson Lake, it is worthwhile to examine information from natural saline lakes. Under natural conditions, a lake in a closed basin with no outflow will eventually become saline. According to Jones and Van Denburgh [15, p. 441], as salts in a closed lake accumulate, the less soluble salts, particularly calcium and magnesium carbonates, precipitate out. Also, some soluble constituents can be trapped in lake sediments [16, p. 10]. The interstitial solutions can be more concentrated than the lake solutions, due to sedimentary insulation or ion filtration [15, p. 445]. Besides the simple precipitation of minerals, such as carbonates and sulfates, under certain chemical conditions, more complex reactions forming silicates are possible [15, p. 443]. In addition to mineral precipitation, organic activity in saline lake sediments can influence chemical processes, particularly carbonate and sulfate species. The chief organic process altering the chemical character of closed lakes is anaerobic decay, which promotes the reduction of sulfur [15,

p. 444]. Precipitates and bacteria in soil voids would tend to reduce permeability.

Summary and Conclusions

Based on an investigation of the soil properties of the compacted soil lining and seepage at Anderson Lake, which was operated as a brine evaporation pond from 1963 to 1976, the following summary and conclusions are made:

1. In March 1982, there was still brine in Anderson Lake with the brine surface about 430 mm below the surface of the salt deposit and about 110 mm below the elevation of the top of the soil-lined area.

2. Although it is not possible to make a direct comparison, from 1962 compaction tests made on soil proposed for lining and the 1982 densities of samples from the lining, it appears that the soil density has increased since lining construction in 1963.

3. The average coefficient of permeability from field tests at three locations in the compacted lining was 1×10^{-6} cm/s, while that from the laboratory tests on samples from the same areas was 3×10^{-6} cm/s.

4. In 1982, the brine level in the salt deposit was dropping slowly, and the seepage rate from the lake was calculated to be about 0.4 mm/day.

5. From calculations of seepage from USGS data during operation of the lake in 1963–1964 and from the drop in brine level during March and April 1982, the seepage rate through the lining appears to have decreased about one order of magnitude from an original average rate of about 3 mm/day.

6. The reasons for the apparent increase in soil density and decrease in seepage are not known at this time, and further study of this would be required.

References

[1] Boegli, W. J., Dahl, M. M., and Remmers, H. E., "Preliminary Study of Solar Ponds for Salinity Control in the Colorado River Basin," Bureau of Reclamation Report REC-ERC-82-19, Denver, CO, Dec. 1982, p. 23.

[2] Hale, W. E., Hughes, L. S., and Cox, E. R., "Possible Improvement of Quality of Water of the Pecos River by Diversion of Brine at Malaga Bend, Eddy County, New Mexico," The Pecos River Commission of New Mexico and Texas, report prepared in cooperation with the U.S. Department of the Interior, Geological Survey, Water Resources Division, Carlsbad, NM, Dec. 1954.

[3] Cox, E. R. and Haven, J. S., "Evaluation of the Queen Lake Depression, Eddy County, New Mexico, as a Storage Basin for Brine," U.S. Department of the Interior, Geological Survey, prepared in cooperation with the Pecos River Commission, Open File Report (unnumbered), Mar. 1981.

[4] "Brine Disposal Area Treatment for Salinity Alleviation at Malaga Bend, Malaga Bend Division, MacMillan-Delta Project, New Mexico," Specifications 500C-127, Bureau of Reclamation, Amarillo, TX, 1963.

[5] Earth Manual, 2nd ed., U.S. Department of the Interior, Bureau of Reclamation, U.S. Government Printing Office, Washington, DC, 1974.

[6] Cox, E. R. and Havens, J. S., "Progress Report on the Malaga Bend Experimental Salinity Alleviation Project, Eddy County, New Mexico," prepared by the U.S. Geological Survey in cooperation with the Pecos River Commission, Carlsbad, NM, Open File Report No. 65-35, Nov. 1965.

[7] Havens, J. S. and Wilkins, D. W., "Experimental Salinity Alleviation at Malaga Bend of the Pecos River, Eddy County, New Mexico," U.S. Geological Survey, Water Resources Investigation 80-4, prepared in cooperation with Pecos River Commission, Carlsbad, NM, Dec. 1979.

[8] Kunkler, J. L., "Evaluation of the Malaga Bend Salinity Alleviation Project, Eddy County, New Mexico," U.S. Geological Survey, Open File Report 800-1111, prepared in cooperation with the Pecos River Commission, Carlsbad, NM, Sept. 1980.

[9] Groundwater Manual, U.S. Department of the Interior, Bureau of Reclamation, U.S. Government Printing Office, Washington, DC, 1977, p. 23.

[10] Drainage Manual, 1st ed., U.S. Department of the Interior, Bureau of Reclamation, U.S. Government Printing Office, Washington, DC, 1978, pp. 67-74.

[11] Blight, G. E., The Civil Engineer in South Africa, Vol. 8, No. 7, Graphic Arts Division of the Council for Scientific and Industrial Research, Pretoria, South Africa, July 1966, pp. 215-221.

[12] Kisch, Michael, "The Theory of Seepage from Clay-Blanketed Reservoirs," Geotechnique, Vol. 9, 1959.

[13] Lee, C. H., "Sealing the Lagoon Lining at Treasure Island with Salt," Transactions of the American Society of Civil Engineers, Vol. 106, Paper 2110, 1941, pp. 577-607.

[14] Turk, L. J. in Proceedings, Fourth Symposium on Salt, Northern Ohio Geological Society, Houston, TX, 1974, pp. 403-406.

[15] Jones, B. F. and Van Denburgh, A. S., "Geochemical Influences on the Chemical Character of Closed Lakes," U.S. Geological Survey, from International Association of Scientific Hydrology, Publication No. 70, Gentbrugge, Belgium, 1966, pp. 435-446.

[16] Langbein, W. B., "Salinity and Hydrology of Closed Lakes," U.S. Geological Survey, Professional Paper 412, U.S. Government Printing Office, Washington, DC, 1961.

Stanley R. Peterson[1] and Glendon W. Gee[1]

Interactions Between Acidic Solutions and Clay Liners: Permeability and Neutralization

REFERENCE: Peterson, S. R. and Gee, G. W., **"Interactions Between Acidic Solutions and Clay Liners: Permeability and Neutralization,"** *Hydraulic Barriers in Soil and Rock, ASTM STP 874,* A. I. Johnson, R. K. Frobel, N. J. Cavalli, and C. B. Pettersson, Eds., American Society for Testing and Materials, Philadelphia, 1985, pp. 229-245.

ABSTRACT: Liner failure, defined as an increase in liner permeability, was not found to be a problem when acidic uranium mill tailings solutions percolated through clay liner materials for periods extending up to three years. Liner materials taken from mill sites in Wyoming decreased in permeability with time in the laboratory columns when permeated with tailings solution. One clay liner decreased in permeability from one half to over two orders of magnitude, depending on the given clay sample and contacting solution. These decreases in permeability were attributed to pore plugging resulting from the precipitation of minerals and solids and to soil particle dispersion.

The clay liner material from Morton Ranch, Wyoming, exhibited a residual buffering capacity that was able to maintain column effluent pH values at higher levels than the influent pH values for extended time (in excess of 30 pore volumes). A likely cause for the elevated pH is the redissolution of iron and aluminum hydrous oxides. Redissolution of iron and aluminum hydrous oxides consumes hydrogen ions.

KEY WORDS: buffering capacity, hydraulic conductivity, particle-size analysis, permeameter, precipitation

Clay materials have been proposed as liners in uranium mill tailings impoundments. However, long-term response of clay to contact with acidic tailings solution is not well understood. Crim et al [1] found measurable increases in permeability in montmorillonite clays subjected to extended contact with an acidic (pH < 1) tailings solution. These increases in permeability varied by more than two orders of magnitude, but tended to increase most dramatically after the pH of the effluent dropped below 4 (from an initial value near 8). Gee et al [2,3] studied the effects of extended contact of

[1]Senior research scientist and staff scientist, respectively, Battelle, Pacific Northwest Laboratory, Richland, WA 99352.

229

acidic tailings solution (pH ~ 2) on the permeability of native clays from Morton Ranch in central Wyoming. Their preliminary results indicated that the permeability of the clays gradually decreased with time during a 16-month period.

This paper summarizes the results from additional permeability tests that were run in flow-through columns packed with clays and percolated with various tailings solutions. The tests were designed to evaluate the effect of long-term contact of acidic uranium mill tailings solution on clay liners.

Materials and Methods

Clay Liner Material

All geologic characterization work on the clay liner materials was done using American Society of Agronomy soil testing procedures described by Black [4,5]. The key physical and chemical characteristics of the test materials are listed in Table 1.

Materials selected for use as clay liners were taken from two uranium mills: Morton Ranch (near Glenrock in central Wyoming) and Lucky Mc (near

TABLE 1—*Characterization of the Morton Ranch and Lucky Mc clay liners.*

Key Characteristics of Test Materials	Morton Ranch Clay Liner	Lucky Mc Clay Liner
Water content, g/g, %, after air drying	4.10	2.95
Particle density, g/cm³	2.72	2.89
Particle size distribution, weight %		
Sand, 50 to 2000 μm	12	7
Silt, 2 to 50 μm	54	45
Clay, <2 μm	34	48
pH of saturated paste	8.2	8.0
Eh of saturated paste, V	+0.406	+0.336
Electrical conductivity of saturated extract, mmhos/cm	0.70	6.5
Liquid limit	43	56
Plasticity index	21	29
Organic matter, weight %	1.44	0.71
CaCO₃, weight %	0.3	4.0
Cation exchange capacity, meq/100 g	31.6	19.4
Water soluble cations, meq/100 g, 2:1 extract		
Potassium	ND[a]	1.25
Sodium	ND	34.24
Calcium	ND	21.23
Magnesium	ND	18.18

[a]ND = not determined.

Riverton in central Wyoming). The Morton Ranch clay liner material analyzed was a composite of sandstone, siltstone, mudstone, and shale. The Unified Soil Classification in ASTM Classification of Soils for Engineering Purposes [D 2487-69 (1975)] of the Morton Ranch clay liner is a clayey silt (CL). The saturated paste of this soil is alkaline, and the organic matter content and the cation exchange capacity are higher than that of the Lucky Mc clay liner (Table 1).

The Lucky Mc clay liner is classified as a fat clay (CH). A distinctive feature of the Lucky Mc clay liner is the relatively high calcium carbonate content of 4% in contrast to the Morton Ranch clay liner, which had a calcium carbonate content of 0.3%. Additional details on the characterization of the Morton Ranch and Lucky Mc clay liners are provided in Gee et al [2] and Erikson and Sherwood [6].

Solutions

Characterization of the original uranium mill tailings solutions along with the components of the prepared synthetic tailings solution is presented in Table 2. Inductively coupled plasma emission spectroscopy was used for macrocation analysis, while graphite furnace atomic absorption was used to analyze trace metals. Anions were determined by ion chromatography and titration, and radionuclides were determined by X-ray and gamma-ray radioanalytical techniques.

Permeability Tests

The method used for determining hydraulic conductivity (herein after referred to as permeability) was a modification of the constant head method. The modifications are described in paragraphs that follow. The permeameter was confined at both ends by filter screens and stationary caps. Flow through the columns was from bottom to top to minimize air bubble entrapment. Pressure, up to 140 000 Pa (1.4 bars), was applied to the influent solution to increase the permeability and accelerate sample collection. The hydraulic head difference for the duration of the permeability tests was 1400 cm water (H_2O).

The liner materials used in the tests were compacted by the sliding-weight tamper ASTM method D 1557-78 in cylindrical columns to densities exceeding 90% of maximum compaction. The test apparatus is shown in Fig. 1. Maximum compaction was determined by the ASTM method D 1557-78, the modified Proctor test.

All columns, with the exception of Column 1, were initially saturated with groundwater. Column 1 was contacted only with synthetic tailings solution. Constant head permeability tests were then run on all columns. Columns 1, 2, and 3 were packed with the Morton Ranch clay liner and leached with syn-

TABLE 2—*Characterization of the Exxon Highland mill, Lucky Mc mill, and synthetic tailings solutions.*

Parameter	Exxon Highland Mill[a]	Lucky Mc mill	Synthetic
	Tailings Solution, mg/L, unless otherwise noted		
Ag	<0.05	<0.05	...
Al	600	1 030	595
As	3.50	19.3	...
B	0.19	1.2	...
Ba	<0.05	0.09	...
Ca	537	600	540
Cd	<0.04	0.28	...
Cr	2.7	2.36	...
Cu	2.3	1.51	...
Fe	2 215	2 780	2 217
K	39.5	156	...
Li	0.9	...	...
Mg	688	1 220	650
Mn	63.5	163	...
Mo	0.35	8.41	...
Na	343	1 630	350
Ni	3.0	...	...
P	30	...	...
Pb	<0.05	0.87	...
Se	0.60	1.61	...
Si	233.5	283	...
Sr	15.7	14.0	...
Th	<1	...	...
U	39.7[b]	...	...
Zn	8.4	16.9	...
Cl	97.1	1 090	103
NO_3	16.5	302.0	15
SO_4	12 850	26 400	13 000
EC, mmhos/cm	18.2	42.0	...
F	4	28	...
pH	1.8	1.2	2.0
Eh, mV	910	698	...
210_{Pb} (pCi/L)	9 701	16 000	...
230_{Th} (pCi/L)	227 286	167 000	...
238_U (pCi/L)	13 216	16 800	...
235_U (pCi/L)	620	789	...
226_{Ra} (pCi/L)	2 252	4 900	...
Totals (pCi/L)	253 075	205 489	...

[a]Tailings solution from the Exxon Highland mill was used because no tailings solutions was available from the Morton Ranch mill.
[b]Radiochemical analysis.

FIG. 1—*Pressurized permeameter setup for measuring permeability of clay liner materials.*

thetic tailings solution. The synthetic tailings solution contained the major constituents of the real uranium mill tailings solution, but was free of radioactivity (Table 2). Columns 4, 5, and 6 were also packed with the Morton Ranch clay liner but leached with the Exxon Highland uranium mill tailings solution. Tailings solution from the Exxon Highland mill was used in Columns 4, 5, and 6 because no tailings solution was available from the Morton Ranch mill. The Exxon Highland mill is within a few kilometers of the Morton Ranch mill site. The Lucky Mc uranium mill tailings solution was used to leach Column 7. This column was packed with the Lucky Mc clay liner.

Results and Discussion

Permeability

The contact time between the solutions and the respective clay liners, the total number of pore volumes of solution that passed through the column during the experiment, and the permeabilities are given in Table 3. The beginning permeabilities were determined by averaging several data points taken over the initial three to ten days of contact time, depending on flow rates. The approximate number of pore volumes to which this corresponds can be found in Table 3. The final permeability achieved in the columns at the end of the specified contact time is also presented.

TABLE 3—*Contact time, pore volumes of effluent, and initial and final permeabilities of the clay liner materials contacted with original and synthetic tailings solution.*

Column	Contact Time, Days	Pore Volumes	Initial Permeability, cm/s	Final Permeability, cm/s
1	1024	34.7	7.0×10^{-9}	1.4×10^{-9}
2	947	21.4	2.8×10^{-8}	7.5×10^{-10}
3	929	13.8	2.2×10^{-9}	8.0×10^{-10}
4	350	0.9	3.5×10^{-10}	flow ceased
5	838	18.7	5.0×10^{-8}	8.1×10^{-11}
6	836	9.4	2.5×10^{-9}	2.5×10^{-10}
7	117	34.2	4.4×10^{-8}	2.4×10^{-8}

The changes in permeability for the Morton Ranch clay liner material contacted with synthetic tailings solution (Columns 1, 2, and 3) are shown in Fig. 2. Columns 1 and 3 experienced a gradual, continuous decrease in permeability. For Column 2 (which was recompacted in place after 50 days of solution contact), the permeability dropped by an order of magnitude after recompaction. Column 2 was recompacted to over 90% of maximum compaction by the same methods that were employed to initially compact the columns. One can conclude from the data that physical manipulation (compaction) has a dramatic effect on permeability, and that physical mechanisms, which optimize density and compaction and minimize channel cracks and fracture flow, will likely play an important role in determining the ultimate permeability of the liner. The nonrecompacted columns (Columns 1 and

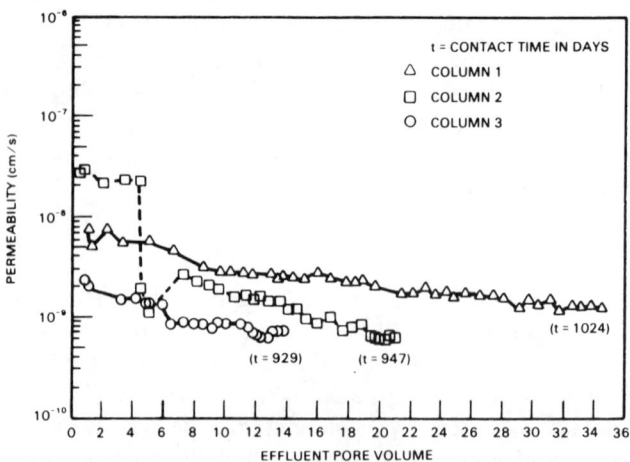

FIG. 2—*Permeability of the Morton Ranch clay liner material, contacted with synthetic tailings solution (Columns 1, 2, and 3), plotted versus pore volumes of effluent.*

3) experienced a decrease in permeability of roughly one half order of magnitude over the ~ 1000-day period.

The effect of the Exxon Highland mill tailings solution on the permeability of the Morton Ranch clay liner (Columns 4, 5, and 6) is shown in Fig. 3.

The permeabilities of Columns 5 and 6 continuously decreased as long as tailings solution was passing through the columns. Flow through Column 4 ceased after nearly one year (350 days) of testing. During this time <1 pore volume of effluent was collected. The initial permeability of this column was very low (3.5×10^{-10} cm/s). The permeabilities of Columns 5 and 6 decreased from one order of magnitude for Column 6 to over two orders of magnitude for Column 5. Differences in permeability between Samples 4, 5, and 6 are attributed to slight differences in packing density. The decreases in permeability of the Morton Ranch clay liner after contact with Exxon Highland mill tailings solution were more dramatic than the decreases observed when the same clay liner was contacted with synthetic tailings solutions. This could be due to the formation of additional precipitates in the Exxon Highland mill tailings solution/clay liner interactions. The precipitation of additional solid phases is likely caused by the presence of soluble constituents in the real tailings solution, which are not encountered in the synthetic tailings solution. An example of such a constituent is potassium, which is necessary for the precipitation of potassium-jarosite $[KFe_3(SO_4)_2(OH)_6]$ and alunite $[KAl_3(SO_4)_2(OH)_6]$. Potassium is found in the Exxon Highland mill tailings solution but not in the synthetic tailings solution. Though potassium would be expected to be found and perhaps leached from the liner material, it evidently was not leached in sufficient quantity to cause appreciable jarosite pre-

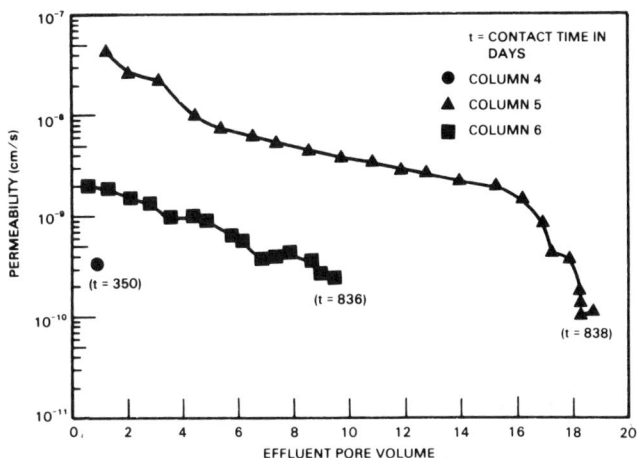

FIG. 3—*Permeability of the Morton Ranch clay liner material, contacted with Highland Mill tailings solution (Columns 4, 5, and 6), plotted versus pore volumes of effluent.*

cipitation in the synthetic tailings experiments. Both potassium-jarosite and alunite were identified, by X-ray diffraction, in the Morton Ranch clay liner material after contact with Exxon Highland mill tailings solution but not after the synthetic tailings solution contacted the same Morton Ranch clay liner [7,8]. The permeabilities observed in the Morton Ranch clay liner material (Columns 1 through 6) are lower than the liner permeability (10^{-7} cm/s) recommended by the U.S. Environmental Protection Agency [9]. The columns packed with the Morton Ranch clay and contacted with Exxon Highland mill tailings solutions (Columns 4 through 6) all had final permeabilities of $<10^{-9}$ cm/s.

Lucky Mc clay leached with the Lucky Mc tailings solution (Column 7) had a final permeability of approximately 2×10^{-8} cm/s, which was a slight decrease from the permeability measured at the beginning of the experiment (Fig. 4).

Mechanisms

Two intertwined mechanisms could help account for the decreases in permeability that were observed when low-pH, high-ionic strength, high-sulfate tailings solutions were allowed to contact clay liners. First, precipitation of minerals due to the increase in pH brought about by the buffering capacity of the soil could result in pore plugging, thus decreasing permeability. Second, soil swelling, dispersion, and deflocculation of clay particles could result in decreases in permeability. Pupisky and Shainberg [10] have shown that dispersion and internal movement of the clay particles can block the conducting

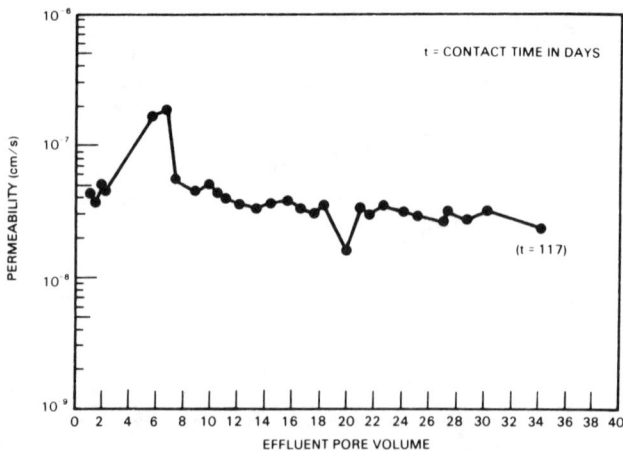

FIG. 4—Permeability of the Lucky Mc clay liner material, contacted with Lucky Mc tailings solution (Column 7), plotted versus pore volumes of effluent.

pores and reduce the permeability of a soil. It will be explained in succeeding paragraphs how these two mechanisms are related.

Many minerals become less soluble as the pH of their aqueous environment is increased. This is explained more fully in Peterson et al [8], but, in general, precipitation of these minerals can occur as the pH rises. This assumes that the given dissolved constituents are sufficiently close to saturation with respect to specific solids at the lower pH values. Dissolution of some soil minerals, such as calcium carbonate, takes place concurrently with the precipitation reactions. Calcium carbonate dissolves when contacted by the acidic solutions, according to the following equation

$$CaCO_3 + 2H^+ \leftrightharpoons Ca^{2+} + H_2O + CO_2 \tag{1}$$

The calcium released into solution can result in the precipitation of gypsum because the tailings solutions tend to be in equilibrium with gypsum initially. Even though, for each mole of calcium carbonate ($CaCO_3$) dissolved, some fraction of a mole of gypsum will precipitate depending on the relative activities of calcium and sulfate; the increase in the pH of the solution will result in the precipitation of many other minerals. A more complete discussion of those minerals expected to precipitate and the stability of various minerals is given elsewhere [6-8].

As tailings solutions move through the clay liner, minerals precipitate, lowering the ionic strength of the solution and changing the relative solution percentage of each element. Using the criteria for dispersion established by Sherard et al [11] indicated that none of the three tailings solutions was initially dispersive. However, as precipitation or adsorption or both preferentially remove nonsodium salts from solution, the tailings solutions can, at some point in the column or soil profile, become dispersive. The nonsodium salts are preferentially removed because of their generally lower solubility and greater affinity for adsorption. The opportunity for a potentially dispersive condition to exist is especially high when the tailings solutions have a high initial sodium content as does the Lucky Mc tailings solution. Some of the effluents from columns leached with the Lucky Mc solution had sodium percentages >40% along with reduced total dissolved solids. The Lucky Mc tailings solution was the only tailings solution that became potentially dispersive as it passed through the column.

In soils containing appreciable amounts of smectite, McNeal et al [12] speculated that the predominant mechanism causing the permeability to decrease was clay swelling. Smectite has been identified in the Morton Ranch clay liner by X-ray diffraction [7]. Soils having larger quantities of 2:1 layer-expanding silicate minerals tend to be more sensitive to high sodium percentages than those containing illite, kaolinite, iron oxides, and amorphous material. The clay-sized fraction of the Morton Ranch clay liner also contains illite and kaolinite along with other minerals [7]. The Lucky Mc clay liner contains eight minerals [8], among which kaolinite, illite, and smectite are found. An inverse relationship was noted between clay content and bulk density and the observed permeability of the soil-liner materials. The precipitation of minerals such as amorphous ferric hydroxide and alunite should help impede any dispersive tendencies. Any dispersion occurring in the Lucky Mc columns involved a short-range migration of particles as no clay particles were visually observed in the effluents.

Neutralization

Effluent pH as a function of time and the number of effluent pore volumes is shown in Figs. 5 through 7. The columns that contained the Morton Ranch clay liner material leached with synthetic tailings solution (Columns 1, 2, and 3) showed disparate pH front breakthroughs (Fig. 5). We define the pH front breakthrough as the point where the effluent from a column drops below a pH value of roughly 6.5. The breakthrough for Columns 1, 2, and 3 ranged between 1 to 7 pore volumes. Columns 1 and 3 had pH values that did not change appreciably until more than three pore volumes had passed through the columns. For Column 2, the pH dropped rapidly after the first pore volume and was below pH 4 after two pore volumes of effluent had been collected. There was visual evidence of cracks and channels along the perimeter of this cell. This was the same cell that was recompacted after ~4 pore volumes had passed through the cell. Recompaction was accompanied by a decrease in the permeability by over one order of magnitude. After recompacting the clay, the pH decreased slowly and in a manner similar to the other columns that were packed with the clay material. After almost three years of

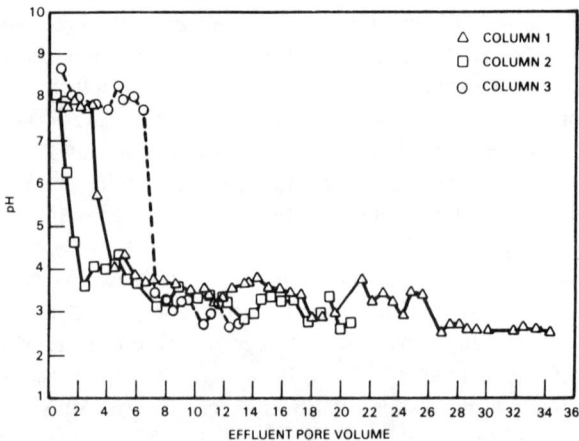

FIG. 5—*Plot of pH versus pore volume for the Morton Ranch clay liner contacted with the synthetic tailings (Columns 1, 2, and 3).*

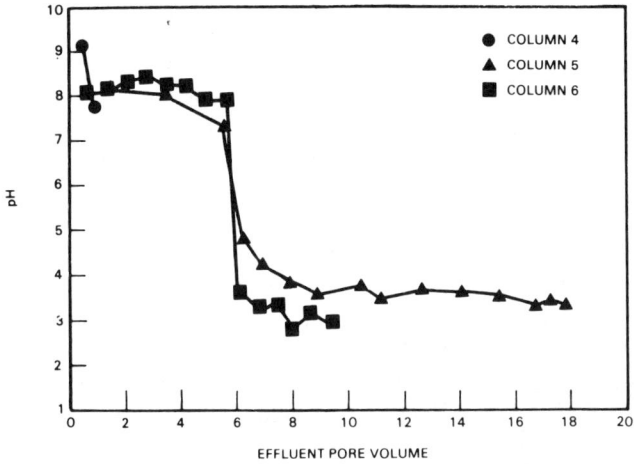

FIG. 6—*Plot of pH versus pore volume for the Morton Ranch clay liner contacted with the Highland Mill tailings solution (Columns 4, 5, and 6).*

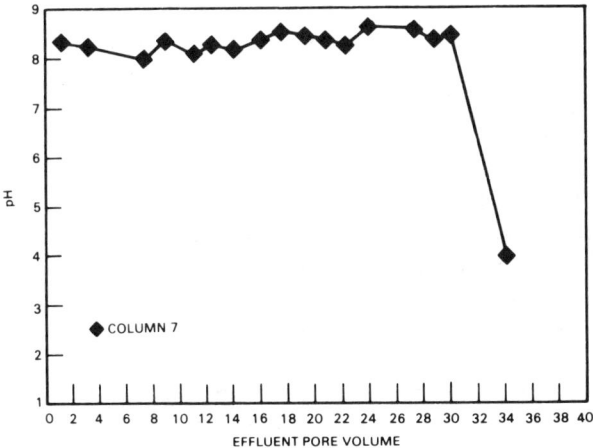

FIG. 7—*Plot of pH versus pore volume for the Lucky Mc clay liner material contacted with the Lucky Mc tailings solution (Column 7).*

leaching, the effluents from Columns 1, 2, and 3 still had not attained influent pH levels, though effluent pH had dropped below 3. This indicates a long-term residual buffering capacity in the liner material.

As indicated in Table 3, only 0.89 pore volumes passed through Column 4 before flow through the column ceased, so the pH of this effluent solution never dropped much below 8. Both Columns 5 and 6 achieved acid breakthrough at roughly six pore volumes. The pH of the effluent from these two columns slowly decreased, after the initial breakthrough, to values near 3 (Fig. 6).

The single column (Column 7) packed with the Lucky Mc clay liner and leached with acidic tailings solution from the Lucky Mc mill reached the pH front breakthrough after slightly over 30 pore volumes had passed through the column (Fig. 7). Analytical chemistry on the effluents of the permeability columns is given elsewhere [8].

Numerous factors (for example, carbonate and layered silicate dissolution and precipitation, aluminum and iron hydrolysis, oxidation of iron, uranium, vanadium, and manganese and the adsorptive capacity of the soil) can potentially affect the pH of the effluent from these column experiments. In many calculations involving the buffering capacity of the soil, it is assumed, as a first approximation, that this buffering capacity is due entirely to the calcium carbonate content of the soil [13,14]. If one makes this assumption and also assumes a neutralization of the influent solution to a pH around 7, then the neutralization process involving $CaCO_3$ can be represented by the following reaction

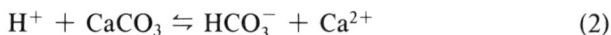

$$H^+ + CaCO_3 \leftrightharpoons HCO_3^- + Ca^{2+} \tag{2}$$

where one mole of $CaCO_3$ will neutralize one mole of hydrogen ions. We are assuming also that the total acidity of the solution comes from the measured activity of the hydrogen ion, which may not be the case, especially in uranium mill tailings solutions, where a large part of the total acidity may arise from species such as iron and aluminum.

Based on the data from Tables 1 and 3, solution data in Table 2 (pH), and Eq 2, it should take 14.2 pore volumes to overcome the calcium carbonate buffering capacity of the Morton Ranch clay liner that was used in Columns 1 through 3. The experimental data show that roughly three pore volumes percolated through Columns 1 and 3 before the pH front breakthrough took place. Columns 5 and 6 did not experience pH breakthrough until about six pore volumes had passed through the columns (compared to the predicted value, based on the just stated assumptions, of 9.3). The column packed with the Lucky Mc clay liner (Column 7) had a predicted pH breakthrough, based on calcium carbonate content, of approximately 60 pore volumes. The actual pH front breakthrough occurred at about 30 pore volumes.

All of the clay liner materials tested exhibited pH front breakthroughs that occurred before the pH front breakthrough calculated by considering only the buffering capacity of the calcium carbonate and the acidity based upon hydrogen ion activity. This earlier-than-calculated breakthrough could partly be due to the release of hydrogen ions through the precipitation of aluminum and iron hydrous oxides as the pH rises.

Several salient points arise when examining the buffering capacity data. First, there appears to be an inherent long-term residual buffering capacity in existence in the Morton Ranch clay liner. The experiments with the Lucky Mc clay liner were not continued long enough to determine if there was a residual buffering capacity present. More research needs to be done to define this mechanism exactly, but there are several possible explanations for the occurrence of the residual buffering capacity. This residual buffering capacity that keeps the effluent pH values from reaching the influent values may be due to layered silicate mineral dissolution that forms silicic acid. Also, in arid environments, relatively insoluble iron oxides sometimes form as coatings on soil minerals. Part of the residual buffering capacity could be due to the formation of these relatively insoluble iron oxides on the surface of calcium carbonate particles and the slow dissolution of these surface coatings as acidic solution flows past them. As these coatings slowly dissolve, more calcium carbonate is gradually made available for buffering of the soil acidity. Another possible mechanism for this residual buffering capacity is the dissolution of minerals that had previously precipitated at higher pH values. For example, the aluminum and iron in the acidic tailings solutions hydrolyze and precipitate as amorphous ferric hydroxide [$Fe(OH)_3$ A (A = amorphous)] and amorphous aluminum hydroxide [$Al(OH)_3$ A], releasing hydrogen ions into solution according to the following equations [15]

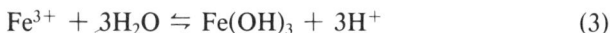

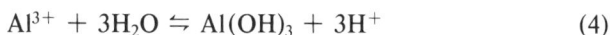

$$Fe^{3+} + 3H_2O \leftrightarrows Fe(OH)_3 + 3H^+ \tag{3}$$

$$Al^{3+} + 3H_2O \leftrightarrows Al(OH)_3 + 3H^+ \tag{4}$$

As the pH drops to lower values and the calcium carbonate dissolves and disappears from the soil, the iron and aluminum hydroxides previously precipitated from solution according to Eqs 3 and 4 become more soluble and begin to dissolve. These dissolutions proceed according to the reverse of Eqs 3 and 4 (right to left) and, in the process, consume hydrogen ions and buffer the liner material against further decreases in pH.

The most dramatic rise in the effluent concentrations of iron and aluminum occurred between the pH values of 4 and 5 [8,15]. Effluent iron and aluminum concentrations increased in the same pH range at which the residual buffering capacity was operable. The marked increase in effluent iron and aluminum concentrations around pH 4 is a partial confirmation of the hypothesis that two of the reactions responsible for the occurrence of the residual buffering capacity are the dissolution of amorphous iron and aluminum hydroxides according to the reverse of Eqs 3 and 4.

Relyea and Martin [16] recently demonstrated that no residual buffering capacity is evidenced when hydrochloric acid was used to titrate geologic materials as opposed to a prevalent residual buffering capacity (below pH 4.5) when acidic tailings leachates containing high concentrations of aluminum were used. Thus, acidic tailings solutions have the capacity to precipitate iron and aluminum and release hydrogen ions into solution as the pH rises, and then may buffer the soil against further decreases in pH by dissolving (and consuming hydrogen ions) once the pH drops to around 4. The release of hydrogen ions into solution, as the pH rises, could help explain why the experimental pH front breakthroughs occurred before the pH breakthrough calculated by considering only the calcium carbonate dissolution reaction.

The second point, which can be gleaned from examining the buffering capacity data, is that the assumption that the total acidity of the tailings solution corresponds to the measured pH and the total buffering capacity of the soil to the calcium carbonate content, according to Eq 2, is insufficient to explain the observed pH front breakthroughs. The best estimator of pH breakthroughs for these columns came from considering the total acidity to be due to the total of the measured pH, aluminum, and iron and the buffering capacity to be due to the calcium carbonate content of the soil plus the soils cation exchange capacity. Using these parameters to calculate a total sediment buffering capacity and total solution acidity, the predictions of the pH front breakthroughs for Columns 1 through 6 matched closely with observed breakthroughs, but were off considerably (predictions too low) for Column 7.

Field Predictions

For a 1-m-thick clay liner material, with a permeability of 10^{-8} cm/s, and assuming a constant pressure head gradient of 10, we calculate that <2 pore volumes of solution would flow through the clay liner in 20 years. This size of head gradient might exist in an evaporation pond that was constantly filled to a depth of 9 m, but would be conservatively high, by a factor of 5 to 10, for most tailings disposal alternatives [17]. Even with the low $CaCO_3$ content (0.3%) of the Morton Ranch clay liner, it would take roughly 65 years, using the assumed values for hydraulic head and permeability, for the acid tailings liquor to overcome the buffering capacity of the liner and for the low pH front to move through the clay liner and into the surrounding geologic media. Using the same hydraulic head and permeability in conjunction with the Lucky Mc clay liner (4.0% $CaCO_3$), we compute a time of ~330 years for the pH front to advance through a 1-m-thick clay liner. These breakthrough times are based on the actual column breakthrough data and not on calculations involving the calcium carbonate content of the soil.

Summary and Conclusions

Liner failure, defined as an increase in permeability, has not been found to be a problem when acidic uranium tailings solutions contacted two different clay liner materials over a period of almost three years. On the contrary, all liner material samples exposed to the harsh environment, produced by contact with acidic tailings solutions, showed a decrease in permeability with time. Reactions that occur, such as chemical precipitation, particle dispersion and subsequent pore plugging, tend to decrease the permeability and thus increase the overall stability of the liner materials.

These tests demonstrate that the materials tested containing >30% clay are able to maintain their permeabilities at the initial values or lower for extended periods when contacted with acidic uranium tailings solutions. The tests also demonstrate that soil compaction procedures, which optimize density and minimize channel cracks and fracture flow, can play an important role in maintaining acceptable permeabilities. The time dependence of the permeability changes was not predictable from the tests run to date.

Data suggest that clay liners can provide a barrier to contaminant migration, not only through their low permeabilities but also by their buffering capacities. The best predictions of pH front breakthroughs were obtained by considering the total acidity due to the total measured pH and aluminum and iron concentrations, and the buffering capacity due to the soil's calcium carbonate content plus cation exchange capacity. The Morton Ranch clay liner exhibited a residual buffering capacity below a pH of around 4, which was able to maintain the effluent pH values at a higher level than the influent pH values for the duration of the test period (≈ 3 years). The redissolution of iron and aluminum amorphous hydrous oxides is considered to be a dominant mechanism accounting for the existence of the residual buffering capacity.

Acknowledgments

This study was funded by the U.S. Nuclear Regulatory Commission as part of the Uranium Recovery Research Program at Pacific Northwest Laboratory. We wish to thank Brian Opitz, Doug Sherwood, Wayne Martin, Michael Dodson, Marvin Mason, Joel Tingey, Ann Campbell, and Cathy Begej for designing and setting up the permeability columns and for performing the chemical analyses. Special thanks is given to Jeff Serne and John Fruchter for their helpful reviews and suggestions.

References

[1] Crim, R. G., Shepherd, T. A., and Nelson, J. D. in *Proceedings*, Second Symposium on Uranium Mill Tailings Management, Civil Engineering Dept., Colorado State University, Fort Collins, CO, 1979, pp. 41–53.

[2] Gee, G. W., Campbell, A. C., Sherwood, D. R., Strickert, R. G., and Phillips, S. J., "Interaction of Uranium Mill Tailings Leachate with Soils and Clay Liners," NUREG/CR-1494, National Technical Information Service, Springfield, VA, 1980.

[3] Gee, G. W., Campbell, A. C., Opitz, B. E., and Sherwood, D. R. in *Proceedings*, Third Symposium on Uranium Mill Tailings Management, Civil Engineering Dept., Colorado State University, Fort Collins, CO, 1980, pp. 333–352.

[4] Black, C. A., *Methods of Soil Analysis. Part 1, Physical and Mineralogical Properties Including Statistics of Measurement and Sampling*, Monograph 9, American Society of Agronomy, 1965.

[5] Black, C. A., *Methods of Soil Analysis.* Part 2, *Chemical and Microbiological Properties*, Monograph 9, American Society of Agronomy, 1965.

[6] Erikson, R. L., and Sherwood, D. R., "Interaction of Acidic Leachate with Soil Minerals at Lucky Mc Pathfinder Mill, Gas Hills, Wyoming," in *Proceedings*, Fifth Symposium on Uranium Mill Tailings Management, Civil Engineering Dept., Colorado State University, Fort Collins, CO, 1982.

[7] Uziemblo, N. H., Erikson, R. L., and Gee, G. W., "Contact of Clay Liner Materials with Acidic Tailings Solutions, I. Mineral Characterization," in *Proceedings*, Fourth Symposium on Uranium Mill Tailings Management, Civil Engineering Dept., Colorado State University, Fort Collins, CO, 1981.

[8] Peterson, S. R., Erikson, R. L., and Gee, G. W., "The Long-Term Stability of Earthen Materials in Contact with Acidic Tailings Solutions," NUREG/CR-2946, PNL-4463, Pacific Northwest Laboratory, Richland, WA, 1982.

[9] U.S. Environmental Protection Agency, "Hazardous Waste, Proposed Guidelines and Regulations and Proposal on Identification and Listing," *Federal Register*, Vol. 43, No. 243, 1978, pp. 58946–59026.

[10] Pupisky, H., and Shainberg, I., *Soil Science Society of America Journal*, Vol. 43, 1979, pp. 429–433.

[11] Sherard, J. L., Dunningan, L. P., and Decker, R. S., *Journal of the Geotechnic Engineering Division*, Vol. 102, 1976, pp. 287–301.

[12] McNeal, B. L., Norvell, W. A., and Coleman, N. T., *Soil Science Society of America, Proceedings*, Vol. 30, 1966, pp. 313–317.

[13] Shepherd, T. A., and Cherry, J. A., "Contaminant Migration in Seepage from Uranium Tailings Impoundments: An Overview," in *Proceedings*, Third Symposium on Uranium Mill Tailings Management, Civil Engineering Department, Colorado State University, Fort Collins, CO, 1980.

[14] Cherry, J. A., Shepherd, T. A., and Morin, K. A., "Chemical Composition and Geochemical Behavior of Contaminated Groundwater at Uranium Tailings Impoundments," presented at the SME-AIME annual meeting, Dallas, TX, 14–18 Feb. 1982.

[15] Peterson, S. R., Felmy, A. R., Serne, R. J., and Gee, G. W., "Predictive Geochemical Modeling of Interactions Between Uranium Mill Tailings Solutions and Sediments In a Flow-Through System," NUREG/CR-3404, National Technical Information Service, Springfield, VA, 1983.

[16] Relyea, J. F., and Martin, W. J., "Evaluation of Inactive Uranium Mill Tailings Sites for Liner Requirements: Characterization and Interaction of Tailings, Soil and Liner Materials," in *Proceedings*, Fifth Symposium on Uranium Mill Tailings Management, Civil Engineering Department, Colorado State University, Fort Collins, CO, 1982.

[17] Nelson, R. W., Reisenauer, A. E., and Gee, G. W., "Model Assessment of Alternatives for Reducing Seepage of Contaminants from Buried Uranium Mill Tailings at the Morton Ranch Site in Central Wyoming," PNL-3378, Pacific Northwest Laboratory, Richland, WA, 1980.

DISCUSSION

W. D. Hammond[1] (written discussion)—For acid leachates, would there be any value in mixing basic material like lime into the clay liner?

S. R. Peterson and G. W. Gee (authors' closure)—Yes. Mixing $Ca(OH)_2$

[1]California Department of Water Resources, Sacramento, CA 95802.

(lime) directly into clay liner material has some distinct advantages over neutralizing an entire waste pile. Cost would be one of the major advantages. When the liner material is low in carbonate, the addition of lime would greatly assist in the neutralization of acid leachate. The resultant seepage water would tend to have low radionuclide (radium, thorium, uranium) and heavy metal (iron, aluminum, arsenic, chromium, vanadium) concentrations because of the elevated pH. Recently, Optiz and Sherwood (1984)[2] demonstrated that in addition to improvement in water quality, there could also be a significant permeability modification because of lime additions to liner materials (laboratory tests showed liner permeability decreased by two orders of magnitude after lime additions). Field tests of lime-treated clay liners for acid leachate control are needed.

Y. Acar (additional closure)—Sodium carbonate is very effective in preventing or reducing the attack of acids on bentonite. Lime would have to be added beforehand, otherwise it will reduce the swelling of the bentonite.

B. S. Beattie[3] (written discussion)—Why were not commercially available oil well grade, American Petroleum Institute (API) 13*a* specification, sodium, bentonite clays utilized in your test?

S. R. Peterson and G. W. Gee (authors' closure)—The clays used were taken from sources local to the two uranium mill sites described in our report. For site operators, major saving in material costs can be realized if the clays are not imported. The local clays, however, must be carefully evaluated to ensure that they have acceptable chemical and physical properties for use as engineered liners. Our laboratory results suggest that for acid tailings solutions, local clays can be used successfully as liner materials.

---

[2]Optiz, B. E. and Sherwood, D. R., "Neutralizing Barrier for Reducing Contaminant Migration from a Uranium Mill Tailings Disposal Pond," in *Management of Uranium Mill Tailings, Low Level and Hazardous Waste Symposium*, Colorado State University, Fort Collins, CO, 1984.

[3]Federal Bentonite, Montgomery, IL 60538.

John A. Mundell[1] *and Bruce Bailey*[2]

The Design and Testing of a Compacted Clay Barrier Layer to Limit Percolation Through Landfill Covers

REFERENCE: Mundell, J. A. and Bailey, B., **"The Design and Testing of a Compacted Clay Barrier Layer to Limit Percolation Through Landfill Covers,"** *Hydraulic Barriers in Soil and Rock, ASTM STP 874*, A. I. Johnson, R. K. Frobel, N. J. Cavalli, and C. B. Pettersson, Eds., American Society for Testing and Materials, Philadelphia, 1985, pp. 246–262.

ABSTRACT: The design and testing of a compacted clay barrier layer to restrict vertical percolation through landfill covers is discussed. General relationships between compaction water content, dry unit weight, and permeability related to changes in soil fabric due to varying compaction conditions are reviewed. Laboratory testing programs to evaluate the degree of imperviousness capable of being achieved in the field for a given soil type are outlined, and a case study of the design and testing of a compacted clay barrier over a landfill is presented. Based on the results of the laboratory testing program prior to construction, it was determined that a design permeability of from 1 to 5×10^{-8} cm/s could be achieved by controlling the minimum dry unit weight to greater than 95% of the standard Proctor dry density and the compaction water content to greater than 1% wet of the line of optimums. Results of laboratory permeability testing on undisturbed ring and block samples taken from the landfill barrier layer indicated that an average permeability of 2×10^{-8} cm/s had been achieved.

KEY WORDS: permeability, seepage, landfill, leachate generation, compacted clay, compaction water content, dry unit weight, compactive effort, line of optimums, laboratory testing, design methods

The incorporation of a properly designed final cover layer system over a completed landfill (Fig. 1) is the most effective method for limiting the amount of moisture percolation through waste materials and provides the first line of defense against the generation of significant quantities of leachate

[1]Environmental research associate, Department of Civil Engineering, University of Notre Dame, Notre Dame, IN 46556.
[2]Vice president—Engineering, ATEC Associates, Dallas TX 75229.

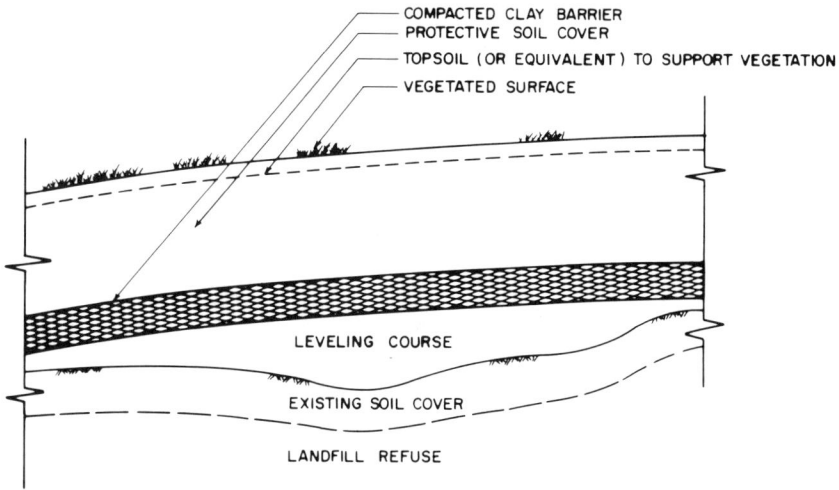

FIG. 1—*Typical landfill cover section.*

from a landfill [1–5]. The primary inhibitor with respect to the reduction of percolation through a soil cover is the compacted clay barrier layer. The overlying protective cover will serve to shield this barrier from the effects of weathering and erosion and will support a healthy vegetated cover. The leveling course beneath the barrier will serve to provide a graded work surface firm enough to permit satisfactory compaction of the barrier layer.

The amount of infiltration into the protective cover layer and any subsequent percolation down through waste materials will be controlled by the characteristics of the surface and the protective cover and by the climatological conditions of the site location. The type and thickness of the cover soil influence percolation by the soil's capacity to store water. The amount of available percolation through a cover is determined by performing water balance calculations [3,6,7] that involve assessing the seasonal variations in precipitation, surface runoff, infiltration, and evapotranspiration and their effect on soil moisture recharge and utilization. During those times of the year when water from precipitation is available in excess of the protective soil cover field moisture capacity after evapotranspiration and surface runoff losses, percolation through the compacted clay barrier can occur.

The seepage model shown in Fig. 2 can then be used to calculate vertical flow through the clay barrier. The total head loss, h, is taken equal to the barrier thickness, L, plus the field capacity of the protective cover (determined from the soil type) divided by its porosity. As illustrated, the seepage rate, q, is directly proportional to the permeability of the clay barrier

$$q = k \frac{h}{L} \tag{1}$$

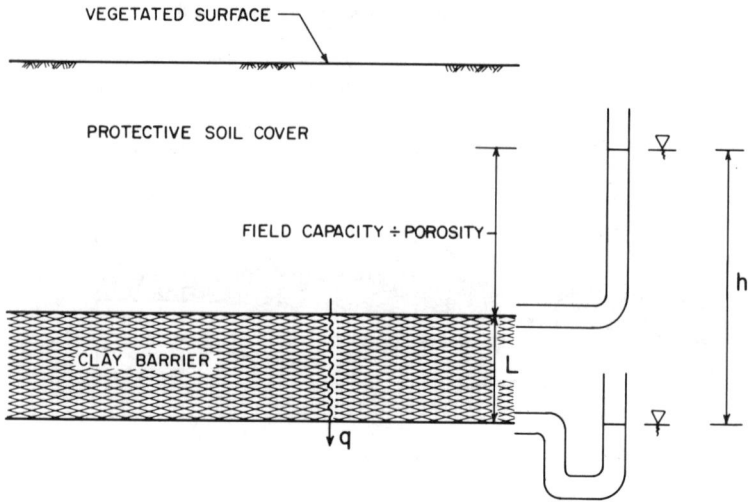

BARRIER SEEPAGE RATE: $q = k \dfrac{h}{L}$ PER UNIT AREA

FIG. 2—*Seepage model through cover.*

The total percolation quantity per year can then be calculated by multiplying the seepage rate by both the landfill area and the annual duration of flow (as determined by the water balance).

To regulate the potential amount of leachate which may be generated due to downward percolation, a design permeability of the compacted clay barrier can be selected which would limit vertical flow through the cover system. The estimated leachate quantities generated could then be used either in the design of a leachate collection system or for comparison with the attenuation capabilities of the underlying or adjacent soils or a combination of the two.

Permeability of Compacted Clay

General relationships between permeability and compacted characteristics (for example, compaction water content, dry unit weight, compactive effort) for fine-grained cohesive soils have appeared in the literature [8–11], and the general trends are shown in Fig. 3. It may be observed from the figure that at each compactive effort, the permeability values decrease dramatically at compaction water contents wet of the optimum water content. In addition, the higher the compactive effort, the lower the permeability value for a given compaction water content. The permeability of soil specimens compacted dry of optimum is generally 10 to 1000 times larger than the permeability of specimens compacted wet of optimum.

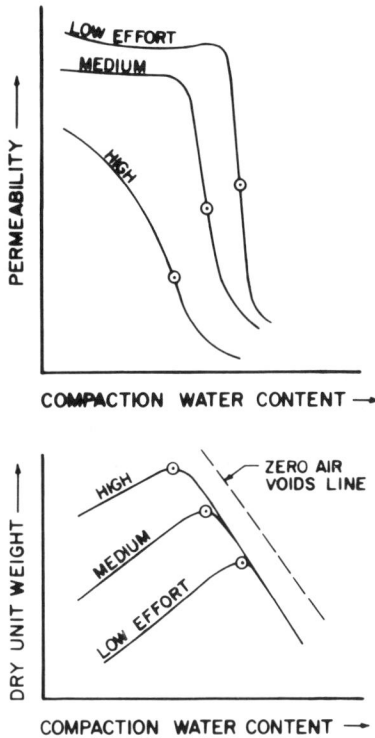

FIG. 3—*Permeability and compaction relationships.*

This behavior has been attributed to changes in the soil fabric and geometry of the porous network at varying water contents and dry unit weights [9,10,13–15]. Lambe [12] postulated that individual clay particles exist in a flocculated state dry of optimum and exhibit a more dispersed structure wet of optimum and that this difference controls the behavior of the soil mass. More recent investigations [9,13,14] have explained the behavior and characteristics of compacted clay using a deformable aggregate soil model in which the individual soil particles group together in agglomerations, and flow is controlled primarily by the size, frequency, and orientation of large interaggregate pores and to a lesser degree by a network of small intraaggregate pores. As the compaction water content increases, the aggregates decrease in strength and easily undergo large deformation during compaction, resulting in a decrease in the size of the interaggregate pore space and, in turn, a decrease in permeability.

Design and Testing Concepts

Several of the problems associated with predicting the field permeability of a compacted clay liner using laboratory prepared samples and permeability tests have been summarized by Daniel [*16*] and are again presented in Table 1. In general, for compacted clay barrier layers over landfills, the two sources of error which appear to be the most common and which yield the greatest magnitude of error on the unconservative side are the first two listed in the table.

If the field compaction water content is lower than that used in the laboratory testing program for the same compactive effort, the field permeability will be much larger than estimated in the laboratory. In addition, if the compactive effort in the field is less than that used in the laboratory, the field permeability will be much higher than expected due to the shift in the optimum water content for the lesser degree of compaction, as shown in Fig. 4.

The effect of landfill settlements also has to be considered in the design of cover systems. Large differential movements, characteristic of landfill surfaces, can contribute to significant cracking of otherwise low-permeability clayey barrier layers. Leonards and Narain [*17*] performed tests which indicated that clayey soils compacted above optimum water content were significantly more flexible and thus more resistant to cracking than if compacted dry of optimum.

Based on the preceding discussion, it is possible to develop a laboratory testing program which is capable of determining the degree of impermeability

TABLE 1—*Sources of error in estimating field permeability of compacted clay liners from laboratory tests.*[a]

Potential Source of Error	Possible Number of Orders of Magnitude of Error	Laboratory Permeability Compared to Actual Field Permeability
1. Compaction at a higher water content in laboratory than in field	1 to 3	low
2. Use of more compactive effort in the laboratory than in the field, resulting in optimum water content higher in field than in laboratory	1 to 3	low
3. Deleterious substances present in the field but not in laboratory samples	1 to 3	low
4. Maximum size of clay chunks smaller in laboratory than in field	1 to 2	low
5. Use of static (impact) compaction rather than kneading compaction to prepare laboratory specimens	0 to 1	high
6. Air in laboratory samples	0 to 1	low
7. Use of excessive hydraulic gradients causing particle migration	0 to 1	low
8. Steady-state seepage not attained	0 to 1	high
9. Sample size too small in laboratory test	0 to 3	low
10. Dessication cracks in field	no data	low

[a]After Daniel [*16*].

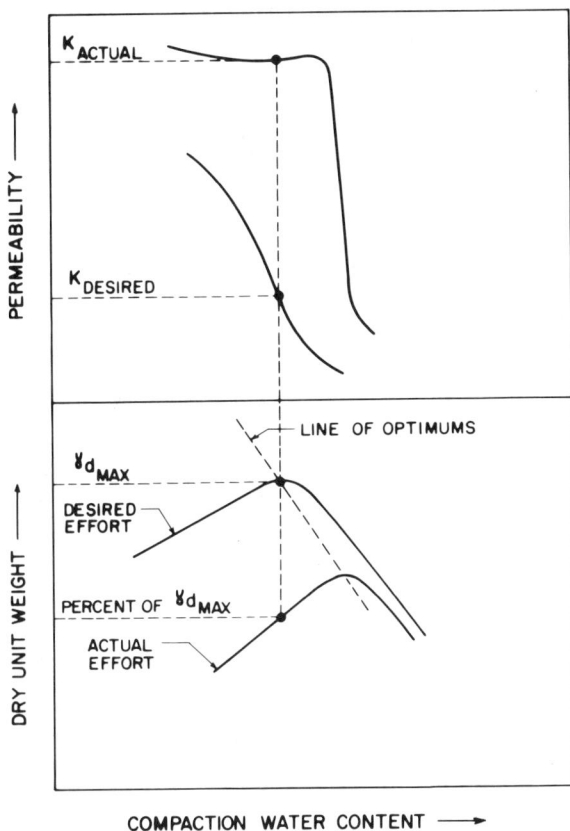

FIG. 4—*Error in estimation of field permeability (after Daniel [16]).*

that can reasonably be achieved in the field for a given fine-grained clayey soil and the range of field conditions (compactive effort and field compaction water content) to achieve a desired design permeability. The initial portion of the testing program includes a complete series of classification tests (for example, natural water content and dry unit weight, Atterberg limits, specific gravity, grain-size analysis) on the range of expected soil types from the proposed borrow area which may be used in the compacted clay barrier layer.

Figures 5 and 6 illustrate a procedure that may be used to define the range of lower permeability values that can exist for a given soil type. As shown, three compactive efforts (high, medium, low) are used to define the possible range of compactive efforts that may be used or achieved in the field. The highest effort curve (C) generally corresponds to the standard Proctor effort [ASTM Test Methods for Moisture-Density Relations of Soils and Soil-Aggregate Mixtures, Using 5.5-lb (2.49-kg) Rammer and 12-in. (304.8-mm)

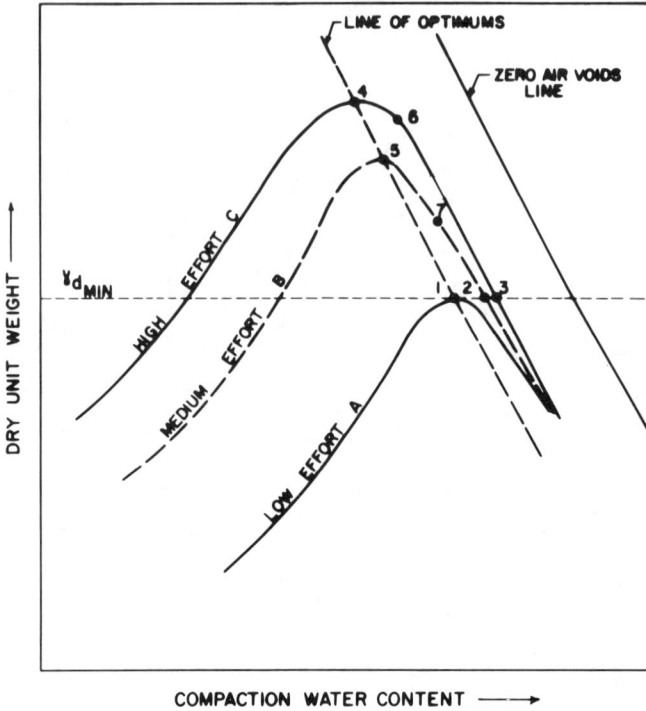

FIG. 5—*Laboratory testing to evaluate field permeability.*

Drop (D 698-78)]. The lowest effort curve (*A*) corresponds to the effort necessary for the maximum dry unit weight to equal the expected minimum dry unit weight requirement specified during the field construction. This may vary between 90 to 95% of the standard Proctor value, depending on the degree of compaction necessary to achieve the design permeability value. Points labeled 1 through 7 correspond to the water content, dry unit weight, and compactive effort conditions considered for the preparation of seven specimens for permeability testing. At Point 1, the minimum dry unit weight is achieved at the lowest selected compactive effort possible with the water content at optimum. The permeability of the specimen prepared at this point will be the greatest of all seven points. Conversely, a specimen at Point 3 compacted at the greatest water content using the highest effort will generally yield the lowest permeability of the seven points.

After the permeability tests have been performed, the resulting values are plotted on a graph of percent wet of compactive effort optimum versus dry unit weight as shown in Fig. 6. From these seven points, the design permeability line may be determined by interpolating between the permeability test results K_1 through K_7. For conditions to the right of the line, the as-

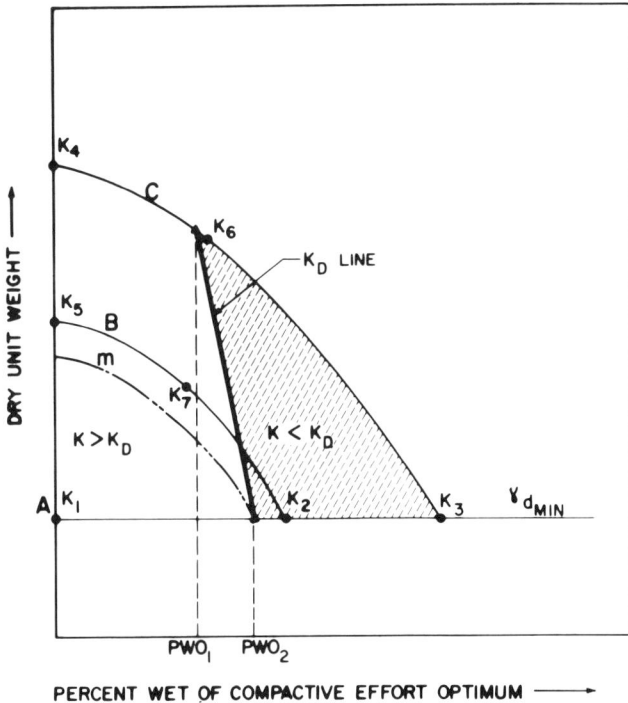

FIG. 6—*Determination of design permeability line.*

compacted permeability of the barrier will be less than the selected design permeability. To the left of the line, the compaction conditions will yield higher (and therefore unsuitable) permeability values. The two end points of the design permeability line, Control Points PWO_1 and PWO_2 (percent wet of compactive effort optimum), provide the control points which define the compaction conditions (shaded area in Fig. 7) necessary to achieve a permeability value less than the selected design permeability.

To illustrate this procedure, test data from Mitchell [11] have been plotted in Fig. 8 showing permeability values obtained for three different compactive efforts, curves A, B, and C, at compaction water contents above the optimum value for a silty clay. These points numerically illustrate how the permeability decreases with increased compactive effort and with increased compaction water content. Curves of equal permeability values (dashed lines) are approximately parallel for this soil.

It may be observed from Fig. 7 that a design permeability may be achieved by specifying a minimum dry unit weight with compaction water contents a given percentage wet of the line of optimums as defined by Control Point PWO_2. The construction criterion then becomes a vertical line through Con-

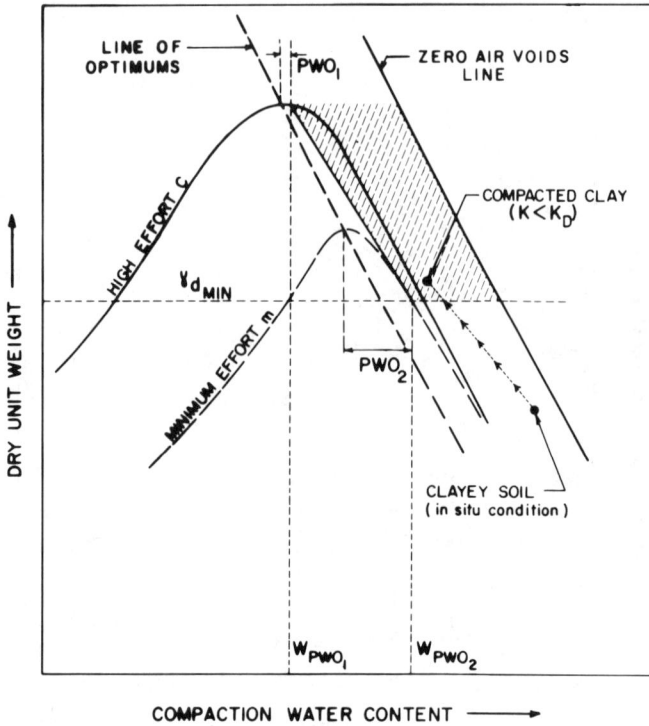

FIG. 7.—*Control area to achieve design permeability.*

trol Point PWO_2 on Fig. 6 rather than the K_D line. If variations in soil types or soil plasticity are indicated from the borrow area testing, this procedure would be repeated for each soil type such that a water content/dry unit weight control area is developed for each soil type as shown in Fig. 9. The control area for a given soil type could be minimally estimated by running at least three permeability tests corresponding to Points K_1, K_2, and K_3 in Fig. 6.

Field Control

To adequately control the construction of a compacted clay barrier to achieve permeability values less than the selected design permeability, a continuous evaluation of the soil used for the barrier layer is necessary as well as careful monitoring of the compacted moisture and density conditions. Subsequent to the laboratory testing program, standard Proctor curves generated for the various soil types evaluated from the proposed borrow area should be available and the compaction criteria necessary to control the permeability of the compacted clay barrier established (see shaded area in Fig. 7). As soil is

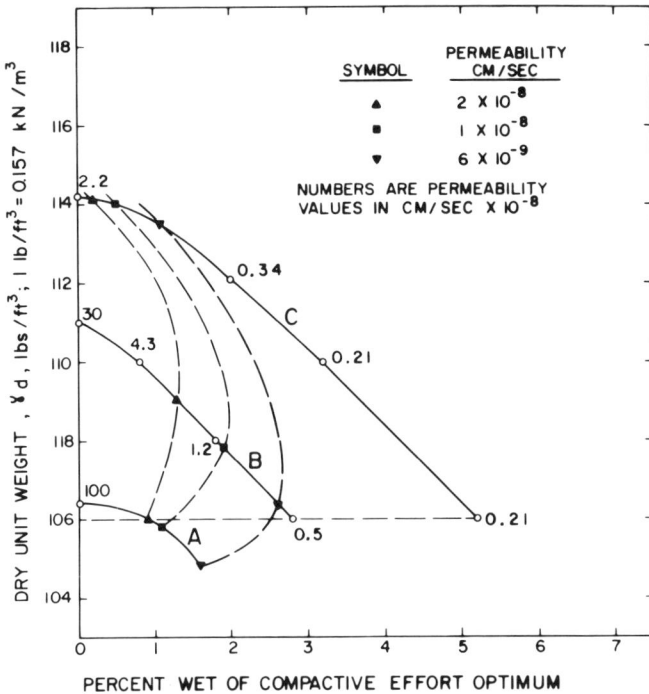

FIG. 8—*Design permeability line concept example (from Mitchell et al [11]).*

brought to the landfill site from the borrow area, the soil is visually classified and compared to the soils previously tested. When differences in soil types are observed, one-point standard Proctor tests and Atterberg limits are run to determine the variability of the soils. As shown in Fig. 9, if an untested soil type falls between two of the soil types tested, the line of optimums for that particular soil type can be estimated by interpolating between the two known soils. A shaded control area (in which the permeability will be less than the design permeability) can then be estimated for that soil type based on the results of the laboratory testing program. It should be emphasized that the water content at which a one-point field compaction test is run should be at or slightly dry of optimum so as to yield a distinct interpretation of maximum dry unit weight and optimum water content for the soil. The wet of optimum family of curves are generally so close together that there is much greater un-certainty in the interpretation in this area.

During placement and compaction of the barrier layer, periodic field den-sity tests (on a random grid pattern) using the sand cone and nuclear mois-ture/density tests can be performed to continuously monitor and maintain control of the compaction process.

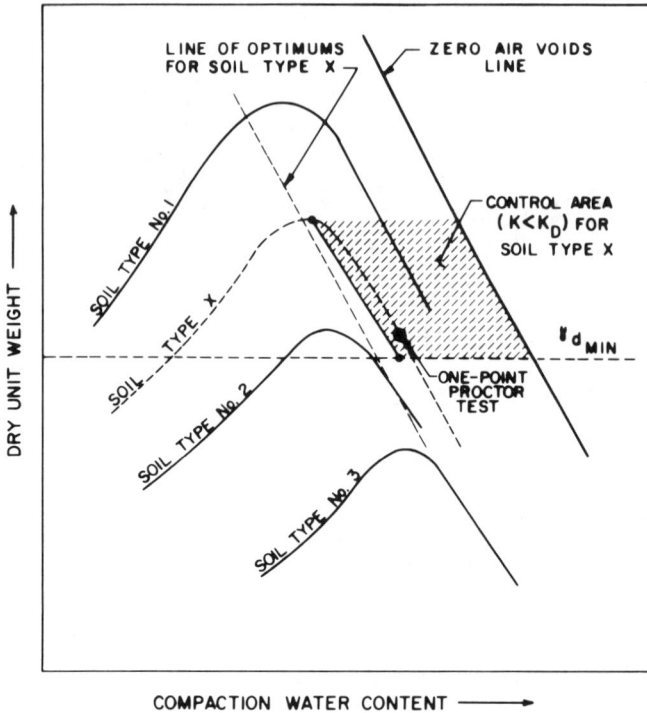

FIG. 9—*Field control area for intermediate soil type.*

Case Study

The closure plan for a waste disposal facility in the midwestern United States included the placement of an engineered final soil cover barrier system over the landfilled area in order to limit vertical percolation and subsequently reduce the potential amount of leachate generated. The landfilled wastes were generally of limited thickness, and the site had been inactive and covered with silty soils for several years. Thus, excessive postconstruction settlements were not expected. Although regulations called for a permeability of 1×10^{-7} cm/s or less for the cover system, it was determined from water balance calculations that an effective barrier design permeability on the order of 2 to 5×10^{-8} cm/s would limit percolation to an acceptable level.

A laboratory testing program was directed toward establishing the character and variability of soils encountered within a nearby borrow area selected as the source for landfill cover materials. A summary of the results of classification tests performed on the range of soils anticipated for use in the compacted clay barrier is shown in Table 2.

Based on these test results, two soil types were selected for extensive perme-

TABLE 2—*Results of general classification testing on borrow soils.*

Parameter	Average	Range
Natural moisture content, %	25.5	17.7 to 29.2
Natural dry unit weight, lb/ft³[a]	98.2	93.8 to 104.6
Atterberg limits		
Liquid limit	38.6	32 to 45
Plastic limit	18.3	17 to 20
Plasticity index	20.3	12 to 28
Clay content, %[b]	23.8	17.5 to 28.5
Specific gravity	2.71	2.68 to 2.75
Unified soil classification	CL[c]	CL

[a]1 lb/ft³ = 0.157 kN/m³.
[b]Less than 0.002 mm.
[c]CL = Inorganic clays of low to medium plasticity.

ability testing: Soil 1, a gray to brown, mottled silty clay judged to be of average plasticity for the cohesive soils in the borrow area and Soil 2, a light brown, mottled, lower plasticity silty clay which was expected to provide an indication of some of the more marginal fine-grained borrow soils to be used in the barrier cover layer. Classification tests for these two soils are shown in Table 3.

As shown in Fig. 10, a series of compaction tests at three different compactive efforts were performed on Soil 1 to establish guidelines for preparing samples for permeability testing and to determine the location of the line of optimums. Samples were compacted in three layers into a standard 101.6-mm (4-in.)-diameter mold with the number of blows per layer varying from 10 (low compactive effect) to 25 (standard Proctor, ASTM D 698). The line of optimums was observed to be very nearly parallel to the zero air voids line, as expected.

Based on the results of these tests, two specimens were prepared at water contents of 19.0 and 20.3%, respectively, and compacted using 14 blows per layer in order to achieve a dry unit weight above 95% of the standard Proctor

TABLE 3—*Results of general classification testing on soil 1 and soil 2.*

Soil No.	Maximum Dry Unit Weight, lb/ft³[a]	Optimum Water Content, %	Atterberg Limits[c]			Particle Size Gradation, %			Specific Gravity
			LL	PL	PI	Sand	Silt	Clay[b]	
1	110.9	16.0	39	18	21	2.4	73.1	24.5	2.69
2	112.3	15.1	32	20	12	3.3	79.2	17.5	2.68

[a]1 lb/ft³ = 0.157 kN/m³.
[b]Less than 0.002 mm.
[c]LL = liquid limit; PL = plastic limit; PI = plasticity index.

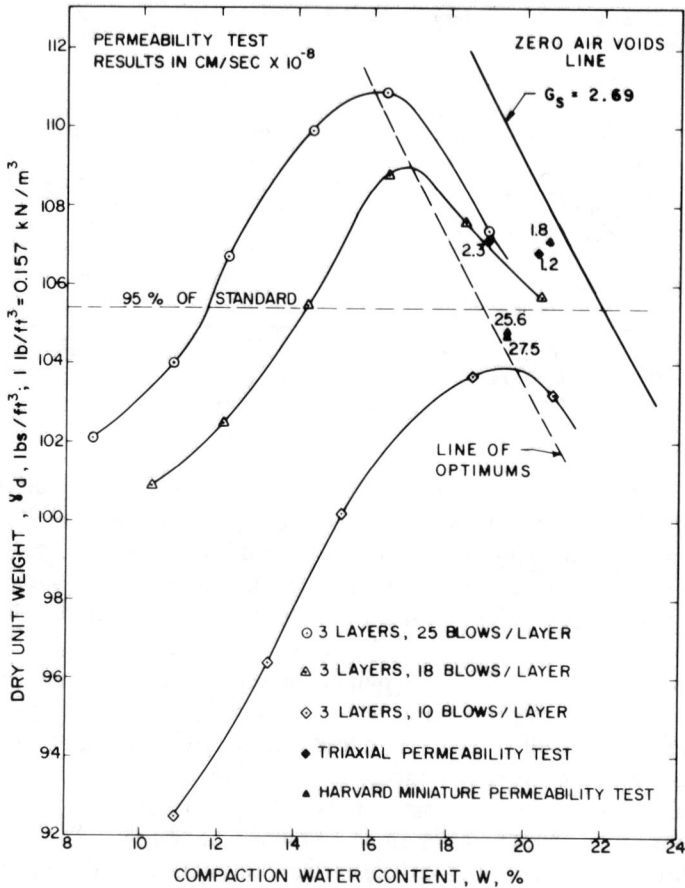

FIG. 10—*Laboratory test results—Soil 1.*

maximum and wet of the line of optimums. These specimens were trimmed from the mold, placed in a triaxial cell, and back-pressure saturated with a confining pressure of 345 to 552 kPa (50 to 80 psi). A constant-head permeability test was then performed with a pressure differential of 20.7 kPa (3 psi) maintained across the specimens to encourage flow at a measurable rate. Another specimen prepared in a similar manner using 12 blows per layer at a water content of 20.6% was trimmed into a Harvard miniature permeameter and a constant-head permeability test was performed. Two additional specimens were compacted directly into Harvard miniature molds (five layers, five blows per layer) at dry unit weights slightly below 95% standard and near the line of optimums.

The results plotted in Fig. 10 indicate that the permeability values ranged from 1.2 to 2.3 × 10⁻⁸ cm/s for specimens with water contents at least 1%

wet of the line of optimums and greater than 95% of standard, and from 25.6 to 27.5 $\times$ 10^{-8} cm/s for specimens with water contents less than 1% wet of the line of optimums and dry unit weights slightly less than 95% of standard.

Similarly, for Soil 2 (Fig. 11), four specimens compacted to 94.0 to 96.5% of standard Proctor with water contents at least 1% wet of the line of optimums yielded permeability value between 3.1 to 4.4 $\times$ 10^{-8} cm/s for permeability tests performed in both the triaxial cell and the Harvard miniature permeameter.

From the results of the limited laboratory testing program, it was decided to compact the natural silty clay borrow soils to at least 95% of the standard Proctor maximum dry unit weight at water contents at least 1% wet of the line of optimums to achieve a coefficient of permeability in the range of 1 to 5 $\times$ 10^{-8} cm/s.

During the construction of the compacted clay barrier layer, field quality control testing and inspection was performed on a continuous basis. Daily activities included determination of the suitability of the borrow soils (by visual classification and Atterberg limits tests), nuclear moisture/density tests to establish the degree of compaction, check field density tests by the sand cone method, and one-point Proctor tests (see Fig. 12) to maintain control of the compaction process with respect to the line of optimums.

To assess the overall degree of imperviousness of the compacted clay bar-

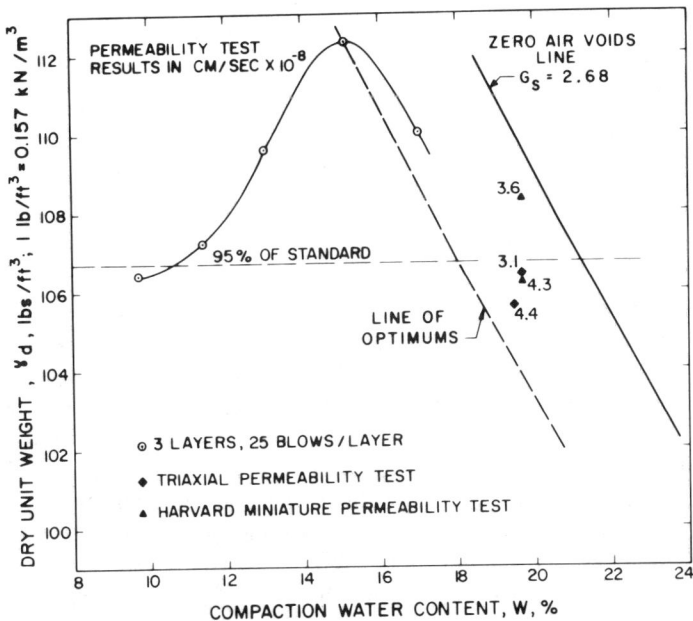

FIG. 11—*Laboratory test results—Soil 2.*

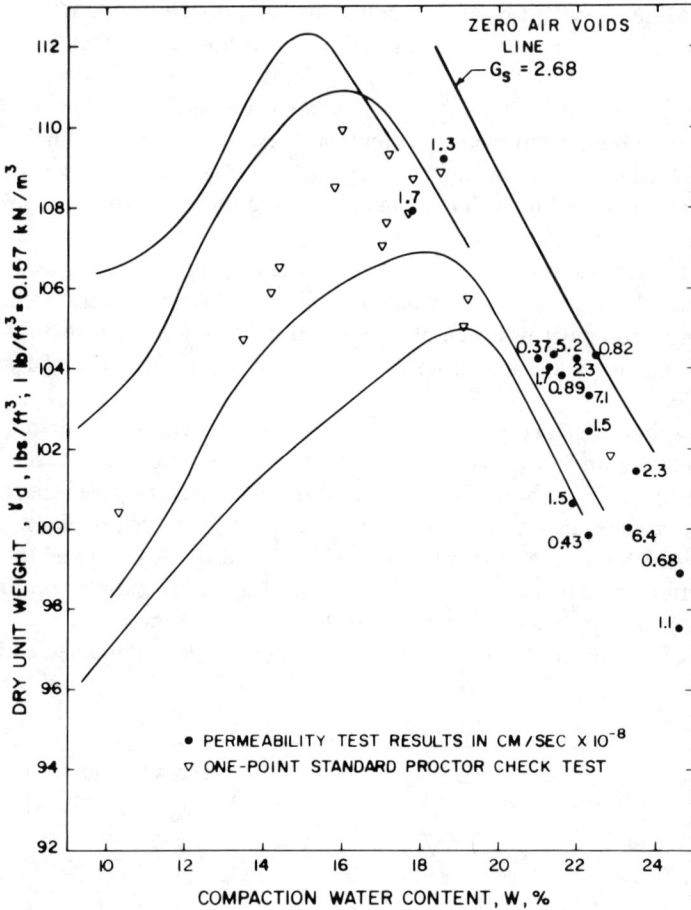

FIG. 12—*Summary of field one-point proctor and as-compacted permeability tests.*

rier layer covering the landfill during construction of the final cover section, the as-compacted permeability of the barrier layer was determined from 18 specimens trimmed from samples of the barrier layer taken from selected random locations across the landfill area. Samples were obtained either by manually driving a 101.6-mm (4-in.)-diameter metal ring sampler into the compacted clay or by cutting large undisturbed block samples from the barrier layer. The results plotted in Fig. 12 indicate that the permeability values from constant-head, back-pressure saturated tests performed in the triaxial cell ranged from 3.7×10^{-9} to 7.1×10^{-8} cm/s with an average value of about 1.8×10^{-8} cm/s. One sample which exhibited a higher permeability value (2.1×10^{-7} cm /s) was found to contain a continuous vertical silt seam and judged to be a localized condition based on additional testing and field inspection.

Conclusions

The generation of leachate quantities from landfills is dependent on the amount of percolation that passes through the final cover layer system. The primary inhibitor with respect to the reduction of percolation through the cover layer is the compacted clay barrier layer. Based on the general relationships between compaction water content, dry unit weight, and permeability for compacted clay soils, a laboratory permeability testing program can be used to determine the field compaction conditions necessary to achieve a design field permeability value for a clay barrier layer. From this testing program, construction specifications can be developed requiring that the clay barrier be compacted to minimum dry unit weight wet of the line of optimums by a given percentage in order to meet the design permeability.

Results from a case study indicate that an average permeability of 2×10^{-8} cm/s had been achieved for a disposal facility by controlling the minimum dry unit weight to greater than 95% of the standard Proctor value and the compaction water content to greater than 1% wet of the line of optimums. Field control procedures included performing classification and one-point Proctor compaction tests to determine the position of the line of optimums so that the compaction moisture content and dry unit weight would fall within the design permeability range.

Acknowledgments

The authors' interest in the permeability of compacted clays has been developed by their involvement in numerous projects relating to sanitary and hazardous waste disposal facilities while employed by ATEC Associates, Inc. The laboratory testing of the soils referenced in the text was performed by Diane Garrison, Gordon Pickett, and Jim Taylor. Audrey Smith and David Haupt contributed to the field control and undisturbed sampling of the compacted clay barrier.

References

[1] Lutton, R. J., Regan, G. L., and Jones, L. W., "Design and Construction of Covers for Solid Waste Landfills," Report EPA-600/2-79-165, U.S. Environmental Protection Agency, Cincinnati, OH, Aug. 1979.

[2] Lutton, R. J., "Evaluating Cover Systems for Solid and Hazardous Waste," EPA SW-867, U.S. Environmental Protection Agency, Cincinnati, OH, Sept. 1980.

[3] Fenn, D. G., Hanley, K. J., and DeGeare, T. V., "Use of the Water Balance Method for Predicting Leachate Generation from Solid Waste Disposal Sites," EPA/530/SW-168, U.S. Environmental Protection Agency, Cincinnati, OH, Oct. 1975.

[4] Perrier, E. R. and Gibson, A. C., "Hydrologic Simulation on Solid Waste Disposal Sites," EPA SW-868, U.S. Environmental Protection Agency, Cincinnati, OH, 1980.

[5] Dass, P. et al, Journal of the Environmental Engineering Division, American Society of Civil Engineers, Vol. 103, 1977, pp. 981–988.

[6] Thornthwaite, C. W. and Mather, J. R., Publications in Climatology, Drexel Institute of Technology, Laboratory of Climatology (Centerton, NJ), Vol. 10, No. 3, 1957, pp. 105–311.

[7] Thornthwaite, C. W. and Associates, *Publications in Climatology*, Drexel Institute of Technology, Laboratory of Climatology (Centerton, NJ), Vol. 17, No. 3, 1964, pp. 419–615.

[8] Bjerrum, L. and Huder, J. in *Proceedings*, Fourth International Conference on Soil Mechanics and Foundation Engineering, London, Vol. 1, 1957, pp. 6–10.

[9] Garcia-Bengochea, I., Lovell, C. W., and Altschaefel, A. G., *Journal of the Geotechnical Engineering Division*, American Society of Civil Engineers, Vol. 105, No. GT7, July 1979, pp. 839–856.

[10] Lambe, T. W. in *Permeability of Soils, ASTM STP 163*, American Society for Testing and Materials, Philadelphia, 1954, pp. 56–67.

[11] Mitchell, J. K., Hooper, D. R., and Campanella, R. G., *Journal of the Soil Mechanics and Foundations Division*, American Society of Civil Engineers, Vol. 91, No. SM4, July 1965, pp. 41–65.

[12] Lambe, T. W., *Journal of the Soil Mechanics and Foundations Division*, American Society of Civil Engineers, Vol. 84, No. SM2, May 1958, pp. 1–34.

[13] Barden, L. and Sides, G. R., *Journal of the Soil Mechanics and Foundations Division*, American Society of Civil Engineers, Vol. 96, No. SM4, July 1970, pp. 1171–1200.

[14] Hodek, R. J., "Mechanisms for the Compaction and Response of Kaolinite," thesis presented to Purdue University, West Lafayette, IN, 1972, in partial fulfillment of the requirements for the degree of doctor of philosophy (see also Joint Highway Research Project Report No. 36, Purdue University, Nov. 1972).

[15] Olsen, H. W., *Clays and Clay Minerals*, Vol. II, 1962, pp. 131–161.

[16] Daniel, D. E. in *Proceedings*, Fourth Symposium on Uranium Mill Tailings Management, Geotechnical Engineering Program, Civil Engineering Dept., Fort Collins, CO, Oct. 1981, pp. 665–676.

[17] Leonards, G. A. and Narain, J., *Journal of the Soil Mechanics and Foundations Division*, American Society of Civil Engineers, Vol. 89, No. SM2, March 1963, pp. 47–98.

James H. Kleppe[1] and Roy E. Olson[2]

Desiccation Cracking of Soil Barriers

REFERENCE: Kleppe, J. H. and Olson, R. E., **"Desiccation Cracking of Soil Barriers,"**
Hydraulic Barriers in Soil and Rock, ASTM STP 874, A. I. Johnson, R. K. Frobel, N. J.
Cavalli, and C. B. Pettersson, Eds., American Society for Testing and Materials, Phila-
delphia, 1985, pp. 263–275.

ABSTRACT: Compacted soil liners have been used to retard leakage of fluids from burial
sites. If allowed to desiccate, such liners may shrink, crack, and lose their integrity. As a
result of the expense and control problems associated with field tests, an initial laboratory
study was made of shrinkage, cracking tendency, and hydraulic conductivity of various
compacted clay/sand mixtures. The study showed that desiccation shrinkage increased
linearly with compaction water content and was unaffected by density. Soaking prior to
desiccation increased strains markedly for specimens compacted dry of optimum. Shrinkage strains greater than 10% should cause serious problems in the field. Clay/sand mixtures were prepared which were crack resistant and which had low hydraulic
conductivities.

KEY WORDS: impermeable liners, clays, compaction, shrinkage, desiccation, cracking,
hydraulic conductivity, permeability

Compacted cohesive soils are often used alone as "impermeable" barriers
or as backups to synthetic liners at sites where wastes are buried in the
ground. Laboratory tests, using compacted samples, may indicate that the
soil is relatively impervious. However, compacted clay liners may fail to function satisfactorily because of

1. Chemical attack from retained fluids.

2. Development of cracks and holes due to differential settlement, penetration by plant roots and animals, freeze-thaw cycles, and desiccation.

This paper is concerned with desiccation cracking. Desiccation cracking
occurs when the compacted soils are exposed to the atmosphere. Pore water
evaporates, causing development of negative pore water pressures in the soil.
The negative pore water pressures cause increases in effective stress and a
consequent reduction in volume. Because the pore water pressure acts in all
directions, the soil tends to shrink in all directions and cracking results. If the

---

[1]Geotechnical engineer, Hart-Crowser and Associates, Seattle, WA 98102.
[2]Professor of civil engineering, University of Texas, Austin, TX 78712.

soil is wetted with water, it will swell and tend to close the cracks, resealing the liner. However, if foreign material is washed into cracks, or material along cracks spalls into cracks upon rewetting, then permanent zones of higher permeability may result. Cyclic wetting and drying could lead to propagation of cracks to greater depths. If the fluid entering a crack is not absorbed by the soil, then the cracks may stay open. With thick layers of soil, the overburden pressures will tend to keep cracks from opening at depth, but compacted clay caps and liners are typically thin, only a metre or two thick. Cracking due to desiccation can, of course, be prevented by avoiding evaporation by watering, sealing, or burying the liner.

Desiccation cracking is a field problem and is probably best addressed using field tests. Unfortunately, field tests are expensive and often involve complex, uncontrolled, and sometimes undefinable boundary conditions. The first stage in an investigation is thus in the laboratory.

Based on the assumption that desiccation cracking results from shrinkage strains, strain measurements were made on 300 specimens of compacted soil as they dried. In addition, about 60 tests were made with flat plate-shaped specimens to observe cracking patterns. Various efforts to minimize shrinkage and cracking also tended to increase the hydraulic conductivity. Accordingly, nearly 60 tests were made of hydraulic conductivity.

If shrinkage was the cause of desiccation cracking, then it seemed reasonable to assume that a soil containing a sufficient amount of sand to have a framework of rigid particles but enough clay to plug up the void spaces between particles would be resistant to desiccation cracking but would still be relatively impervious. Accordingly, tests were performed using several clays and clay/sand mixtures.

Based on the tests, a rationale for design of cohesive soil barriers to minimize desiccation cracking was developed.

Soils Used

The clays used in this study included Elgin fire clay (Liquid Limit (LL) = 61%, Plasticity Index (PI) = 41%), Taylor marl (LL = 75%, PI = 49%), and Wyoming bentonite (LL = 564%, PI = 515%). The sands were Los Alamos silty sand (sieve size and percent passing: No. 10—100%, No. 20—70%, No. 40—57%, No. 100—42%, No. 200—30%) and a fine to coarse clean sand (No. 10—91%, No. 20—67%, No. 40—35%, No. 100—4%).

Experimental Procedure

Shrinkage Tests

Tests with pure clay generally involved use of small specimens because of the large number of specimens to be prepared and also to avoid cracking. The specimens were originally 38 mm (1.5 in.) in diameter and 76 mm (3.0 in.) high and were compacted using ten layers and three to nine blows per layer of a 1-kg (2.2-lb) hammer falling 305 mm (12 in.). In general, densities achieved using this method of compaction were greater than using the standard Proctor method of compaction [ASTM Test Methods for Moisture-Density Relations of Soils and Soil-Aggregate Mixtures, Using 5.5-lb (2.49-kg) Rammer and 12-in. (304.8-mm) Drop (D 698-78)].

When sand was added to the clay, a Proctor mold was used (102 mm inside diameter by 116 mm high) together with standard Proctor compaction procedures to form the test specimens.

Permeability tests were performed on specimens prepared by compacting the soil into steel rings used in consolidation tests. The rings were 82 mm inside diameter (3.25 in.) by 25 mm (1.0 in.) high and the soil was compacted in five layers using 7 to 14 blows per layer from a 1-kg hammer falling 305 mm.

Samples compacted in the 38-mm inside diameter molds and in the proctor molds were removed from the molds and allowed to air dry with time-dependent measurements of height, diameter, and weight. All samples were oven dried at 115°C after air drying.

Cracking Tests

In the field, the compacted clay liner would be wide and thin and the tendency for lateral shrinkage would be partially restrained by supporting soils which would be less desiccated. To simulate such conditions, samples were compacted in flat plates, with horizontal dimensions of 150 by 300 mm (6 by 12 in.) and a thickness of 50 mm (2 in.). Specimens were compacted on plastic plates, and thin nails were pressed 13 mm (0.5 in.) into the bottom of the specimens through holes in the base plates to simulate the effects of a nonshrinking, adhering, base layer. Edges of the bars were clamped to prevent curling and waxed so desiccation could occur only through the flat faces. The specimens were subjected to cyclic wetting and drying, and dimensions and weights were recorded as functions of time. Wetting was accomplished by submergence.

Determination of the densities of the specimens, with usual accuracy, was not feasible because the apparatus was heavy compared to the soil weight, specimen dimensions were somewhat irregular, and soil was lost during cycles of wetting and drying. A few special tests indicated that the moisture-density curves of bar samples were of the usual shape, but the optimum moisture content was higher than the standard Proctor optimum for both Taylor marl/sand and Elgin Fire clay/sand mixtures by an amount ranging from 1% for samples with 88% sand to 2½% for pure clay. No data were collected to show the effects of compaction procedures on soil fabric.

Permeability Tests

Specimens compacted into consolidation rings were permeated with upwards flowing de-aired tap water at gradients ranging from 11 to 39. These tests were not intended to simulate a field condition where coarser material might enter open cracks.

Measurements of Shrinkage Strains

Tests with Pure Clay

Dry densities and the shrinkage strains for compacted Taylor marl are shown in Fig. 1 (data for Elgin fire clay were similar). The volumetric strains were taken as the sum of the three principal strains. The data indicate that the specimens were nearly isotropic. Shrinkage increased linearly with compaction water content. Shrinkage strains presumably develop because of increases in effective stress as the pore water pressure decreases during desiccation. For specimens compacted at low water contents, the degree of saturation is low and the developing negative pore water pressures have little impact on effective stress. At high water contents the high degree of satura-

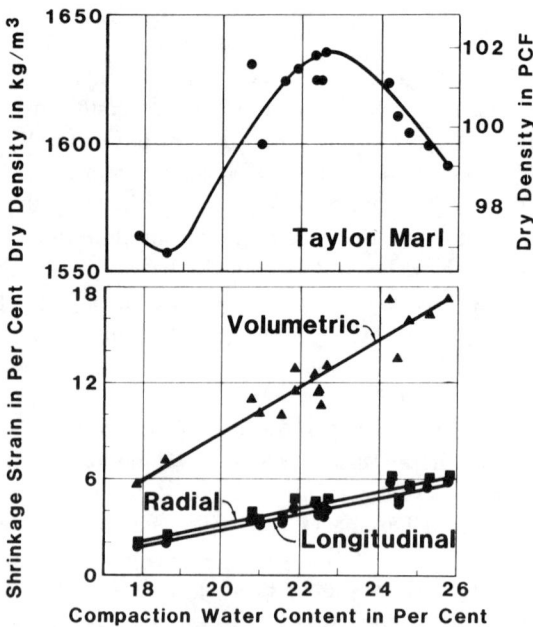

FIG. 1—*Moisture-density relations and shrinkage strains for specimens of compacted Taylor marl.*

tion leads to substantial changes in effective stress during desiccation. The data can be interpreted in terms of Bishop's x-factor [1,2].

An apparent lack of an effect of density on shrinkage strains was observed. To investigate effects of density further, samples of Taylor marl and Elgin fire clay were prepared in the 38-mm diameter molds using 3, 6, and 9 blows/layer. In spite of significant differences in density, there was a unique relationship between shrinkage strain and water content, as shown in Fig. 2 for Taylor marl.

To determine if higher degrees of saturation would lead to more shrinkage, specimens were compacted in pairs in consolidation rings and one specimen was allowed to air dry at once, whereas the other was first saturated at constant volume by flushing with de-aired water and then was also air dried. As shown in Fig. 3, shrinkage does indeed increase when the soil is initially saturated.

Tests with Clay-Sand Mixtures

Specimens of Elgin fire clay and Taylor marl were mixed with the clean sand, compacted in Proctor molds, extruded, and desiccated. Volumetric

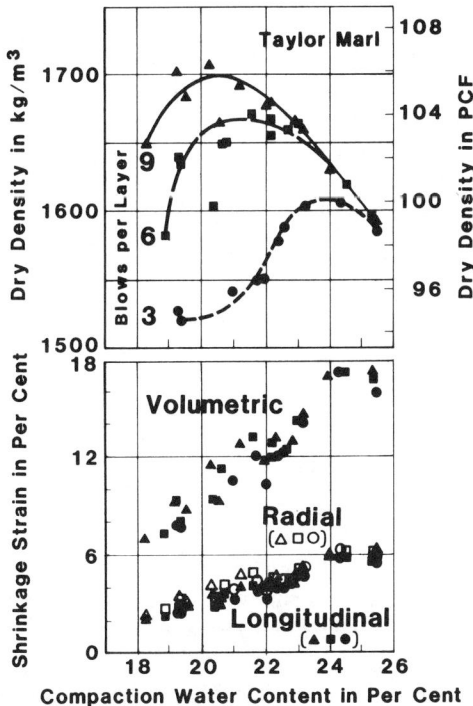

FIG. 2—*Moisture-density relations and shrinkage strains for specimens of Taylor marl prepared using a range of compactive energy.*

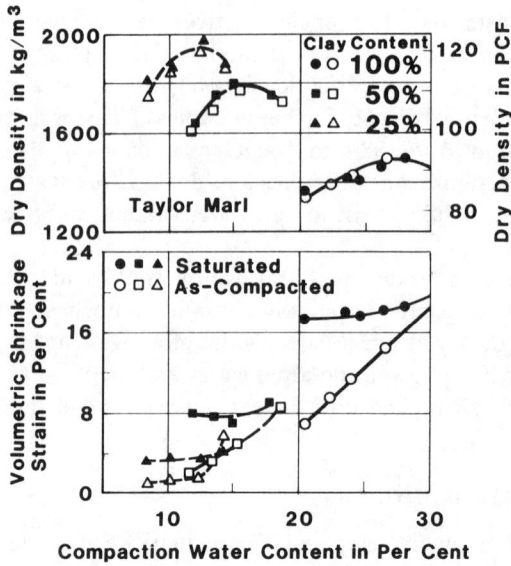

FIG. 3—*Moisture-density relations and shrinkage strains for saturated and as-compacted specimens of Taylor marl and sand.*

shrinkage strains are shown in Fig. 4 for Taylor marl containing 50 to 88% clean sand. Similar results were obtained for Elgin fire clay. The presence of substantial amounts of sand reduced shrinkage greatly.

A similar series of tests was performed with mixtures of Los Alamos tuff (silty sand) and bentonite. Indications from the data (Fig. 5) are that marked reductions in shrinkage occur as the sand fraction increases.

Cracking Tests with Clay Plates

Drying after Compaction

The first series of tests was performed using samples that were air dried immediately after compaction. Dimensions of specimens were measured at a number of locations, allowing gross volumetric strains to be measured. However, there was no way to correct for the volume of open cracks, so the gross strains were smaller than for the shrinkage tests on circular specimens.

The number of visible cracks maximized after about one day of drying. Thereafter there was a tendency for some cracks to grow and others to close up.

Samples of pure Taylor marl and Elgin fire clay all cracked. For samples compacted wet of optimum, the cracks were severe; typically there was one large crack across the sample with a width generally of 5 to 15 mm. The Elgin

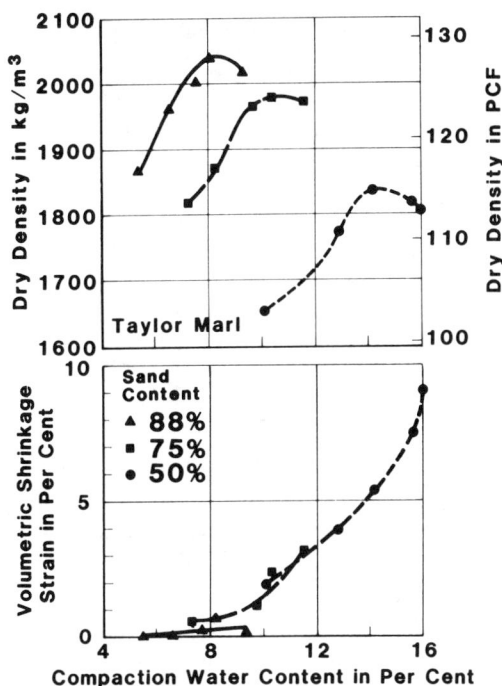

FIG. 4—*Moisture-density relations and shrinkage strains for specimens of Taylor marl and sand.*

fire clay had deep, narrow cracks for samples prepared dry of optimum but the Taylor marl had only narrow, shallow cracks dry of optimum. None of the samples of either Taylor clay or Elgin fire clay containing sand underwent any detectable cracking (samples were 50 to 88% sand).

All specimens of bentonite and tuff underwent severe cracking when compacted wet of optimum. Samples compacted dry of optimum underwent a degree of cracking ranging from moderate (50% sand) to minor (75% sand) to indetectable (88% sand).

Drying after Soaking

Soaking the specimens prior to drying led to a marked increase in cracking. All specimens of pure Taylor clay and Elgin fire clay cracked severely (cracks were 12 to 25 mm wide and full depth of 50 mm). For both clays, cracking was still severe with 50% clean sand added (cracks 5 to 11 mm wide and 30 to 50 mm deep) but diminished as sand content increased, and all cracks disappeared when the specimens were 88% sand.

For bentonite-tuff mixtures, samples with 50% tuff cracked and distorted

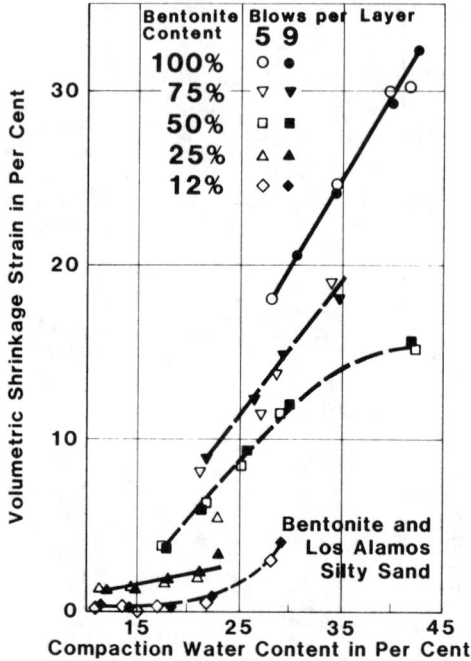

FIG. 5—*Influence of compaction water content, compactive energy, and percent Los Alamos silty sand on the volumetric shrinkage strains of bentonite.*

severely after one day of drying, and that series of tests was discontinued. Samples with 75% tuff cracked clear through, and cracks were 10 to 15 mm wide. At 88% tuff the main cracks were 3 to 5 mm wide and 10 to 30 mm deep. For all bentonite-sand mixtures there were two sets of cracks. One set was shallow and polygonal, and the second set was deep and tended to traverse the whole specimen.

Attempts to relate the severity of cracking for plate-shaped samples and shrinkage in cylindrical specimens was successful only qualitatively. Part of the difficulty was probably due to differences between the two types of specimens, due to differing compaction energies, differing boundary conditions during compaction and desiccation, and doubtless other effects as well. Part of the problem is also involved with the difficulty of assigning a quantitative value to cracking. As an initial effort, a cracking severity number was defined as indicated in Table 1 and was plotted against the maximum-volumetric strain in a more or less equivalent cylindrical sample prepared in the 82-mm-diameter by 25-mm-high rings (Fig. 6). If a cracking severity number of 3 is assumed to indicate probable problems in the field and 4 indicates a certain problem, then volumetric strains greater than about 5% indicate potential problems and greater than 10% indicate almost certain problems in the field.

TABLE 1—*Scale of cracking severity.*

Number	Description
0	no cracking
1	minor cracking; cracks up to 1 mm wide and 5 mm deep
2	moderate cracking; cracks open 3 to 10 mm and up to 30 mm deep
3	major cracking; more than 10 mm wide or penetrating through the specimens or both
4	severe cracking, more than 20 mm wide; would probably penetrate to substantial depth in the field

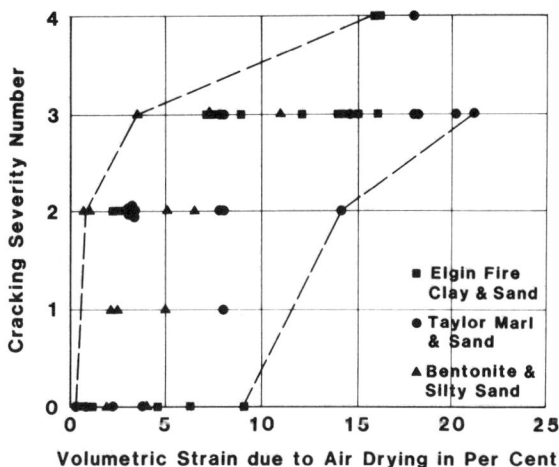

FIG. 6—*Influence of volumetric strain due to air drying on the severity of cracking.*

Hydraulic Conductivity

The tests just reported show that addition of substantial amounts of sand to a clay will minimize shrinkage and cracking, but the function of the clay is to retard fluid flow and the addition of an excessive amount of sand should cause increased hydraulic conductivity. To determine if the hydraulic conductivities were increasing significantly, specimens were compacted into metal rings and immediately permeated with de-aired tap water. The measured values of hydraulic conductivity are thus for uncracked specimens.

Measured hydraulic conductivities are presented in Figs. 7a, 7b, 7c, and 7d. The data indicate that large increases in hydraulic conductivity may be expected for compaction at water contents below optimum and for soil containing large amounts of sand. For soil mixtures of clay and sand only (no

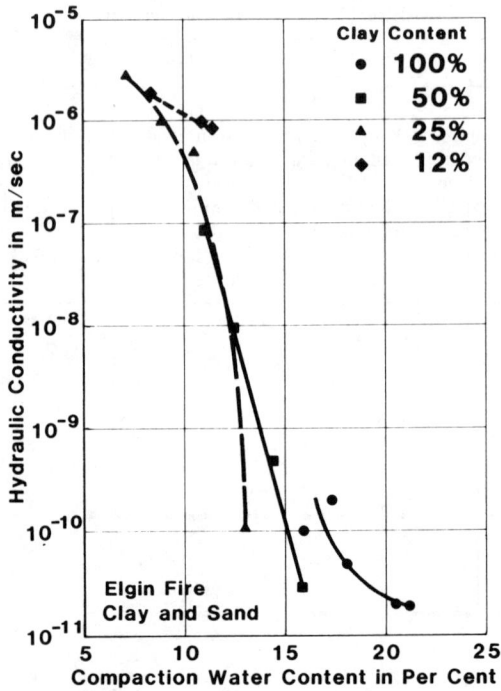

FIG. 7a—*Hydraulic conductivity for specimens of Elgin fire clay and sand, compacted with seven blows per layer.*

silt), values of hydraulic conductivity less than 1.0×10^{-9} m/s were difficult to obtain at percentages of 25% clay or less.

Discussion

Based on all of the data, it appears that there are ranges in particle size and compaction water content that represent adequately low saturated hydraulic conductivity but may or may not minimize the probability of cracking due to desiccation. For the Taylor marl/sand mixtures with 50 and 75% sand, un-soaked samples did not crack, soaked samples cracked moderately, and hydraulic conductivities were reasonably low. Samples of Taylor clay with no added sand were relatively impervious but cracked badly if soaked before drying. Samples with 88% sand did not crack regardless of compaction water content, but were probably too pervious.

As observed in the Los Alamos silty sand/bentonite samples, it appears that a soil with 50 to 75% sand, a broad range in grain sizes, and enough silt and clay to plug up the voids would be ideal in terms of crack resistance and

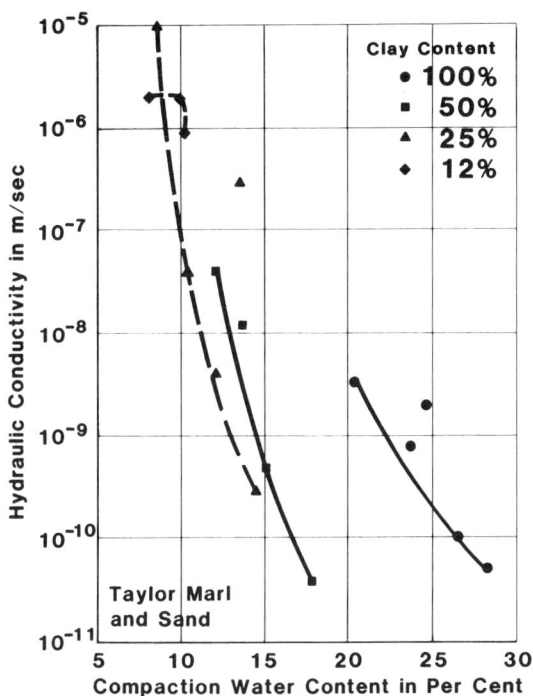

FIG. 7b—*Hydraulic conductivity for specimens of Taylor marl and sand, compacted with seven blows per layer.*

low hydraulic conductivity. Such a material may require artificial blending on site, and thus natural soils meeting the criteria would be preferred.

Summary and Conclusions

Shrinkage tests were performed using more than 300 specimens of compacted clay and clay-sand mixtures. Cracking tests were performed on 60 plate-shaped specimens, and permeability tests were performed on 56 specimens.

Specimens of the original clays were relatively impervious in their as-compacted condition, especially if compacted wet of optimum, but they tended to crack if desiccated from their as-compacted condition. Cracking became much more severe when samples were first soaked and then dried. Addition of sand reduced the amount of drying shrinkage and the tendency to crack but also increased the permeability, especially for dry-side compaction.

Volumetric shrinkage strains less than about 5% led to relatively minor cracking, and values exceeding 10% generally led to severe cracks.

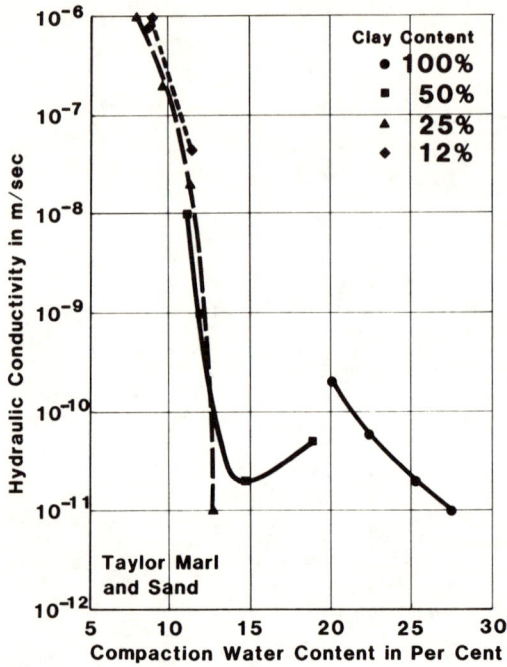

FIG. 7c—*Hydraulic conductivity for specimens of Taylor marl and sand, compacted with 14 blows per layer.*

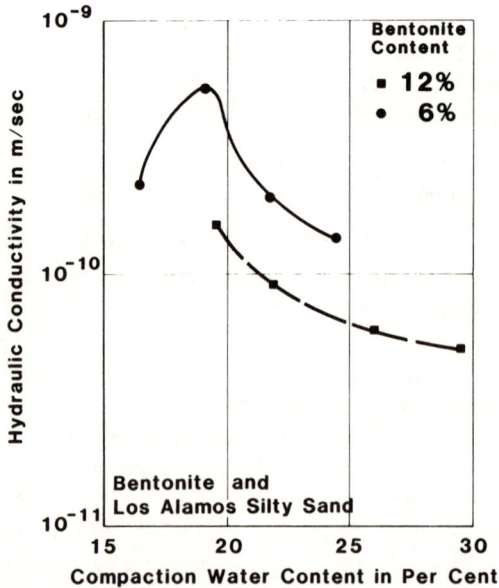

FIG. 7d—*Hydraulic conductivity for specimens of bentonite and Los Alamos silty sand, compacted with 14 blows per layer.*

Shrinkage strains were essentially linear functions of compaction water content and, surprisingly, did not depend on dry density for the range in compaction procedures used here.

Based on the data reported here, it appears possible to find natural soils or to prepare soil blends that will be highly crack resistant but still relatively impervious, but careful control of compaction moisture content is required.

Acknowledgments

This investigation began under the sponsorship of the Los Alamos Scientific Laboratory with Dr. Merlin L. Wheeler acting as the contract officer. Detailed results were included in the thesis by Kleppe [*3*].

References

[*1*] Bishop, A. W., Alpan, I., Blight, G. E., and Donald, I. B. in *Proceedings*, American Society for Civil Engineers Research Conference on the Shear Strength of Cohesive Soils, ASCE, New York, pp. 503-532.

[*2*] Bishop, A. W. and Donald, I. B. in *Proceedings*, Fifth International Conference on Soil Mechanics and Foundation Engineering, Vol. 1, Dunod, Paris, pp. 13-22.

[*3*] Kleppe, J. H., "Desiccation Cracking of Cohesive Soil," M.S. thesis, University of Texas, Austin, 1981.

Steven R. Day[1] and David E. Daniel[2]

Field Permeability Test for Clay Liners

REFERENCE: Day, S. R. and Daniel, D. E., **"Field Permeability Test for Clay Liners,"** *Hydraulic Barriers in Soil and Rock, ASTM STP 874,* A. I. Johnson, R. K. Frobel, N. J. Cavalli, and C. B. Pettersson, Eds., American Society for Testing and Materials, Philadelphia, 1985, pp. 276–288.

ABSTRACT: A method of measuring the hydraulic conductivity of compacted clay liners in the field using single-ring infiltrometers has been developed. It is assumed that the ring has a diameter that is no less than the thickness of the clay liner and that the clay liner is underlain by a freely draining material with negligible suction. Finite element analyses were performed to develop correction factors that account for horizontal seepage for cases in which the ring infiltrometer is partially embedded into the liner. The correction factors were developed for a range in diameter of the ring and for ratios of horizontal to vertical hydraulic conductivity of 1, 10, and 100. Laboratory experiments were conducted to verify the results of finite-element analyses, but the laboratory results showed considerable scatter and were successful only in demonstrating that the finite-element results show the proper trends. Finally, the test method was tried in the field on a full-sized clay liner in which the actual hydraulic conductivity of the entire liner could be calculated from the known rate of leakage. The hydraulic conductivity measured in the infiltration test agreed almost perfectly with the computed overall hydraulic conductivity of the entire liner. It is concluded that the single-ring infiltration test can be used to measure the hydraulic conductivity of clay liners, although it is difficult to measure hydraulic conductivities that are substantially lower than 1×10^{-7} cm/s. In addition, the field tests may take several weeks to complete.

KEY WORDS: permeability, hydraulic conductivity, infiltration, clay, compacted clay, clay liner, infiltration test, single-ring infiltrometer, seepage, case history, permeability test, field test

Compacted clay liners are widely used as hydraulic barriers for water retention reservoirs, solid waste landfills, sludge ponds, waste lagoons, and other types of impoundments. The principal requirement of clay liners is low hydraulic conductivity. In most instances, the hydraulic conductivity (k) of the clay liner is evaluated by compacting samples in the laboratory and then mea-

[1]Project engineer, Geo-Con, Inc., Pittsburgh, PA 15235.
[2]Assistant professor, Department of Civil Engineering, University of Texas, Austin, TX 78712.

suring k with a suitable permeameter. However, there is increasing recognition that laboratory permeability tests tend to underestimate k [1], even for compacted clay [2]. Field permeability tests offer many advantages over laboratory tests, but methods of performing field permeability tests on clay liners are ill-defined and there has been very little experience with such tests.

The study described in this paper was conducted to develop a relatively large-scale permeability test for clay liners. It was desired to maximize the volume of soil permeated in the field to improve the probability that the measured hydraulic conductivity will be representative of a large section of the full-sized liner. Single-ring infiltrometers with various configurations were evaluated. Finite-element analyses were conducted to determine the appropriate shape and depth factors. Bench-scale experiments were performed in an attempt to verify the analytical results. Finally, the method was applied to an actual case history.

Background

Infiltration tests of various sizes and shapes have been used by agricultural researchers and practitioners since the early 1900s [3]. Engineers have used infiltration tests to evaluate the suitability of septic tank absorption fields and for a variety of other purposes. There are several examples of applications of ring infiltrometers in the literature [3-10], but none involve clay liners.

In general, infiltration tests fall into three categories. The first involves drilling or digging a hole and filling the hole with water (Fig. 1a). The rate of water seepage is measured and reported as the "percolation rate." This test is of limited use for clay liners because the interpretation of test results is open to question, the volume of soil tested is small, and the rate of percolation would be too low to measure with most compacted clays.

The second type of device is termed a "single-ring infiltrometer" and is depicted in Fig. 1b. If the diameter of the ring is much larger than the thickness of the liner, the test involves essentially one-dimensional vertical seepage through a relatively large volume of the liner. The problem of measuring low flow rates still exists.

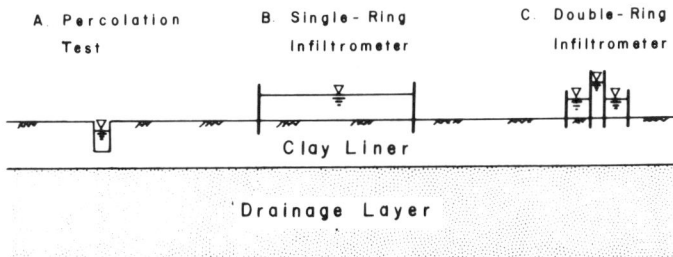

FIG. 1—*Types of field permeability tests for clay liners.*

The third type of test is the "double-ring infiltrometer" (Fig. 1c). The diameter of the outer ring is usually less than 50 cm and the diameter of the inner ring seldom exceeds 20 cm. The rings are normally driven into the ground. The purpose of the outer ring is to promote one-dimensional seepage in the soil located directly beneath the inner ring. It is assumed that one-dimensional flow exists, and, if the boundary conditions are known, the flow rate from the inner ring can be used to compute k. An ASTM Standard, Method for Infiltration Rate of Soils in Field Using Double-ring Infiltrometers (D 3385-75), exists for the double-ring test, but the standard is not intended to apply to "heavy clay soils." In addition, the assumption of one-dimensional seepage beneath the inner ring is open to question, and a poor seal is sometimes obtained when the rings are driven into the soil.

Of the three types of tests, the large-diameter, single-ring infiltration test appears to be the least ambiguous to interpret and the test that is most readily adapted to testing a relatively large volume of soil. Accordingly, this method was selected for further evaluation.

Single-Ring Infiltration Test

There are actually several ways to construct a single-ring infiltrometer. One way is to embed a ring in a circular trench and then backfill the trench with a mixture of excavated soil and bentonite (Fig. 2a). The ring can be constructed from a segment of metal or plastic pipe, from a strip of sheet metal that is sealed where the two ends of the strip meet, from prefabricated livestock watering tanks, or from other materials. The ring can be also constructed from steel drums, which can be particularly advantageous for measuring low flow rates if a small diameter tube is attached to the top of the barrel (Fig. 2b) such that the air/water interface is located inside the small tube.

The test can be also performed on an even larger scale by constructing earthen dikes as indicated in Fig. 2c.

One-dimensional seepage can be developed by extending the ring or seal through the full thickness of the liner (Fig. 3a). However, the excavation for the ring would be tedious and time consuming (particularly if the liner is thick), and the seal around the ring would have to be nearly flawless. If the ring is partially embedded, as shown in Fig. 3b, installation is simplified. In addition, if the backfilled trench leaks, the leak will moisten the soil next to the ring and the presence of a leak should be obvious. On the negative side, if the ring is partially embedded, there will be a horizontal component to the flow, and this "lateral spreading" of water (Fig. 3b) must be taken into account in the calculation of hydraulic conductivity.

Calculation of Hydraulic Conductivity for a Partially Penetrating Single Ring

The rate of flow out of a partially penetrating, single-ring infiltrometer will be controlled by the geometrical parameters shown in Fig. 4, the vertical (k_y)

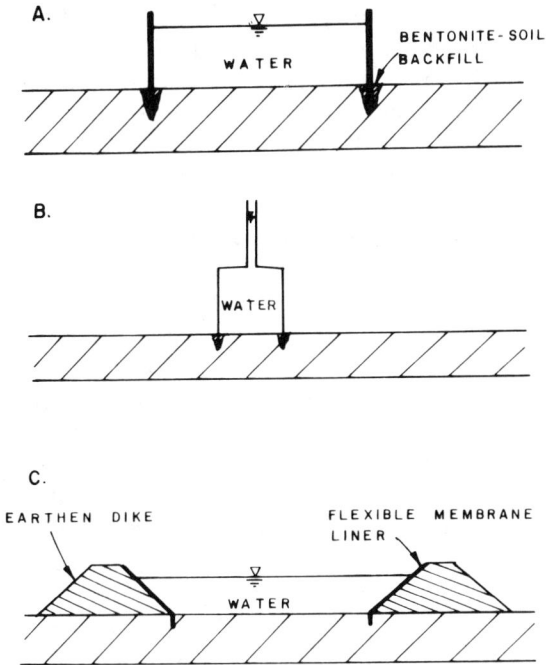

FIG. 2—*Alternative designs for a single-ring infiltrometer used for a permeability test on a compacted clay liner.*

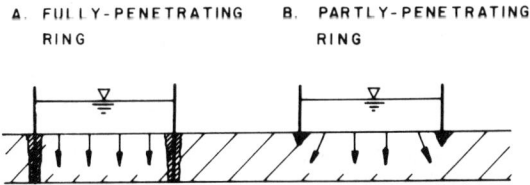

FIG. 3—*Fully and partly penetrating single-ring infiltrometers.*

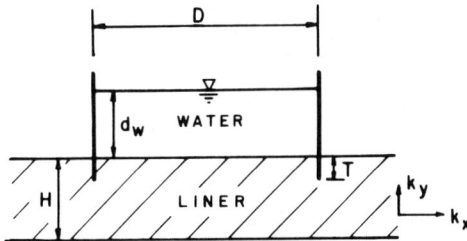

FIG. 4—*Definition of terms used in calculating the hydraulic conductivity from the leakage rate in a ring infiltration test.*

and horizontal (k_x) hydraulic conductivities of the soil, and the boundary conditions. The horizontal extent of the liner can be assumed to be infinite. If the ring is covered to eliminate evaporation of water from inside the ring, the relevant boundary conditions are those at the top and base of the liner. At the top of the liner inside the ring the pressure head is equal to the depth of water (d_w). A freely draining material, which might be part of a leak detection system, usually underlies the liner. The pore water pressure at the base of the liner was assumed to be zero in this study. Capillary stresses probably exist outside the zone of seepage, but within the zone of seepage the assumption of zero pore water pressure at the base of the liner seems reasonable. The flux across the upper surface of the liner outside of the ring was also assumed to be zero. Additional analyses were performed assuming the pore water pressure was zero on the top of the liner outside the ring, but the results were practically the same.

A finite-element code was used to compute rates of inflow over a range of test conditions [11]. It was assumed that one would measure the rate of flow from the ring per unit time (Q), multiply that value by some factor (F), and compute hydraulic conductivity in the vertical direction (k_y) from Darcy's law as follows

$$k_y = \frac{FQ}{iA} \tag{1}$$

where, for the assumptions discussed previously, i is the hydraulic gradient defined as follows

$$i = \frac{d_w + H}{H} \tag{2}$$

and A is the area of the inside of the ring

$$A = \frac{\pi D^2}{4} \tag{3}$$

If seepage is one-dimensional, then F is unity. If lateral spreading occurs, the rate of flow from the ring will exceed the value for one-dimensional seepage, all other factors being equal, in which case F must be less than unity in order to compute the proper k from the measured Q. The factor F will always be between 0 and 1.

A finite element grid with 350 linear quadrilaterals and 396 nodal points was established (Fig. 5). The parameters that were varied were the embedment ratio (T/H), the width ratio (D/H), and the conductivity ratio (k_x/k_y). Steady state seepage was assumed, and the liner was assumed to be homoge-

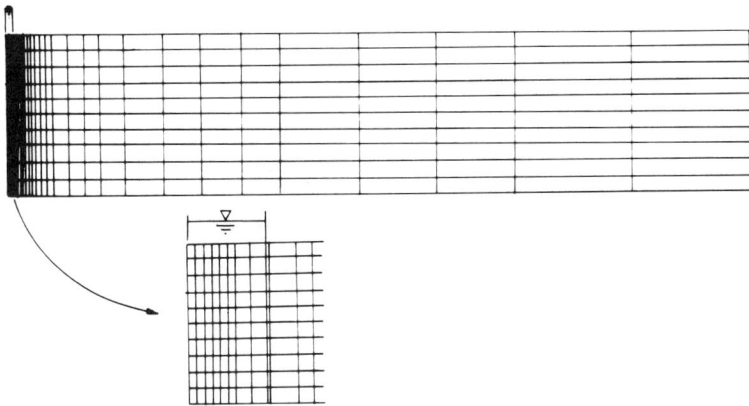

FIG. 5—*Finite-element grid used to compute* F. *The thickness of the grid was adjusted to obtain different values of* D/H.

neous. The factor F was computed for conductivity ratios of 1, 10, and 100 and is plotted in Fig. 6.

The values plotted in Fig. 6 may be inappropriate if (1) there is significant evaporation from the surface of the liner, (2) the liner is not homogeneous, (3) there is a significant suction at the base of the liner in the zone of seepage, (4) the underdrain is not freely draining, (5) the degree of saturation varies significantly within the zone of seepage, or (6) steady state conditions do not exist. The theory suggested here for reduction of the test data is obviously not perfect but nevertheless provides a good starting point for development of the test.

In order to use a partially penetrating ring, one must either know k_x/k_y or one must be able to eliminate the conductivity ratio as a variable. It is possible, at least in principle, to perform a series of tests at constant T/H, but variable D/H, and to determine both k_x and k_y. However, such an approach is impractical because there would be substantial variation in k from one location to another and because the constraints of time and budget would almost certainly preclude multiple tests at each test site. The more practical approach is to assume that k_x/k_y is unity, or if it is suspected that the conductivity ratio is substantially in excess of 1, to perform laboratory permeability tests on undisturbed specimens to determine k_x/k_y. If the embedment ratio (T/H) is greater than 0.3 and the width ratio (D/H) exceeds 5, the Factor F is relatively insensitive to the value of k_x/k_y. (The Factor F varies only by a factor of less than 2 for k_x/k_y equal to 1 or 100—a factor of 2 is much less than an order of magnitude and is not considered particularly significant in hydraulic conductivity testing.) In such cases, there is no point in going to great lengths to determine the conductivity ratio.

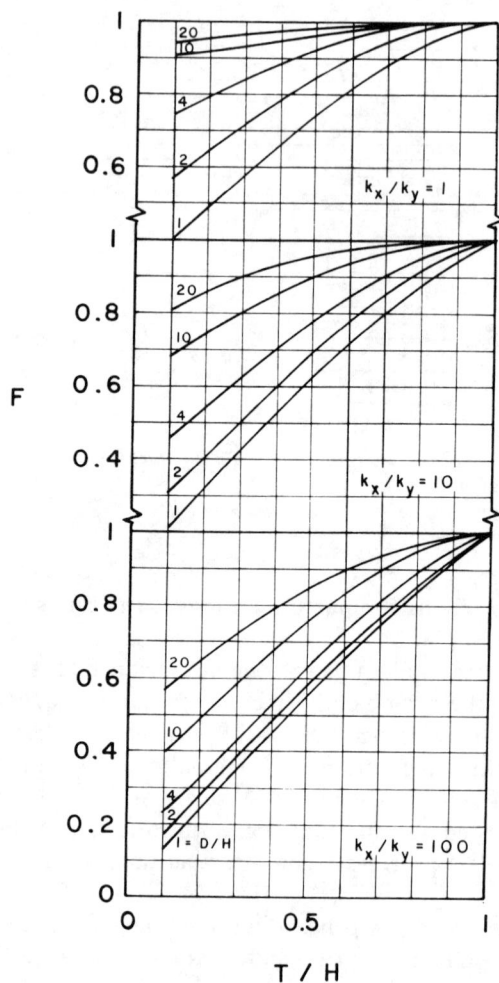

FIG. 6—*Factor* F *plotted as a function of width ratio* (D/H), *embedment ratio* (T/H), *and conductivity ratio* (k_x/k_y).

Laboratory Experiments

Because the calculation of the Factor F involved use of several simplifying assumptions, it was desirable to verify at least some of the numbers through direct measurement. Accordingly, a model tank was constructed in the laboratory. A schematic drawing of the tank is shown in Fig. 7. A 7.5-cm-thick layer of sand was placed in the tank and was underlain by a drainage system. The outflow was collected, and the rate of outflow determined. Rings of various diameters were embedded to various depths in the layer of sand. Water

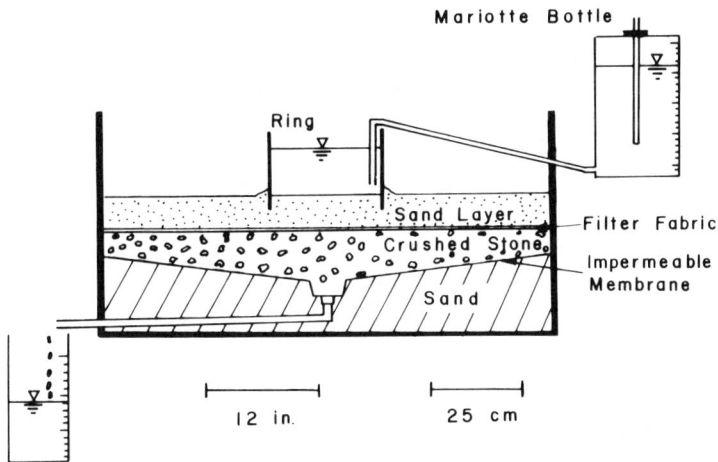

FIG. 7—*Schematic drawing of model tests.*

was placed inside the ring, and the depth of water was kept constant with a Mariotte bottle [1]. The Mariotte bottle was also used to measure the rate of inflow. The maximum rate of flow through the Mariotte bottle and tubing was checked to be certain that head losses in the hydraulic system did not introduce significant error. Similar checks were made for the outflow collection system. Each test was continued until the rates of inflow and outflow were equal and steady with respect to time.

The experiments were conducted using a uniform fine sand with a mean grain diameter of 0.4 mm. The sand was placed in 2-cm-thick lifts and compacted with a drop weight falling on a rectangular foot that was moved across the surface of a lift. The hydraulic conductivity of the soil was measured with a fully penetrating ring and was 1×10^{-2} cm/s. Additional tests were performed by adding 5% by dry weight of very fine silica sand to the original sand. The second sand blanket was constructed just like the first one. The finer sand had a hydraulic conductivity of 1×10^{-4} cm/s. It was assumed that k_x/k_y was unity for both sands.

The rings used for the tests were constructed from plastic or steel pipes and had sharpened edges machined on one end. The rings were pushed or driven into the soil, and then the soil next to the ring was pressed against the ring using hand pressure. A small berm of liner material was pressed into place around the outside of the installed ring. The berm helped to minimize piping and stabilized the ring in place.

The results of the experiments are shown in Fig. 8 for D/H of 1, 2, and 4. In nearly all cases, the measured Factor F was less than the calculated value. Some of the measured F's for $D/H = 2$ are anomalously low. The fact that the measured F's were lower than expected means that the rates of flow were higher than anticipated. Part of the high flow rate was due to upward seepage just outside the ring. The upward seepage even caused piping and quick conditions in several tests with low embedment ratios. Leakage along the ring-soil contact may have also accounted for the high flow rate.

Attempts were made to adjust the theoretical curves to bring them into better agreement with computed curves. These attempts led to several anomalies and generally produced unsatisfactory results. Reconsideration was given to the discrepancies shown in Fig. 8. The ratio of measured F to calculated F, which will be called C, was determined for each test (including results not

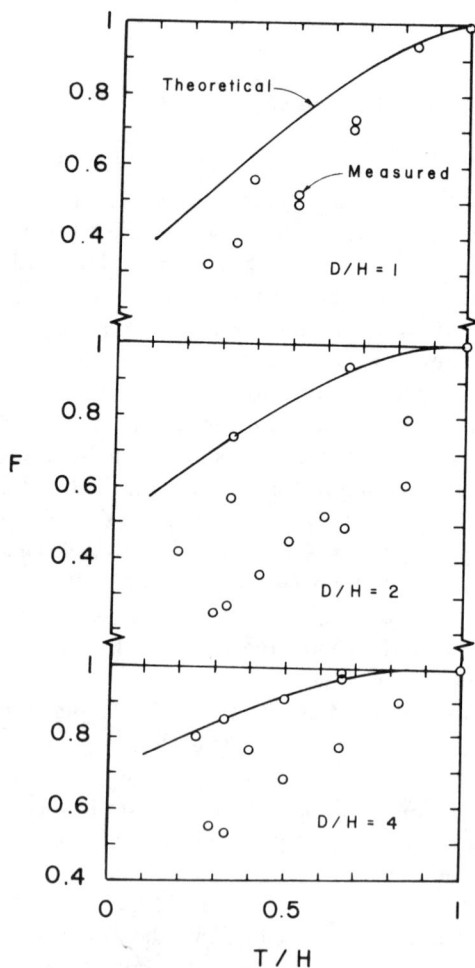

FIG. 8—*Results of model tests.*

plotted in Fig. 8). The mean ratio was 0.80, the range was 0.35 to 1.00, and the standard deviation was 0.19. The letter C may be used as follows

$$C = \frac{F_{measured}}{F_{calculated}} = \frac{k_{measured}}{k_{calculated}} \qquad (4)$$

and

$$k_y = \frac{CFQ}{iA} \qquad (5)$$

To be conservative, one would use $C = 1.0$. However, a value of $C = 0.8$ is in better agreement with the laboratory data. As more data are developed, particularly from larger-scale tests on clay, C can be adjusted accordingly.

Case History

A pond was constructed at a site in central Texas by excavating 1.8 m below the surrounding ground surface and then lining the bottom and the sides of the excavation with 30 cm (12 in.) of compacted clay (Fig. 9) that was constructed in two lifts. The pond covered 8094 m^2 (2 acres) and was designed to hold fresh water pumped from a local water well. Subsoils at the site consist of 1.5 m (5 ft) of clay underlain by silty and clayey gravel. Clay removed from the excavation was used to construct the liner.

Attempts were made to fill the pond, but leakage was so great that the pond could not be filled to a depth of more than 0.3 to 0.6 m (1 to 2 ft). The design water depth was 1.2 m (4 ft). Attempts to fill the pond continued for three months; the rate of pumpage and water level in the pond remained essentially constant during this period. From the known rate of pumpage and the estimated rate of evaporation, the average rate of leakage through the liner was calculated to be between 1.3 and 1.7 L/s (20 to 30 gal/min). The average

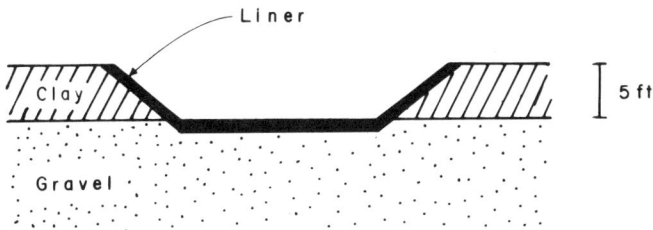

FIG. 9—*Pond at site in central Texas.*

hydraulic conductivity of the liner was computed from the leakage rate and was found to be between 2×10^{-5} and 5×10^{-5} cm/s, which was much higher than anticipated.

A field permeability test was performed by draining the pond and installing a 2.4-m (8-ft) diameter ring. The ring was constructed from sheet metal. The walls of the ring were embedded approximately 8 cm (3 in.) into the liner in a circular trench that was backfilled with a mixture of soil and bentonite to provide a seal. About 15 cm (6 in.) of tap water were placed and later maintained in the ring, and then the ring was covered with a sheet of plastic to prevent evaporation. Water levels inside the ring were monitored for four weeks; the rate of leakage after the first week remained steady and was 1 cm/day.

The soil was assumed to be homogeneous and isotropic. Values of $D/H = 8$ and $T/H = 0.25$ were used, for which case $F = 0.9$. It was assumed that $C = 1$. With these assumptions and a measured water loss of 1 cm/day, the hydraulic conductivity was computed and found to be 4×10^{-5} cm/s, which is nearly identical to the average value that was back-calculated from the field leakage rate for the entire liner. In this case, the field permeability test seemed to work well. The high hydraulic conductivity was thought to have been caused by desiccation cracks or some other type of hydraulic defect. Laboratory permeability tests on undisturbed samples of the liner showed much lower hydraulic conductivities [2].

The liner was ultimately repaired by excavating and recompacting the clay. The repair lowered the hydraulic conductivity by a factor of 10, which enabled the ponds to function successfully.

If the hydraulic conductivity of the liner had been less than approximately 1×10^{-7} cm/s, the rate of flow would have been too low to measure accurately. For low flow rates, a scheme such as the one suggested in Fig. 2b would be required.

Conclusions

A field permeability test for clay liners employing a single-ring, partially penetrating infiltrometer has been developed and employed on an actual clay liner. The equation to be used in determining the hydraulic conductivity for vertical flow (k_y) is Eq 5. The factor F is plotted in Fig. 6 and accounts for partial embedment, variable width of the ring, and variable ratio of horizontal to vertical k. The important assumptions are that the liner is homogeneous, the degree of saturation within the zone of seepage is constant, the ring is properly sealed in the liner, and the flow is steady with respect to time. The method was used successfully on a full-sized liner with relatively high hydraulic conductivity.

There are several important limitations to the method. In order to achieve steady state seepage, the test may have to last for several days or in some cases several weeks. The soil in the zone of seepage beneath the ring may contain entrapped air and therefore may not be fully saturated. One must have at least an approximate idea of the hydraulic conductivity ratio k_x/k_y. The clay liner must be homogeneous or the method may not apply. Finally, it is difficult to measure a hydraulic conductivity that is less than about 1×10^{-7} cm/s because the flow rate will be very small. The method does show good promise for verifying low hydraulic conductivity even though one might have difficulty in determining just how low k is.

Further experience with the method is needed before it would be recommended to practicing engineers, geologists, or earth scientists. However, the increasing body of data that shows that hydraulic conductivities measured in the laboratory may be in error by several orders of magnitude [2] provides the motivation to continue developing and refining field permeability tests of this type.

Acknowledgments

John Bowders performed the literature review on ring infiltrometers. James Long and Steve Trautwein helped with the field permeability test.

Funding for this investigation was provided partly by the U.S. Environmental Protection Agency under cooperative agreement CR-810165-01-0. This paper has not been subjected to the agency's peer and administrative review and, therefore, does not necessarily reflect the view of the agency and no official endorsement should be inferred.

References

[1] Olson, R. E. and Daniel, D. E. in *Permeability and Groundwater Contaminant Transport, ASTM STP 746*, American Society for Testing and Materials, Philadelphia, 1981, pp. 18-64.

[2] Daniel, D. E., *Journal of Geotechnical Engineering*, American Society of Civil Engineers, Vol. 110, No. SM2, Feb. 1984, pp. 285-300.

[3] Lewis, M. R., *Transactions*, American Geophysical Union, Vol. 18, 1937, pp. 361-368.

[4] Kohnke, H., *Proceedings*, Soil Science Society of America, Vol. 3, No. 4, 1938, pp. 296-303.

[5] Schiff, L., *Transactions*, American Geophysical Union, Vol. 34, 1953, pp. 257-266.

[6] Aronovici, V. S., *Proceedings*, Soil Science Society of America, Vol. 19, No. 1, 1955, pp. 1-6.

[7] Burgy, R. H. and Luthin, F. N., *Transactions*, American Geophysical Union, Vol. 37, No. 2, 1956, pp. 189-191.

[8] Schiff, L., *Transactions*, American Geophysical Union, Vol. 38, No. 2, 1957, pp. 260-261.

[9] Swartzendruber, D. and Olson, T. C., *Proceedings*, Soil Science Society of America, Vol. 25, No. 1, pp. 5-8.

[10] Bouwer, H., *Proceedings*, Soil Science Society of America, Vol. 25, No. 6, 1961, pp. 334-339.

[11] Daniel, D. E., "Moisture Movement in Soils in the Vicinity of Waste Disposal Sites," Ph.D. dissertation, University of Texas, Austin, 1980.

DISCUSSION

R. J. Hodek[1] (written discussion)—Please explain the installation technique used for single-ring field test.

S. R. Day and D. E. Daniel (authors' closure)—The installation of a ring infiltrometer into a clay liner may be compared to the trimming of a soil sample into a consolidation ring. Care and patience are the primary attributes required. The installation procedure found most effective was discovered by trial and error. After a location was chosen, an impression of the cutting edge on the ground was made. The ring, once set, was maintained level and round. The ring was then pushed by hand and driven with a hammer into the liner. A board was used to cushion the blow of the hammer and protect the ring. Frequent checks were required to maintain the ring level. When the driving became difficult or threatened to damage the ring, a narrow trench was excavated along the entire outside of the ring wall. The trench was not extended below the bottom level of the ring. After the trench was excavated, the ring drove with less resistance. This procedure continued until the desired depth of embedment was reached. The trench was then backfilled with a mixture of clay and bentonite. The backfill was carefully compacted into 2.5 to 5-cm (1 to 2-in.) lifts using a board or a tamping iron. A berm of native soil was compacted around and up against the ring to minimize piping.

I used 208-L (55-gal) steel barrels and 570-L (150-gal) stock tanks, modified for testing purposes, to serve as the single-ring infiltrometer. The steel barrels were modified by cutting off the bottom of the barrel and sharpening the lip at the open end. The stock tanks were modified by cutting off the top lip. This provided a ring with one open end and a thin-walled [less than 1.3 mm (0.05 in.)] cutting edge. Both the barrels and the stock tanks retained one closed end to minimize evaporation. A small opening in the top of the infiltrometer allowed measurements to be made but made it difficult to inspect the inside edge of the ring after it was embedded.

I believe that the installation could be simplified and the performance improved if a specially constructed single-ring infiltrometer was developed. The improved ring would have a removable lid (for ease of installation), which could be sealed onto the ring to minimize evaporation and permit the use of a standpipe. The standpipe would allow finer resolution and permit the use of higher hydraulic heads. This should decrease testing time and permit the timely measurement of hydraulic conductivity below 1×10^{-7} cm/s.

---

[1]Michigan Technological University, Houghton, MI.

Kevin D. Vesperman,[1] *Tuncer B. Edil,*[2] *and*
Paul M. Berthouex[2]

Permeability of Fly Ash and Fly Ash-Sand Mixtures

REFERENCE: Vesperman, K. D., Edil, T. B., and Berthouex, P. M., **"Permeability of Fly Ash and Fly Ash-Sand Mixtures,"** *Hydraulic Barriers in Soil and Rock, ASTM STP 874,* A. I. Johnson, R. K. Frobel, N. J. Cavalli, and C. B. Pettersson, Eds., American Society for Testing and Materials, Philadelphia, 1985, pp. 289-298.

ABSTRACT: In an experimental program aimed at evaluating the potential of fly ash for use in liners, permeability tests were performed on two pozzolanic fly ashes from western U.S. coal sources compacted at varying densities and mixed in varying percentages with a quartz sand. Permeability testing was performed in a falling head, flexible membrane, triaxial cell permeameter. A back pressure of 380 kPa and a hydraulic gradient of 10 to 20 were applied to the permeability specimens.

The two pozzolanic fly ashes exhibit different degrees of "self-cementation," which causes a dramatic influence on the resulting permeability values. The highly self-cementing fly ash exhibits nearly a full three orders of magnitude lower permeability than the less self-cementing fly ash. Increasing density, while not as important as the coal source, decreases permeability in both types of fly ash on the order of about half a magnitude. Moisture content is also an important factor since it controls the extent of cementation as well as compaction density. Increasing percentages of fly ash in the mixtures decreases the permeability up to a limiting value when the volume of fly ash and water added exceeds the available pore space between the sand grains; for instance, the mixture of 40% self-cementing fly ash and 60% sand exhibits essentially the same permeability as the 100% fly ash specimen.

The study indicates that the addition of pozzolanic fly ash to an otherwise highly permeable soil results in a dramatic reduction in permeability. The "self-cementing" fly ash exhibits 10^3 to 10^4 folds of permeability reduction with final permeabilities of less than 10^{-6} mm/s and seems quite suitable for liner applications.

KEY WORDS: permeability, soil barriers, soil liners, permeability test, fly ash, fly ash-sand mixtures, hydraulic conductivity, compaction, pozzolan cementing

Over 60% of the electric power generated in the United States is produced from coal combustion, which produces over 70 million metric tons of coal ash

[1]Environmental engineer, Wisconsin Power & Light Co., Madison, WI 53701.
[2]Professor of civil and environmental engineering, University of Wisconsin, Madison, WI 53706.

per year. Fly ash, the lighter, smaller particles carried in the flue gas, is collected by electrostatic precipitators, filter bag houses, venturi-type scrubbers, etc. The fly ash removed by electrostatic precipitators and filter bag houses is frequently handled "dry" and stored for disposal or utilization in storage silos. This dry storage maintains the pozzolanic capabilities of the fly ash [1].

Disposal of fly ash in an environmentally sound manner must be based on an understanding of the physical and chemical nature of the waste. Obviously, the rate of discharge of chemical substances leached from landfilled fly ash depends on several factors: (1) the flow pattern through the waste matrix, that is, pore flow versus crack flow; (2) the rate of flow through the pore spaces; and (3) the leaching characteristics of the waste, etc. Therefore, testing the permeability of the fly ash is an important step in estimating the rate of flow through the pore spaces and, subsequently, the discharge of leachable constituents.

Fly ash-stabilized soil may have potential for use as a liner material at fly ash or scrubber sludge landfill sites or both, nonhazardous waste lagoons—for example, manure pits, wastewater treatment lagoons, etc.—and hazardous waste facilities either alone or in combination with geomembranes. The potential for each of these applications will require, as an initial first step, an understanding of the permeability of the fly ash and fly ash and soil mixtures and what factors affect permeability.

This paper deals with a testing program designed to provide a better understanding of (1) how placement conditions may affect the permeability of fly ash and fly ash-sand mixtures, (2) the percentage of fly ash necessary to develop low permeability liner materials, and (3) the effect of the coal source on the just-mentioned facts. All of these factors are critical in the design of fly ash landfills and in the use of fly ash as a soil cement and its subsequent use as a liner material.

Testing Program

The major objective of the program was to test the permeability of varying percentage mixtures of fly ash and sand mixtures compacted at varying densities. Previous testing of fly ash-stabilized soils indicated that moisture content and compaction effort are the major factors affecting density [2]. Based on this, the experimental design was established as indicated by Table 1. The more extensive testing of the Colstrip fly ash was designed to elucidate the effect that landfill placement conditions may have on the permeability. On the other hand, testing of the Belle Ayr ash was designed to test placement at optimum and worst case conditions.

Materials Tested

The two fly ash materials tested were both pozzolanic in nature but from two distinct western U.S. coal sources. The first fly ash is a self-cementing ash

TABLE 1—*Test design.*

	Belle Ayr Fly Ash		Colstrip Fly Ash
	100%	40%	100%
Optimum moisture content, standard compaction effort	x	x	x
Optimum moisture content, low compaction effort			x
Less than optimum moisture, standard compaction effort			x
Less than optimum moisture, low compaction effort	x	x	x
Slurry; no compaction			x

from the Belle Ayr mine near Butte, Wyoming; the second fly ash is a self-cementing ash from the Colstrip mine near Colstrip, Montana. Although it is considered self-cementing, the Colstrip fly ash is not as pozzolanic (see pozzolanic index in Table 2). Analysis of the major mineral constituents reveals a strong similarity between the ashes except for calcium oxide content (see Table 2). This is generally considered a major factor in the self-cementitious properties of coal fly ash [3].

Generally, the major difference in the two fly ashes tested is the source of the mined coal. The coals arrive at the steam electric generating station via unit train and are burned in identical 525 MWE tangentially fired boilers. Both fly ashes are collected in hot-side electrostatic precipitators and stored

TABLE 2—*Chemical and physical analysis of the fly ash.*

Mineral/Parameter	Colstrip Fly Ash	Belle Ayr Fly Ash
SiO_2	45.5%	34.8%
Al_2O_3	21.0%	16.0%
TiO_2	0.9%	1.0%
Fe_2O_3	8.0%	5.8%
CaO	16.1%	27.8%
MgO	4.4%	6.5%
K_2O	0.3%	0.4%
Na_2O	0.7%	1.8%
SO_3	1.0%	3.1%
P_2O_5	0.4%	0.9%
Undetermined	1.7%	1.0%
Specific gravity	2.39	2.71
Fineness (percent retained on no. 325 sieve)	24.8%	11.5%
Compressive strength		
With Portland Cement, Percent of control, 28 days	92.4%	107.1%
With Lime, MPa, 7 days	6.3	10.4

in concrete silos. The major differences in the collection of the fly ashes are: (1) the Colstrip flue gas is chemically treated to enhance particulate removal efficiency, and (2) for the Colstrip fly ash, only the first row of the precipitators is collected and stored in the silos. The first row of the precipitator collects the majority of the fly ash (50 to 70%) and the larger particles [4]. The second, third, and fourth rows of the precipitator collect the finer fractions of the fly ash, and this fly ash is currently sluiced to an ash disposal basin.

The soil used in preparing the fly ash-soil mixtures was a commercial fine quartz sand with a specific gravity of 2.66. The sand is mined near Portage, Wisconsin, and will be simply referred to as a sand.

Testing Apparatus and Procedure

Specimen Preparation

All specimens were prepared by compaction in a Harvard compaction mold. The "standard" compaction effort for this study consisted of 25 blows per layer with five layers per specimen. Blows were delivered by a spring device calibrated to 89 N per blow. The "low" compaction effort was established at the 44.5 N spring setting and 10 blows per layer.

To standardize the preparation, the specimens were prepared and stored for seven days in a humidity-controlled environment at room temperature. Specimens were then trimmed, measured, weighed, and placed into the permeameter. Once optimum moisture content was determined, specimens were prepared according to the test design matrix (Table 1). The less than optimum moisture content was established based on the compaction moisture-density relationship at water contents that would achieve measurably different densities.

Permeability Testing

Permeability of specimens was measured by a falling head test utilizing a flexible membrane in a triaxial cell. The general procedure followed was as outlined by Gordon and Forrest [5]. Distilled de-aired water was used as the permeant. Saturation of specimens was performed by applying a nominal hydraulic gradient for 24 h prior to permeability testing. A back pressure of 380 kPa was then established in both burettes to remove any dissolved air. A gradient of 20 to 1 for the low permeability materials ($k < 10^{-6}$ mm/s) and 10 to 1 for the high permeability materials ($k > 10^{-6}$ mm/s) was then established. Repetitive testing of permeability was performed either until approximately one pore volume of permeant was passed through the specimen or, in the lower permeability specimens, at least one half of a pore volume was passed through the specimen. Intermediate measurements of the heads at various time increments were taken, as well as the initial and final values. This allowed calculation of permeability throughout the testing using the standard falling head permeability equation.

Test Results

Moisture-Density Relationship

The compaction behavior of the fly ash was considerably different than most soil materials. For example, the optimum moisture contents in the 100% fly ash mixtures were only a few percentage points less than the saturation moisture content. Also, specimens compacted on the wet side of the optimum usually liquified during compaction and were difficult to prepare. Presumably, the spherical shape of the fly ash particles and the lubrication of the moisture combined to create the unstable conditions.

As the fly ash percentage in the mixture decreased, the density of the mixture generally increased and the optimum moisture content decreased (Fig. 1). Due to the particle size difference, the fly ash particles presumably would fill the voids between the sand particles and create the denser mixture. The density increase, apparently, continued until the sand particles would be touching and the void space would not be filled completely by fly ash (see 10% Belle Ayr fly ash curve). Similar to clay's effect on sand, the 10% fly ash-sand mixture behaved more like pure sand than like a mixture.

Permeability of Belle Ayr Fly Ash and Fly Ash-Sand Mixtures

Specimens compacted at the optimum moisture content and at the "standard" compaction effort exhibited the lowest permeability. The 100 and the

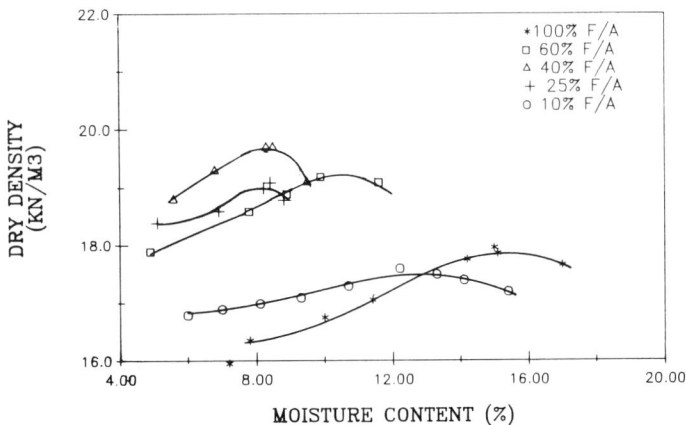

FIG. 1—Dry density versus moisture content, Belle Ayr fly ash/sand mixtures.

40% fly ash-sand specimens exhibited essentially identical permeabilities (Fig. 2) in the range of 10^{-7} or 10^{-8} mm/s. Specimens were also prepared at less than optimum moisture content and with approximately two fifths the energy input, which resulted in a less dense specimen. These specimens exhibited higher initial permeabilities, that is, 10^{-6} and 10^{-5} mm/s for the 100 and 40% fly ash mixtures, respectively. As testing continued, the permeability of the less dense 100% fly ash specimen, however, continued to decrease to levels similar to the denser 100% fly ash specimens.

As the percentage of Belle Ayr fly ash increased in the mixture, the permeability decreased up to a limiting point at 40% fly ash (Fig. 3). Again, the 40% fly ash-sand mixture and the 100% fly ash mixture exhibited essentially identical permeability. Also, a marked decrease in permeability was noted between the 25 and 40% fly ash specimens (see Table 3).

Permeability of Colstrip Fly Ash

Generally, the denser the specimen, the lower the permeability. Specimens compacted at the optimum moisture content exhibited approximately half an order of magnitude lower initial permeabilities (Fig. 4) than those compacted at the lower moisture contents. The level of compaction effort had less of an effect on the density of the specimen than did moisture content (Table 3). Similarly, compaction effort had less of an effect on permeability than did moisture content (see Fig. 4).

Although not as dense, the "slurry" specimen exhibited the lowest permeability of all Colstrip fly ash specimens. The "slurry" specimen density of 13.8 kN/m³ was similar to the low moisture content specimens density of 13.8 and

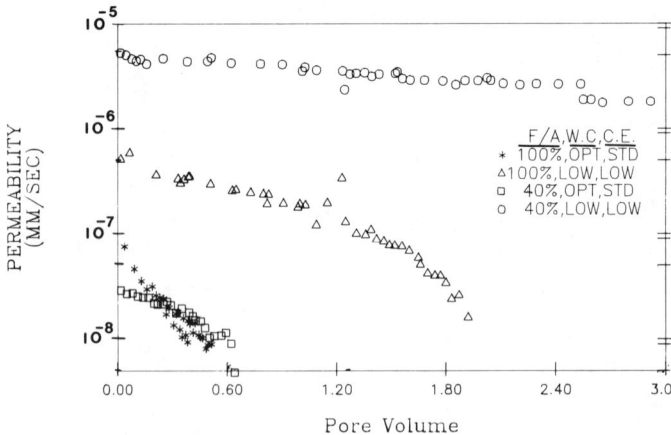

FIG. 2—*Permeability of Belle Ayr fly ash/sand mixtures (F/A = Fly Ash, W.C. = Water Content, C.E. = Compaction Effort).*

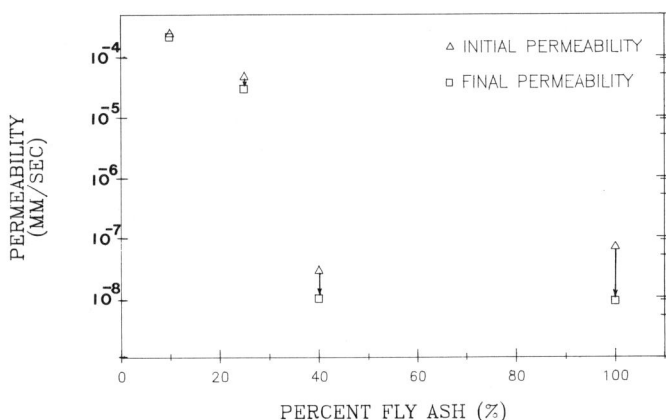

FIG. 3—*Permeability versus percent fly ash, Belle Ayr fly ash/sand mixtures.*

14.1 kN/m^3. Even though it was less dense, the permeability of the slurry specimen was a half an order of magnitude lower than all other specimens (Fig. 3).

Permeability of Belle Ayr Versus Colstrip Fly Ash

Even though both are self-cementitious, the Belle Ayr fly ash exhibited considerably lower permeabilities than the Colstrip fly ash. As indicated in Fig. 2, the 100% Belle Ayr fly ash specimen at optimum moisture content had permeabilities between 10^{-7} and 10^{-8} mm/s. The Colstrip specimens at optimum moisture content (See Fig. 4) had permeabilities between 5.0×10^{-4} and 1.0×10^{-5} mm/s.

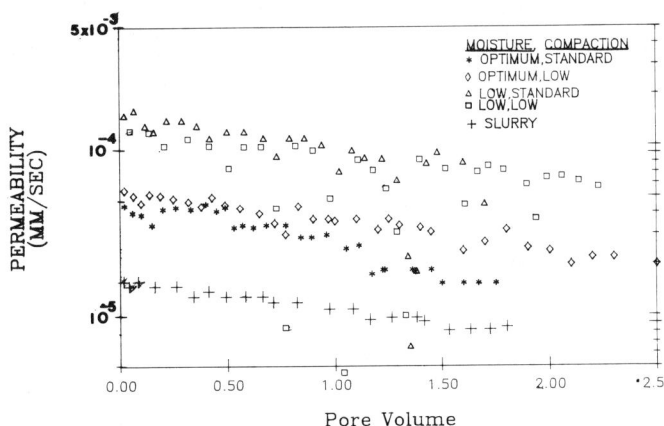

FIG. 4—*Permeability of Colstrip fly ash.*

TABLE 3—*Dry density of permeability specimens.*

	Dry Density, kN/m³		
	Belle Ayr Fly Ash		Colstrip Fly Ash
	100%	40%	100%
Optimum moisture content, standard compaction effort	18.2	19.5	14.8
Optimum moisture content, low compaction effort			14.3
Less than optimum content, standard compaction effort			14.1
Less than optimum content, low compaction effort	16.2	18.5	13.8
Slurry; no compaction			13.8

Discussion of Results

Both fly ashes were self-cementitious, but the permeabilities of the specimens were significantly different. This suggests that another major factor affects permeability that involves the pozzolanic reaction products. Presumably the final pozzolanic reaction products are affected by coal source, combustion process, collection system, etc. Furthermore, these results indicate that specifying a western U.S. coal source or a self-cementitious ash is not enough to ensure the low permeability of a landfilled fly ash or a fly ash liner.

In the permeability testing on these self-cementitious fly ash specimens, the density did have a measurable effect on the permeability. Of the two factors influencing density that were tested in this study, the moisture content at compaction had a greater effect on the specimens' density and, subsequently, on the permeability. The compaction effort, on the other hand, seemed to have only a minor effect on the specimen density; however, the effect on permeability was not noticeable.

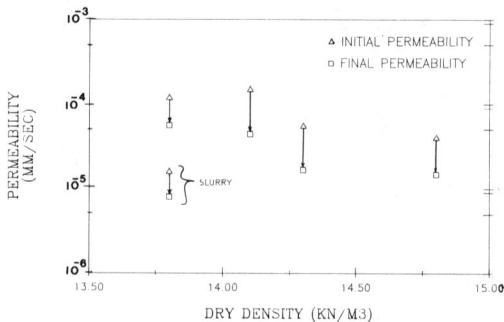

FIG. 5—*Permeability versus dry density, Colstrip fly ash.*

The density, as controlled by moisture content and the level of compaction effort, had a minor effect on permeability (half an order of magnitude change in permeability) when compared to the effect of the coal source. This has several significant design implications. For example, the design of a landfill should recognize the greater importance of coal source versus placement conditions. However, placement conditions that affect density are important in the design of a low permeability liner where a half an order of magnitude change in permeability is critical.

As the fly ash content increases in the Belle Ayr fly ash-sand mixtures, the permeability decreases up to a limiting point. At the 40% fly ash-60% sand level, the permeability of the specimen was essentially identical to the 100% fly ash specimen. Presumably the pore space remaining after the pozzolanic reactions is equivalent in both the 40% fly ash and the 100% fly ash specimens. Between the 25 and 40% fly ash levels tested in this study, the optimum exists where a low permeability specimen is created with a minimum of fly ash use.

Conclusions

Based on the permeability tests on two types of fly ash and the mixtures of one fly ash with a sand, the following conclusions are advanced:

1. The two pozzolanic fly ashes exhibit different degrees of "self-cementation," which causes a dramatic influence on the resulting permeability values. The highly self-cementing fly ash exhibits nearly a full three orders of magnitude lower permeability than the less self-cementing fly ash.

2. Increasing density, while not as important as the coal source, decreases on the order of about half a magnitude.

3. Moisture content is also an important factor since it controls the extent of cementation as well as compaction density.

4. Increasing the percentage of fly ash in the mixtures decreases the permeability up to a limiting value when the volume of fly ash and water added exceeds the available pore space between the sand grains. For instance, the mixture of 40% self-cementing fly ash and 60% sand exhibits essentially the same permeability as the 100% fly ash specimen.

The study indicates that addition of pozzolanic fly ash to an otherwise highly permeable soil results in a dramatic reduction of permeability. The "self-cementing" fly ash exhibits 10^3 to 10^4 folds of permeability reduction with final permeabilities of less than 10^{-6} mm/s and seems quite suitable for liner applications.

Acknowledgment

Funding for this work was provided by the Wisconsin Power & Light Co., Madison, Wisconsin. Special recognition is extended to Alan Erickson, who performed the laboratory work.

References

[1] Bahor, M. P., "Coal Ash Disposal Manual," EPRI Report CS-2059, Electric Power Research Institute, Palo Alto, CA, Oct. 1981.
[2] Parker, D. G., and Thornton, S. I., "Permeability of Fly Ash and Fly Ash Stabilized Soils," FHWA/RD-M-0356, Federal Highway Administration, Washington, DC, 1979.
[3] Terrel, R. L., Epps, J. A., Barenberg, E. J., Mitchell, J. K., and Thompson, M. R., "Soil Stabilization in Pavement Structures," FHWA-1P-80-2, Federal Highway Administration, Washington, DC, Oct. 1979.
[4] Gallaer, C. A., "Electrical Precipitator Reference Manual," EPRI CS-2809, Electric Power Research Institute, Palo Alto, CA, Jan. 1983.
[5] Gordon, B. B. and Forrest, M. in *Permeability and Groundwater Contaminant Transport, ASTM STP 746*, American Society for Testing and Materials, Philadelphia, PA, Sept. 1981, pp. 101–120.

Wayne S. Adaska[1]

Soil-Cement Liners

REFERENCE: Adaska, W. S., **"Soil-Cement Liners,"** *Hydraulic Barriers in Soil and Rock, ASTM STP 874*, A. I. Johnson, R. K. Frobel, N. J. Cavalli, and C. B. Pettersson, Eds., American Society for Testing and Materials, Philadelphia, 1985, pp. 299–313.

ABSTRACT: For over 30 years soil-cement has been effectively used to line lakes, reservoirs, ditches, and irrigation ponds. More recently, wastewater treatment lagoons, sludge drying and ash settling ponds, and coal storage areas have been lined with soil-cement. In addition to its low permeability, soil-cement also provides a reliable method of slope protection.

Numerous laboratory permeability tests have been performed on various soil-cement mixtures. One study included the addition of lime and fly ash to the mixture. Research has been also conducted on shrinkage cracking in soil-cement and its effect on total seepage.

A large-scale field test to determine seepage through a stairstep constructed soil-cement facing was conducted by the Bureau of Reclamation at the Lubbock Regulating Reservoir in Texas. The study indicated the soil-cement to be quite impervious and that most of the seepage occurred through either shrinkage cracks or along the horizontal contact planes. It also showed an overall decrease in seepage with time.

The U.S. Environmental Protection Agency (EPA) has reported laboratory test results on the compatability of soil-cement to various hazardous and other industrial wastes. The tests concluded that exposure to acids such as from electroplating sludges and acidic steel-pickling wastes should be avoided; however, for toxic pesticide formations, oil refinery sludge, toxic pharmaceutical wastes, and rubber and plastic wastes, soil-cement should perform satisfactorily. For caustic petroleum sludges, compatability tests should be conducted on the specific combination of soil-cement and waste.

This paper will include basic information on soil-cement as a liner material and describe in detail the research work on permeability and compatability testing, which has been briefly mentioned. The design, construction, and performance of some unique soil-cement-lined projects will be also presented as well as information on a new composite soil-cement/synthetic membrane liner system that meets the latest U.S. EPA requirements for containing hazardous wastes.

KEY WORDS: soil-cement, impermeable linings, seepage control, slope protection, hazardous waste impoundments, wastewater treatment lagoons

For almost 50 years engineers and contractors have been using soil-cement for paving roads, streets, airports, and parking areas. During most of that

[1]Senior energy and water resources engineer, Energy and Water Resources Dept., Portland Cement Association, Skokie, IL 60077.

time, and particularly in the last 30 years, further developments in technology have led to many other uses, such as linings for reservoirs, channels and lagoons, and slope protection for dams and other embankments.

This paper includes basic information on soil-cement as a lining material and describes research work that has been conducted on permeability and waste compatibility. The design construction and performance of some unique soil-cement-lined projects are also presented as well as information on a new composite soil-cement/synthetic membrane lining system that can be used for the containment of hazardous wastes.

Physical Properties

Soil-cement is a highly compacted mixture of soil, portland cement, and water. As the cement hydrates it hardens into a strong, durable, low permeable material.

Sandy soils containing at least 5% but no more than 35% silt and clay (material passing the No. 200 [75-μm] sieve) are generally used. For channels, ditches, and other applications exposed to debris-carrying, rapid-flowing water [velocities greater than 2 m/s (8 ft/s)], greater abrasion resistance can be obtained where the soil contains at least 20% gravel (material retained on a No. 4 [4.75-mm] sieve) [1]. Fine textured soils such as clays usually are difficult to pulverize and require more cement for satisfactory hardening as do poorly graded granular soils, which have no material passing the No. 200 (75-μm) sieve. Cement contents for typical soil-cement mixtures are generally in the range of 7 to 12% by weight of dry soil.

The durability of soil-cement has been most significantly demonstrated by the satisfactory performance of a 33-year-old test section built by the U.S. Bureau of Reclamation (Fig. 1). Located on the southeast shore of the Bonny Reservoir in eastern Colorado, this test section has been exposed to severe wave action and more than 100 freeze-thaw cycles per year. The test section is 7 m (24 ft) high and approximately 105 m (350 ft) long. It was placed in a series of 15-cm (6-in.)-thick by 2-m (7 ft)-wide successive horizontal layers adjacent to the 2:1 upstream embankment slope. Cement contents were either 10 or 12% by weight depending on the gradation of the sandy soils used [2].

For sludge-drying and wastewater treatment lagoons, ash settling ponds, and coal-storage yards where a firm, puncture-resistant surface is required, liner strength is an important consideration. The liner must be capable of supporting front-end loaders, tractors, dozers, or other equipment that may operate directly on top of the soil-cement.

Similar to concrete, soil-cement continues to gain strength with age. Typical seven-day compressive strengths range from 3.5 to 7 MPa (500 to 1000 psi), while at 28 days the compressive strengths are 5.5 to 10 MPa (800 to 1500 psi). Cores taken from the Bonny Reservoir test section after ten years showed an average compressive strength of 17 MPa (2500 psi).

FIG. 1—*Soil-cement test section, Bonny Reservoir, Colorado. After 29 years.*

The permeability of almost all soil-cement is an inverse function of the cement content, that is, an increase in the cement content decreases permeability. This relationship is illustrated in Fig. 2, which presents permeability test results on a number of sandy soils. Exceptions are some very fine-grained soils that become more permeable with an increase in the cement content. Seldom, however, would these fine-grained soils be utilized for soil-cement liners because of the difficulty in processing and their already low permeability.

Permeability tests have been also conducted in which 2% lime and varying percentages of fly ash were added to soil-cement mixtures. The results indicated that permeabilities were further reduced with optimum mixtures of lime and fly ash, and in a number of cases permeabilities less than 10^{-9} cm/s (10^{-3} ft/year) were recorded [4].

As soil-cement dries it shrinks, causing shrinkage cracks to form. These cracks, however, are very narrow. Field studies have shown that the majority of crack openings are less than 1.6 mm ($^1/_{16}$ in.) wide and spaced for granular soils between 3 to 6 m (10 to 20 ft) apart [5]. An analysis based on a conserva-

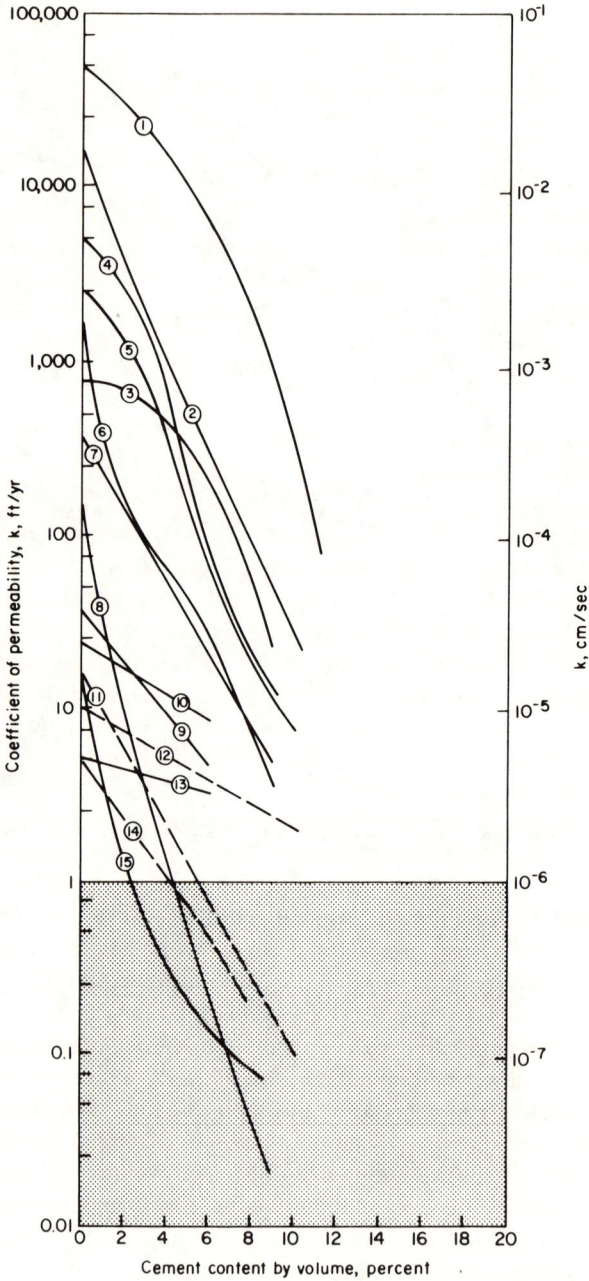

FIG. 2—*Permeability of cement-treated sandy soils. Encircled numbers identify soils described in Ref 3.*

tive assumption of crack width and spacing indicates that seepage increased less than 10% [1]. From a practical standpoint, this is insignificant considering the low permeability of soil-cement. In addition, sediment along the bottom of reservoirs and lagoons tends to fill these cracks.

Reservoirs

Many years ago engineers experimented with soil-cement for lining water storage reservoirs. In 1945 soil-cement was used to line the bottom of a 45-mL (12-million-gal) reservoir in Port Isabel, Texas. This reservoir had been constructed initially with concrete-faced side slopes but no bottom lining. Leakage was so excessive, however, that steel petroleum storage tanks nearby were floated. The soil-cement was mixed in place, using 12% cement by volume, and compacted to a thickness of 10 cm (4 in.).

Lake Cahuilla, a terminal regulating reservoir for the Coachella Valley County Water District irrigation system in the Mojave Desert of southern California, was completely lined with soil-cement in 1969. Designed by the Bureau of Reclamation, the 546 300 m² (135-acre) reservoir bottom has a 15-cm (6-in.)-thick soil-cement lining while the sand embankments forming the reservoir are faced with 0.6 m (2 ft) of soil-cement, measured normal to the slope. In addition to conserving water for irrigation, the lake provides recreational benefits such as swimming, boating, fishing, and picnicking.

Since these early projects, more than 300 water control applications utilizing in excess of 8.4 × 10⁶ m³ (11 million yd³) of soil-cement have been constructed. Applications include slope protection and seepage control for water storage reservoirs, earth dams, highway embankments, settling basins, wastewater treatment lagoons, spillways, channels, and even coastal structures. Information on location, engineer, contractor, volume of soil-cement, cement content, and in some cases cost for many of these projects are described in Ref 6.

Channels and Drainageways

Lining channels and drainageways with soil-cement provides erosion protection and minimizes water loss due to seepage.

Soil-cement linings 15 cm (6 in.) thick have been used for flumes carrying cooling water to and from several power plants. Figure 3 illustrates such a flume, constructed in 1971 at Florida Power & Light Co.'s power plant in Sanford, Florida. The soil-cement containing 10.5% cement by volume was central plant mixed and compacted by a dozer equipped with street pads operating up and down the 3:1 side slopes.

In 1976, following the study of alternatives, Pima County Highway Department engineers chose soil-cement to line the Harrison Hills Drainageway in Tucson, Arizona. The 750-m (2500-ft)-long drainageway, located in a resi-

FIG. 3—*Cooling water flume at power plant in Sanford, Florida.*

dential area, has a 7-m (24-ft)-wide bottom with 3:1 side slopes. To protect against erosion from flow velocities of 3 to 3.5 m/s (10 to 12 ft/s), the side slopes and bottom were lined with 30 cm (12 in.) of soil-cement. Following the fall 1983 flooding, little to no damage was reported for this and numerous other soil-cement slope protected projects in the area.

Sludge-Drying and Wastewater Treatment Lagoons

Recently, the City of Phoenix, Arizona, expanded its 91st Avenue Wastewater Treatment Plant from 34 to 45×10^4 m³/day [90 to 120 million gal per day (mgd)]. Included in this expansion was 404 700 m² (100 acres) of soil-cement-lined sludge drying lagoons. Phoenix's sunny days and low humidity provide ideal conditions for large scale use of drying lagoons.

Prior to plant expansion, existing 30 by 180-m (100 by 600-ft) sludge-drying lagoons were unlined and constructed of native soil compacted to 95% of maximum density. During normal mixing and hauling operations, the saturated native soil became soft and unstable. Front-end loaders, which turned the sludge over in order to accelerate the drying process, and trucks, which hauled the dry sludge out, experienced considerable difficulty in maneuvering along the bottom of the lagoons. Frequently, this heavy equipment would get buried up to its axles.

It was essential that the existing and newly proposed sludge lagoons provide a suitable working surface for operating equipment. The following three alternatives were considered: (1) native soil compacted to 100% of maximum density; (2) a 13-cm (5-in.) asphalt liner over a gravel base; and (3) a 15-cm (6-in.) soil-cement liner with no special base material. A 60 by 80-m (200 by 600-ft) test lagoon was constructed for each alternate. The lagoons were exposed to actual operating conditions for a period of several months.

Analysis of the test program concluded that the unlined lagoon, although compacted to a higher density, provided little improvement over existing conditions. Turning the sludge over in the lagoon with no bottom liner took between two to three h, whereas sludge in the lined lagoons could be turned in about 30 m. Although both the asphalt-lined and soil-cement-lined lagoons performed satisfactorily, soil-cement was chosen because its cost was estimated to be less than one half that of the asphalt alternative [7].

Soil-cement linings are also being used for storage and treatment of wastewater, both municipal and industrial. Outstanding examples are the seven-basin wastewater treatment system of Muskegon County, Michigan, and a similar system at Taft, California.

The Muskegon facility is designed to treat an average flow of 15.9×10^4 m³/day (42 mgd) of municipal and industrial wastewater. It consists of three (8-acre) 32 376 m² aerated lagoons, completely lined with (8 in.) 20 cm of soil-cement; a similar size, fully lined settling lagoon; and two 3 399 480 m² (840-acre) storage basins with soil-cement–faced banks. A total of 107 210 m³ (140 220 yd³) of soil-cement was placed. Constructed in 1972, the ponds are still performing well with little or no maintenance.

In 1973, soil-cement was used as erosion and seepage control for the wastewater treatment lagoons owned and operated by the Taft Heights and Ford City Sanitary District in southern California. The facility is designed to treat an average flow of 4.5×10^3 m³/day (1.2 mgd). It consists of two aerated lagoons, each about 8094 m² (2 acres) (Fig. 4). There are also two settling lagoons. A holding lagoon, 70 m (230 ft) wide, 122 m (4000 ft) long, and 3 m (10 ft) deep, was added in 1978.

Approximately 11 087 m³ (14 500 yd³) of soil-cement was placed on the bottom and on the 3:1 side slopes of these lagoons. The bottom was lined with 15 cm (6 in.) while the side slopes received a 30-cm (12-in.) facing of soil-cement. Cement contents by volume were 10% for the bottom and 12% for the side slopes. To date, all ponds are performing well.

FIG. 4—*Aerated lagoons at Taft Heights–Ford City Sanitary District, Taft, California.*

Sanitary and Hazardous Waste Landfills

A major factor in considering a soil-cement liner for sanitary or hazardous waste landfill sites is its compatibility with stored wastes. The U.S. EPA sponsored laboratory tests to evaluate the suitability of a number of liner materials exposed to various wastes. The results of these tests are reported in Ref 8.

The tests indicated that after one year of exposure to leachate from municipal solid wastes, the soil-cement hardened considerably and cored like portland cement concrete. In addition, it became less permeable during the exposure period.

The soil-cement was also exposed to various hazardous wastes including toxic pesticide formulations, oil refinery sludges, toxic pharmaceutical wastes, and rubber and plastic wastes. Results showed that for these hazardous wastes no seepage had occurred through soil-cement following two and one-half years of exposure. After 625 days of exposure to these wastes, the compressive strength of the soil-cement exceeded the compressive strength of similar soil-cement that had not been exposed to the wastes.

Soil-cement was not exposed to acid wastes. It was rated "fair" in containing caustic petroleum sludges, indicating that the specific combination of soil-cement and waste should be tested. Other wastes being considered for containment should be also tested for compatibility with soil-cement.

Design and Testing

Established American Society for Testing and Materials (ASTM) and Portland Cement Association (PCA) test methods are used to determine the required cement content, optimum moisture content, and density to which the mixture should be compacted.

The optimum moisture content and maximum density are determined by ASTM Method for Moisture-Density Relations of Soil-Cement Mixtures (D 558-82). The required cement content for durability is then established by performing wet-dry (ASTM Wetting-and-Drying Tests of Compacted Soil-Cement Mixtures [D 559-82]) and freeze-thaw (ASTM Freezing-and-Thawing Tests of Compacted Soil-Cement Mixtures [D 560-82]) tests on representative specimens and applying the PCA weight loss criteria [9]. For soil-cement continually submerged, the cement content established from the PCA criteria should be adequate for durability. An additional 2 percentage points of cement is used for slope protection and other areas requiring increased durability.

For slopes exposed to moderate to severe wave action (effective fetch greater than 305 m [1000 ft]) or debris-carrying, rapid flowing water, the soil-cement should be designed 45 to 60 cm (18 to 24 in.) thick measured perpendicular to the slope. It should be placed in horizontal layers 2.1 to 2.7 m (7 to 9 ft) wide and 15.2 to 23 cm (6 to 9 in.) thick (Fig. 5). Details on designing this type of stairstep slope protection may be found in Ref 3.

For less severe applications such as small reservoirs, ponds, or lagoons, the slope protection may consist of a 15 to 23 cm (6 to 9 in.) thick single layer of soil-cement placed parallel to the slope face. This method is often referred to as plating.

Side slopes of 3:1 or flatter are preferred to facilitate spreading and compacting the soil-cement directly on the slope. To control seepage, the bottom lining generally consists of a single 15 to 20 cm (6 to 8-in.)-thick layer of soil-cement.

Shrinkage cracking is to be expected in soil-cement. To minimize shrinkage cracking, the following methods may be used:

1. *Keeping the soil-cement moist.* A continuous water cure is very effective, followed by an application of an asphaltic emulsion or other type of curing compound. Filling a soil-cement-lined reservoir or lagoon with water soon after completion will minimize volume change and largely eliminate shrinkage cracks which, if formed, will be tightly closed hairline cracks.

2. *Applying a bituminous or other type of sealing material over the soil-cement approximately one month after completion.* The one-month delay al-

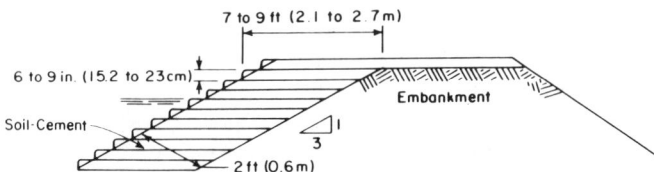

FIG. 5—*Typical design for soil-cement stairstep slope protection.*

lows sufficient time for most of the shrinkage cracks to occur. This method was used for the 46 816 m² (56 000 yd²) of soil-cement paved coal-pile storage pads at the McIntosh Power Plant in Lakeland, Florida. Two coats of liquid asphalt were applied over the completed soil-cement. The first coat was applied at 1.7 L/m² (0.4 gal per yd²) immediately after final compaction and served as a cure coat. In an attempt to seal any hairline shrinkage cracks that may have developed, a second coat at 0.8 L/m² (0.2 gal/yd²) was applied 30 days later [10].

3. *Minimizing the initial moisture content of the soil-cement.* Shrinkage is dependent upon the quantity of water lost. Since only a small quantity of water is needed for cement hydration, the higher the initial water content, the more shrinkage will result. In Switzerland, where noncracked soil-cement is quite common, the quantity of water in the soil-cement at time of compaction averages approximately 70% of the maximum degree of saturation [11].

Construction

Soil-cement may either be mixed in place using a traveling mixing machine or central plant mixed. The mixed-in-place method may be used for single layer bottom lining or plating of flat slopes (4:1 or less). The central-plant method of mixing is recommended for multiple layer and steeper slope construction because it provides more accurate and uniform mixing than the mixed-in-place technique. Also, mixing equipment is difficult to operate on steep slopes.

Soil-cement may be also mixed in place on a horizontal surface known as a "working table." The mixed soil-cement is then hauled and placed on the slope as would be done if the soil-cement was central plant mixed. Often the "working table" is located at or near the toe of the slope to minimize haul distance.

Proper compaction is one of the fundamental requirements of soil-cement construction. If the subgrade is soft and cannot properly support the compaction equipment, adequate density will not be obtained. Therefore, soft areas should be located and corrected before processing begins. On horizontal surfaces soft areas can be generally detected by observing a depression or "pumping action" under the wheels of the motor grader as it shapes the area prior to soil-cement processing.

Shallow, soft, wet subgrade areas can be usually stabilized by aerating and recompacting the soil. When deep unstable areas are encountered, it is usually necessary to remove the underlying wet or unsuitable soil and replace it with stable material.

Slopes that are to receive soil-cement should be firm and shaped to the required lines and grades prior to soil-cement placement. Cut slopes may have to be compacted prior to receiving the soil-cement mixture. Proper preparation of the subgrade will ensure that the soil-cement is of uniform and specified thickness and that adequate density can be achieved with minimal compactive effort.

A number of methods have been used for spreading soil-cement. Aggregate spreaders often pushed by dozers can place premixed soil-cement uniformly to specified widths and thicknesses. These spreaders are especially effective in placing multiple layers of soil-cement.

On relatively low embankments, soil-cement may be placed on the slope using a dozer that pushes the soil-cement up from the toe of the slope. Where water or insufficient room is present at the bottom of the slope, the dozer may have to place the soil-cement by working cross slope. It is important to have a reliable and accurate method for controlling layer thickness to ensure uniformity.

On an 8-m (27-ft)-high embankment for the spillway inlet channel at Palmetto Bend Dam in Texas, soil-cement was placed using an aggregate spreader working from the top of slope down. Central-plant-mixed soil-cement was hauled and dumped into a storage bin at the bottom of the embankment. A front-end loader then fed the spreader as it traveled down the 3:1 slope (Fig. 6). Much less effort was required to place the single 30-cm (12-in.)-thick layer of soil-cement working from top to bottom than if the soil-cement had to be pushed up the long slope.

Compaction of soil-cement may be achieved by any equipment that will satisfy minimum density requirements and provide uniform compaction

FIG. 6—*Placing soil-cement with aggregate spreader, Palmetto Bend Dam, Texas.*

throughout. Tamping, pneumatic-tire, steel-wheel or vibrator rollers or compactors may be used depending on soil type and job conditions. Generally vibratory steel-wheel rollers are recommended for soil-cement made of granular materials. On slopes steeper than 4:1, a traveling "deadman" on top of the embankment can support or assist compaction equipment operating up and down (Fig. 7) or cross slope.

Adequate compaction has been also achieved using dozers. Multiple, overlapping passes are generally necessary to obtain the required density. To minimize surface tearing, cut grousers or street pads should be used.

Regardless of the construction method used, the completed soil-cement must be protected from drying out for a seven-day hydration period. Moist earth, application of water by fog spray, or bituminous or other sealing material can be used for curing, except that no bituminous or other membrane-curing material should be applied to surfaces that will be in contact with succeeding layers of soil-cement.

FIG. 7—*Compacting soil-cement with vibratory sheepsfoot roller.*

Suggested specifications along with procedures for constructing, testing, and inspecting soil-cement are thoroughly described in publications available from the Portland Cement Association [*12–16*].

Soil-Cement/Synthetic Membrane Liner

A composite liner consisting of a synthetic membrane placed between two layers of compacted soil-cement has been recently developed. It is especially applicable for lining hazardous waste and other impoundments where maximum seepage protection is required. Some inherent characteristics of the composite liner are:

1. *It resists degradation*. As previously mentioned, soil-cement is compatible with many hazardous wastes. Synthetic membranes are available that will resist degradation to a broad range of wastes.

2. *It is strong and durable*. The bottom soil-cement layer provides a firm, smooth surface for the synthetic membrane. It also protects the membrane from damage due to burrowing animals. The top layer of soil-cement provides a hardened surface to support maintenance and operating equipment. It protects the membrane from damage due to burrowing animals, vandals, operating equipment, and weathering.

3. *It provides additional impoundment storage volume over thicker liner systems*. A synthetic membrane placed between two 15 cm (6 in.) thick layers of compacted soil-cement is thinner than clay liners 1.5 to 3.0 m (5 to 10 ft) thick or synthetic membranes that often require both a base material for the membrane and a cover material over the membrane. The base and cover material together could total 0.9 m (3 ft) or more.

The construction feasibility of the composite liner was demonstrated in the fall of 1983 at a test section near Apalachin, New York (Fig. 8). To simulate a portion of a waste impoundment, the composite liner was placed 13 m (44 ft) wide along a horizontal length of 9 m (30 ft) and along a 3:1 side slope length of 12 m (40 ft). Both a 30-mm and 40-mm high density polyethylene (HDPE)

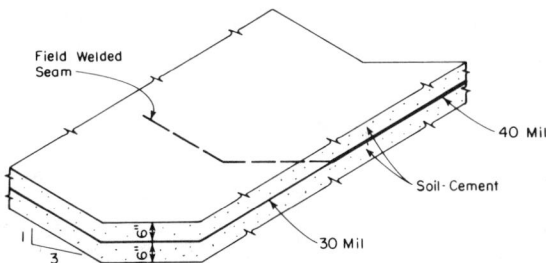

FIG. 8—*Drawing of the composite liner test section, Apalachin, New York.*

membrane were placed adjacent to each other on the bottom 15-cm (6-in.) compacted layer of soil-cement and were field welded. Aggregates ranging in size from 0.6 cm (¼ in.) to 2 cm (¾ in.) were sprinkled beneath a portion of each membrane prior to the placement and compaction of the top layer of soil-cement. A premixed 15-cm (6-in.) layer of soil-cement was then placed over the HDPE membrane by use of a small dozer (Fig. 9). Compaction was accomplished using a smooth, steel drum vibratory roller. After compaction, the top layer of soil-cement was removed at several locations to inspect the membrane for signs of damage. Observations of each membrane showed that there was no damage to either membrane or seam, and both were found to be puncture-resistant to the placement and compaction of soil-cement even with 2-cm (¾-in.) aggregate placed under them. The test proved the ease and reliability of constructing the composite liner.

Conclusions

More than 300 water resource projects over the past 30 years have utilized soil-cement. Applications include slope protection and seepage control for lakes, channels, and sludge-drying and wastewater treatment lagoons. In addition to its low permeability, soil-cement provides a firm working surface to support maintenance and operating equipment. For hazardous waste and other impoundments where maximum seepage protection is required, a composite liner consisting of a synthetic membrane placed between two layers of soil-cement has been developed.

FIG. 9—*Spreading soil-cement on membrane at 3:1 slope, Apalachin, New York.*

References

[1] Nussbaum, P. J. and Colley, B. E., "Dam Construction and Facing with Soil-Cement," No. RD010W, Portland Cement Association, Skokie, IL, 1971.

[2] Coffey, C. T. and Jones, C. W., "10-year Test of Soil-Cement Slope Protection for Embankments," Earth Laboratory Report No. EM-630, Div. of Engineering Laboratories, U.S. Bureau of Reclamation, U.S. Dept. of the Interior, Denver, CO, 1961.

[3] "Soil-Cement Slope Protection for Embankments: Planning and Design," No. IS173W, Portland Cement Association, Skokie, IL, 1975.

[4] Nussbaum, P. J., "Permeability of Soil-Cement, Lime, and Fly Ash Mixes," Portland Cement Association, Skokie, IL, 1979.

[5] "Soil Stabilization with Portland Cement," Highway Research Board Bulletin 292, 1961, National Academy of Sciences—National Research Council, Washington, DC.

[6] "Soil-Cement in Energy and Water Resources," No. SR238W, Portland Cement Association, Skokie, IL, 1981.

[7] "Sludge-Drying Lagoons," No. PL153W, Portland Cement Association, Skokie, IL, 1982.

[8] "Lining of Waste Impoundment and Disposal Facilities," No. SW-870, Office of Solid Waste and Emergency Resources, Washington DC, Mar. 1983.

[9] "Soil-Cement for Water Control: Laboratory Tests," No. IS166W, Portland Cement Association, Skokie, IL, 1976.

[10] "Coal-Handling and Storage Facilities," No. PL152W, Portland Cement Association, Skokie, IL, 1983.

[11] Jonker, C., "Subgrade Improvement and Soil-Cement," in *Proceedings*, International Symposium on Concrete Roads, London, 13–15 Sept. 1982.

[12] "Suggested Specifications for Soil-Cement Linings for Lakes, Reservoirs, Lagoons," No. IS186W, Portland Cement Association, Skokie, IL, 1975.

[13] "Soil-Cement Slope Protection for Embankments: Construction," No. IS167W, Portland Cement Association, Skokie, IL, 1975.

[14] "Soil-Cement Construction Handbook," No. EB003S, Portland Cement Association, Skokie, IL, 1979.

[15] "Soil-Cement Slope Protection for Embankments: Field Inspection and Control," No. IS168W, Portland Cement Association, Skokie, IL, 1976.

[16] "Soil-Cement Inspectors Manual," No. PA050S, Portland Cement Association, Skokie, IL, 1980.

General Discussion

GENERAL DISCUSSION

J. Daemen[1] *(written discussion)*—Would any author comment on the possibilities of reducing test time requirements by some type of accelerated testing? What acceleration methods are available/reliable?

D. Anderson,[2] *W. Crawley,*[3] *and D. Zabcik*[4] *(closure)*—The main method used to decrease testing time has been to increase the hydraulic gradient by increasing the pressure head. Increased pressure may have one of the following effects:

1. It may decrease permeability by causing increased shear forces along pore walls, thereby causing an increase in particle migration and subsequent clogging of pores.

2. It may increase permeability if the migrating particles pipe out of the clay liner rather than clog the pores as just mentioned. Some would also argue that there is a threshold gradient below which flow will not occur in some smaller pores. In this case also, a higher gradient would result in a higher permeability.

Another way that has been suggested to shorten testing time is to increase the concentration of leachate constituents. This is, however, inadvisable because many of the effects constituents have on permeability are concentration dependent.

Other inadvisable methods for shortening time include increased temperature, decreased specimen thickness, and decreased clay content. It is necessary to pass at least two pore volumes of leachate through a barrier to establish how that leachate will affect permeability. With permeabilities less than 1×10^{-7} cm s^{-1} and hydraulic gradients no higher than 40, a test should be expected to last at least six months.

Y. Acar[5] *(closure)*—The API water loss test which we used in our experiments can be used to simulate the effects of salts on slurries. It is a short-term, 24-h test, which gives a good indication as to what the effect would be on bentonite.

Wes Holtz[6] *(written discussion)*—What is the effect of calcium ions when cement is added to sodium bentonite slurry mixes?

[1]University of Arizona, Tucson, AZ.
[2,3,4]K. W. Brown & Associates, Inc., College Station, TX 77840.
[5]Assistant professor, Department of Civil Engineering, Louisiana State University, Baton Rouge, LA 70803.
[6]Woodward-Clyde Consultants, Englewood, CO 80111.

Y. Acar (closure)—Calcium ions will replace sodium ions, causing a reduction in swelling of the bentonite and thus a drop in viscosity and an increase in water loss. Because the slurry thickens as a result of the addition of fines, the bentonite will not drop out of suspension, and, while the permeability of the filter cake increases, it does not become totally permeable. By adding sodium carbonate or a sodium phosphate, the effect can be reduced.

A. I. Johnson[7] (written discussion)—I have some concern that the effects of the temperature and the density of permeating fluids are not entering into calculations of the permeability reported from laboratory tests. Permeabilities frequently are not converted to a standard temperature, for example, and thus cannot be converted easily to field temperatures, such as that of the groundwater.

S. R. Day[8] and D. E. Daniel[9] (closure)—It is true that the hydraulic conductivity (k) is a function of both the fluid and the porous medium. Another measure of flow through porous media is often used in other disciplines known as the intrinsic or absolute permeability (K). These variables are related by the properties of the fluid.

$$k = \frac{K g \rho}{\mu} \qquad (1)$$

where

k = hydraulic conductivity (L/T),
K = intrinsic permeability (L^2),
ρ = density of the fluid (M/L^3),
g = acceleration due to gravity (L/T^2), and
μ = dynamic viscosity of the fluid (M/LT).

It should be remembered that the temperature of interest is the temperature of the fluid flowing through the liner, not the air temperature.

It is possible to correct k to a standard temperature (say 20°C) by forming the following ratio from Eq 1.

$$\frac{k_{20}}{k_T} = \frac{\rho_{20}\mu_T}{\rho_T\mu_{20}} \qquad (2)$$

The magnitude of this correction for water is illustrated in the following table.

[7]Woodward-Clyde Consultants, Englewood, CO 80111.
[8]Civil engineer, GEOCON, Inc., Pittsburgh, PA 15235.
[9]Assistant professor, Department of Civil Engineering, University of Texas, Austin, TX 78712.

Temperature, °C	k_{20}/k_T
0	1.78
20	1.00
40	0.66

The relative importance of this correction should be weighted against the relative accuracy of the total testing program.

In field testing clay liners with water, it is generally not practical to consider this relatively minor correction factor. Typically, there are other sources of error (for example, resolution, representative sampling, soil variability, etc.) that are more critical. When testing with permeants other than water it may be meaningful to convert to absolute permeability. An example of this use of absolute permeability can be found in this publication in the paper by Acar, Field, and Scott.

Y. Acar (closure)—We have not yet been successful enough to obtain a good agreement between the hydraulic conductivity measured in the laboratory and in the field. Differences of orders of magnitude are possible due to different fabric generated by the laboratory and field compaction schemes and the field macrofabric variations that are not accounted for in a laboratory compaction test. In fact, we have not yet established whether the microfabric generated in the field is comparable to the fabric in the laboratory. Consequently, although it is a lot more proper to report absolute values of hydraulic conductivity, such corrections are not expected to decrease the variation between laboratory and field values and hence would not justify the use of the laboratory values for *in situ* conditions.

S. A. Gill[10] *and B. R. Christopher*[11] *(closure)*—Indeed, the effects of temperature and density of the permeating fluid should be entered into calculations, especially with permeating fluids other than water. On permeants such as solvents, drilling fluids, and even water that contains high quantities of contaminants, density determinations of the permeating fluids should be made. Ideally, tests should be performed in laboratories under controlled "standard" conditions. Unfortunately, standard conditions have not been defined for soil testing laboratories. It is felt that comparisons of results between laboratories would be more valid if tests were performed in more controlled environments as opposed to converting laboratory data to standard temperature. In our case study, all tests were performed in a temperature-controlled room maintained at $22\pm2°C$ ($71\pm2°F$). Under these conditions, temperature has an insignificant effect on test results.

I. Johnson (closure)—I appreciate the comments of the authors. The problem exists, however, that permeabilities are important to many problems be-

---

[10]Chief engineer, STS Consultants, Northbrook, IL 60062.
[11]STS Consultants, Northbrook, IL 60062.

sides that of leakage through clay liners at waste ponds or flow of relatively pure groundwater and the range of temperatures or densities related to those situations. Temperatures can be quite high when considering flow problems involved with solar ponds or geothermal energy, for example. I recommend that the permeabilities always be reported to a standard of 20°C. In any case, the least that should be done is to report the temperature of the permeant as used in the test. Permeabilities in my laboratory were performed in a constant temperature room, but I still reported the values at a standard temperature. Furthermore, density also can be an important factor when one considers the wide variety of fluids that are of interest to subsurface flow problems today. Because investigators in soil and rock problems have dealt for so many years primarily with the movement of relatively pure groundwater through the pores at a narrow range of temperatures, there is a tendency to forget the effects of viscosity (temperature) and density. There also is a tendency among investigators to automatically assume that something must be wrong with the laboratory test or sample when a difference between laboratory and field permeability data occurs. One should look at both methods for possible errors or difficulties, remembering that many factors also affect field tests.

J. A. Mundell[12] *(written discussion)*—(1) Many laboratory results have been shown on the effect of various permeants on the permeability of compacted clay. What effect does the state of compaction (molding moisture content, compaction energy) have on these effects? (2) Have the authors taken this into account in their studies? (3) Results presented at this symposium indicate that the results of various permeants on the permeability of lower plasticity silty clay may be much less than for highly plastic, bentonite clays. What are the authors' experiences with regard to this?

Y. Acar (closure)—(1) To this author's knowledge, no specific study exists that scrutinizes the effect of different permeants on the hydraulic conductivity of samples compacted at different molding water contents and compaction energies. However, when water is the permeant, the effect of such mechanical variables on hydraulic conductivity is well established. Studies by various investigators indicate that while the clay fabric is not very stable at the dry side of the optimum water content, hydraulic conductivity could be orders of magnitude greater than that obtained at the wet side of optimum water content. As a consequence, it is expected that soils compacted at or slightly above the optimum water content would offer the most stable fabric. As to the effect of compaction energy, previous experience with earth dams indicates that soils compacted at efforts greater than the standard compaction effort tend to be brittle in deformation behavior and more susceptible to desiccation cracking when subjected to any volume decreases. Consequently, standard Proctor efforts and optimum moisture contents were used in our studies.

(2) This author believes that it is necessary to restrict the activity of the soil used in the liners for shallow land waste disposal facilities where the postcon-

[12]Environmental research associate, Department of Civil Engineering, University of Notre Dame, Notre Dame, IN 46556.

structional pore fluid chemistry is expected to be significantly different than the initial molding fluid. Various results presented at this symposium emphasize the necessity to bring criteria that provide a control on not only the mechanical variables such as the compactive effort, the type of compaction, and the molding water content but also on the compositional variables such as the clod size, the activity of the soil, and the chemical characteristics of the leachate generated.

(3) High swelling bentonites will work very well to prevent the breakthrough of certain leachates such as those of municipal sanitary landfills. Strong leachates, like certain industrial ones, will have a more deleterious effect on these types of bentonites versus silty clay because the quantity of fines is much lower. Consequently, when there is a reduction in the size of the double layer, the porosity is increased and therefore the permeability also. This will not happen to any large degree with silty clays because the pores are filled with fines rather than with swollen clay particles. It is important that the materials are tested in the lab for compatibility with the leachate prior to installation. Bentonites are often much cheaper than such silty clays.

C. Williams[13] *(written discussion)*—The audience is led to believe that soil liners have few problems with most leachates. When are acids and bases a problem? The pH range? Organic comment rations? Soil types and minerology?

D. Anderson,[8] *W. Crawley,*[9] *and D. Zabcik*[10] *(closure)*—Dilute aqueous leachates with a near neutral pH have not been shown to affect the permeability of compacted clay liners or soil-bentonite mixtures. Both acids and bases may increase or decrease permeability as follows:

1. Dissolution of soil particles or soil binding agents by strong acids or bases will release particle fragments for migration.

2. If particle migration continues for an extended time, the pores will enlarge and, consequently, the permeability will increase.

3. If the pores are too small for passage of the migrating particles, the pores will become clogged and, consequently, the permeability will decrease.

Acids or bases can be a problem where the clay liner or slurry trench is adjacent to soils with pores large enough to allow for the migration of clay particles.

Y. Acar (closure)—Acids and bases are a problem for bentonite at a pH lower than 2 and higher than 11. Organic materials are a big problem if they are pure, which means they will compete with water, and if sinkers (higher specific gravity than water) are at a relatively high concentration (more than 20 ppm). For more information, check "General Classification of Wastes," IMCLAY Binder, International Minerals and Chemical Corp., Mundelein, Illinois (Industry Group). Acids and bases are also a problem if the soils contain such soluble minerals as gypsum, creating pore spaces upon dissolution.

[13]McBride-Ratcliffe & Associates, Houston, TX 77040.

Subject Index

Author Index